THRIPS

their biology, ecology and
economic importance

Corn thrips (*Limothrips cerealium*) (a Shell photograph).

THRIPS

their biology, ecology and economic importance

TREVOR LEWIS

Department of Entomology
Rothamsted Experimental Station
Harpenden, Hertfordshire, England

1973

ACADEMIC PRESS LONDON AND NEW YORK
A Subsidiary of Harcourt Brace Jovanovich, Publishers

ACADEMIC PRESS INC. (LONDON) LTD.
24/28 Oval Road,
London NW1

United States Edition published by
ACADEMIC PRESS INC.
111 Fifth Avenue
New York, New York 10003

Library of Congress Catalog Card Number: 72–12273
ISBN: 0 12 447160 9

Printed in Great Britain by
Butler & Tanner Ltd, Frome and London

PREFACE

The main purpose of this book is to complement the hitherto largely descriptive works on the order Thysanoptera, by presenting thrips as living animals, stressing the behaviour of individuals and populations, their varied and complex relationships with plants, other animals and the physical components of their environment, their abundance in undisturbed and in cultivated habitats, and their economic importance as pests and beneficial insects.

A considerable literature on thrips has accumulated over the last one hundred years. Unfortunately much of this information is fragmentary, scattered through obscure pamphlets, journals and bulletins, and accessible only with difficulty. Also, many observations and impressions collected personally from entomologists, ecologists and farmers remain unrecorded. Thus, a subsidiary reason for writing this book is to collect and arrange these many valuable snippets of published and unpublished information and combine them with major studies before the task becomes too daunting. In so doing, many of the earlier observations on thrips have been re-interpreted in the light of modern ecological concepts derived from studies on other groups of insects.

To encourage general entomologists, ecologists and agriculturalists to use the book, non-specialist terminology has been used wherever possible, and to stimulate a wide academic and applied interest illustrative examples have been chosen from a great variety of countries, habitats and crops. Following a brief introduction to the structure and classification of adult thrips, the text is divided into four sections. The first describes the biology and behaviour of individuals, the second, the techniques needed to study field populations, a hitherto sparsely documented topic. Section III deals in detail with thrips ecology, especially survival in unfavourable weather, migration and dispersal, population dynamics and community structure, and finally Section IV gives a survey of the economic importance of the group.

The current generally accepted nomenclature for thrips has been used throughout, rather than using the names given in the original papers consulted. There is, however, an Appendix listing the authorities for all specific names mentioned, and synonyms of the more common species.

I hope that the book will stimulate further research on thrips by providing a compact source of basic information about them as well as exciting interest in their varied and fascinating habits and way of life.

Rothamsted Experimental Station   TREVOR LEWIS
Harpenden
June 1973

ACKNOWLEDGMENTS

I have had help from many people while writing this book, but I wish especially to thank my wife, Margaret E. Lewis, who has helped patiently in all the stages of its preparation, with translations, and on whom has fallen the burden of checking errors in grammar and literary style. Mr. L. A. Mound of the British Museum (Nat. Hist.) also has my special thanks for invaluable advice on nomenclature and encouragement over several years.

I am grateful to those who have read and criticized parts of the book: Dr. N. Waloff, Dr. F. D. Bennett, Dr. A. J. Cockbain, Mr. D. E. Evans, Mr. B. R. Pitkin, Dr. H. W. Simmonds and Dr. L. R. Taylor, although in fairness to them I alone accept responsibility for the views expressed. I acknowledge the debt of stimulating discussions with many entomologists, ecologists and farmers; to Prof. D. K. McE. Kevan who introduced me to thrips, to Prof. O .W. Richards who guided my interest in the early years, and the kindnesses received from Prof. T. N. Ananthakrishnan, Dr. C. G. Johnson, Dr. G. D. Morison and Mr. K. Sakimura.

Elizabeth Haines Dip.AD. carefully re-drew or made original drawings of insects and plants (with the exception of Figs. 8, 16, 17, and 81), Mr. H. H. Franklin made the inventive drawings of apparatus; many people, acknowledged in the legends, provided photographs; Mr. F. D. Cowland also helped with photography; Mrs. Tyra Bacon, Miss Monica Williams and Miss Roseanne Adam typed the manuscript and the latter, with Miss J. B. Sherrard, checked the references; Mrs. A. Kay checked grammatical points in the text; Mr. H. Lewis catalogued reprints; Mrs. Norah J. Gray, Miss Katherine Voorhoeve, Mr. J. Frerichs, Dr. E. Judenko, and Mr. K. Nakusono translated papers; to all of them I tender my sincere thanks.

I thank the many individuals who have allowed me to re-draw or modify diagrams from their papers and to use unpublished data, and the following for permission to use diagrams and photographs:
The publishers: Collins, London; Commonwealth Agricultural Bureau; John Gresham and Co., London; Geest and Portig KG, Leipzig; Gustav Fisher Verlag, Stuttgart; North Holland Publishing Company,

Amsterdam; VEB Gustav Fischer Verlag, Jena; Verlag Paul Parey, Hamburg; W. Junk, The Hague.

The company: British South Africa Co., Southern Rhodesia.

The Councils, Editors and Publishers associated with the following scientific journals: *Acta Botanica Neerlandica*, *Annals of Applied Biology*, *Annales Société Entomologique de France*, *Atti dell Accademia pontificia dei Nuovi Lincei*, *Biologiske Meddelelser*, *Bulletin of Entomological Research*, *Bulletin of Illinois State Natural History Survey*, *Canadian Entomologist*, *Entomologia Experimentalis et Applicata*, *Hilgardia*, *Indian Journal of Entomology*, *Journal of Animal Ecology*, *Journal of the Council of Scientific and Industrial Research Australia*, *Journal of Economic Entomology*, *Milling*, *Mitteilungen der Schweizerische Anstalt für das forstliche Versuchswesen*, *Mitteilungen der Schweizerischen entomologischen Gesellschaft*, *Oikos*, *Parasitiology*, *Pedobiologica*, *Proceedings of the Royal Entomological Society of London*, *Psyche*, *Redia*, *Report on Cacao Research*, *I.C.T.A. Trinidad*, *Transactions of the Royal Entomological Society of London*, *Zeitschrift für angewandte Entomologie*, *Zoologické Listy—folia Zoologica*.

"Some climb the stubble, others try the power,
And compass of their wings, by many a flight,
From herb to herb and onward draw, as if
They wished to join the reapers in their talk;
Some less ambitious, less aspiring, crawl
In silence, on the thread-bare soil, they seek
No higher ground."

Thomas Francis—Harvest Day, 1859.

Beware of the Thrips!
It constantly sips
The milk, or the juice, of the wheat-ear;
It flies in your face,
And stings at the place,
And never appears as a treat here.

Yes! get rid of Thrips!
It crawls to the pips
(Or stones) of good olives and peaches;
Though tiny as specks,
That tiny things vex
As deeply as big things, it teaches.

Jennet Humphreys—circ. 1910

To:

MARGARET, HEATHER and ROGER

CONTENTS

SECTION IV. ECONOMIC IMPORTANCE

1 Introduction

Geographical and Historical Background

Thrips are diminutive insects, individually so small that they easily pass unnoticed, yet collectively spectacular when they appear in vast numbers. They are widespread throughout the world, with a preponderance of tropical species, many temperate ones, and even a few species extending to arctic regions. Their habitats range through forests, grasslands, scrub, desert, most cultivated crops and gardens, and they include phytophagous and carnivorous species, gall-makers and inquilines as well as themselves being prey for other arthropods and vertebrates.

Most laymen would probably dismiss an individual thrips as a mere dark or pale speck, and even many entomologists group the 5,000 or so known species (zur Strassen, 1960) under the single all-embracing term "thrips" or Thysanoptera. In fact, the order comprises an array of beautiful and diverse creatures, either brightly coloured or sombrely shaded, often intricately sculptured, and occasionally of bizarre shapes and proportions. But the most striking feature, common to all but a minority of species, is their delicately fringed wings which give the order its scientific name, Thysanoptera, derived from two Greek words meaning a fringe, and a wing (Fig. 1). Their common name, thrips, is also derived from the Greek, meaning a "wood louse". Other names are bladderfeet (Blasenfüsse), Physopoda or Vesitarses, because of their peculiar terminal tarsal segment; fringewings; and thunderflies or stormflies, names attributable to the frequent appearance of some species in stormy weather. This variety of form and structure provides a basis for the taxonomic and systematic classification of the group, which has been largely documented since the year 1900, though many revisions of the early descriptive work are necessary and much taxonomic research remains to be done.

Compared with many groups of insects, few early entomologists specialized in studying thrips. Perhaps this was partly because many species can only be found by diligent searching, and captured specimens need careful preparation and mounting for microscopic examination; perhaps also because more obvious crop pests were easier to study.

B

However, since they were recognized as economically important insects, first as pests damaging crops directly in many parts of the world (Marsham, 1796; Curtis, 1860; Osborn, 1888; Froggatt, 1906), then as potential predators of other crop pests (Walsh, 1864, 1867a, b), as vectors of plant

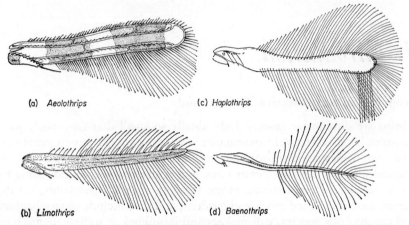

(a) Aeolothrips   (c) Haplothrips

(b) Limothrips   (d) Baenothrips

FIG. 1. Examples of wings from four sub-families of thrips showing their diverse shapes and pigmentation. (a) Evenly broad, blunt tip, cross-veined, banded (Aeolothripinae: *Aeolothrips*); (b) narrow, pointed tip, no cross-veins, tinted (Thripinae: *Limothrips*); (c) unevenly broad, no veins, almost colourless (Phlaeothripinae: *Haplothrips*); (d) slightly sinuous and slender, single pigmented vein (Urothripinae: *Baenothrips*) (original).

virus and bacterial diseases (Pittman, 1927; Hansen, 1929; Bald and Samuel, 1931) and finally as agents of weed control (Simmonds, 1933) they have received progressively more attention.

General Structure and Classification

Although detailed morphological descriptions and systematics are beyond the scope of this book, a brief introduction to them is necessary to appreciate much in the following chapters.

As an order, the thrips are the smallest winged insects, ranging from about 0·5 to 14 mm in length. Most large species are tropical; the common temperate species are usually no more than 1 to 2 mm. Living specimens, seen with the naked eye, are slender, often deeply pigmented and shiny. The head is distinct, and the division between thorax and abdomen sometimes distinguishable, but when at rest the four wings are kept folded flat along the back and are only visible as pale or silver strips glinting in the light. In the sub-order Terebrantia the wings lie parallel to each other but in the sub-order Tubulifera they overlap so that only one is completely visible (cf Figs 2a, c).

Sub order TEREBRANTIA

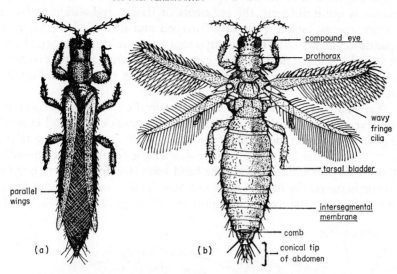

parallel wings

(a)

compound eye
prothorax
wavy fringe cilia
tarsal bladder
intersegmental membrane
comb
conical tip of abdomen

(b)

Sub order TUBULIFERA

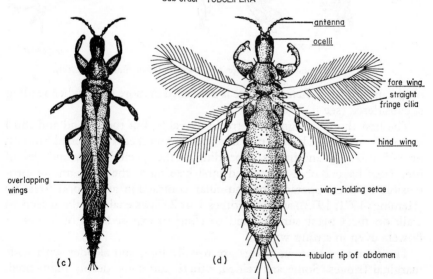

overlapping wings

(c)

antenna
ocelli
fore wing
straight fringe cilia
hind wing
wing-holding setae
tubular tip of abdomen

(d)

Fig. 2. Living (left-hand side) and mounted (right-hand side) thrips of the two sub-orders compared. Note the setae and cuticular structure revealed in the slightly flattened mounted specimens, contrasting with the elongated, relatively smooth silhouette of living insects. Features common to both sub-orders are underlined (original).

The appearance of mounted specimens (see Appendix 2) under magnification is quite different; the segments of thorax and abdomen become distinct, the abdomen is slightly flattened and distended to display the separate segments with their component parts and intersegmental membranes, the wings and legs are extended, and often a complex array of spines, setae, tubercles and cuticular wrinkles is revealed (cf Figs 2a, c with b, d).

The head bears a pair of 4 to 9, but usually 7 or 8-segmented antennae. These are inserted at the front between prominent compound eyes with large convex facets; most adults also have 3 ocelli arranged in a triangle on top of the head. But the most characteristic feature is the specialized mouthcone protruding beneath the head and often appearing to originate near or between the base of the front legs. This is best seen in side view (Fig. 3). Because the economic importance of thrips is so closely dependent

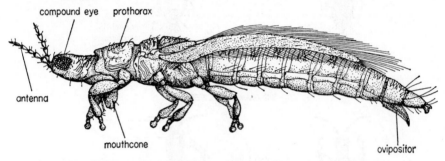

FIG. 3. Side view of *Anaphothrips sandersoni* (after Stannard, 1968).

on the mouthparts, their structure is described in more detail when feeding is considered (see p. 36).

The first thoracic segment is freely movable, but the second and third segments are fixed. The legs may be slender or markedly thickened, smooth or with tubercles and hooks, depending on the species' habits and way of life. They have 1 or 2-segmented tarsi bearing at the tip a unique protrusible bladder, operated by muscular contraction and blood pressure (Heming, 1971, 1972), and sometimes 1 or 2 claws enabling the insects to walk on most plant surfaces and to cling to exposed tips of leaves or flowers even in strong winds.

The four membranous wings are usually long and slender with wide marginal fringes. Some are veined, others not; some deeply pigmented, others merely tinted (Fig. 1). The length of the wings relative to the body often differs between groups, species and sexes; both sexes may have long-winged (macropterous) or short-winged (brachypterous) forms; within some species male and female have wings of different length and there may even be polymorphic forms of the same sex present in a single

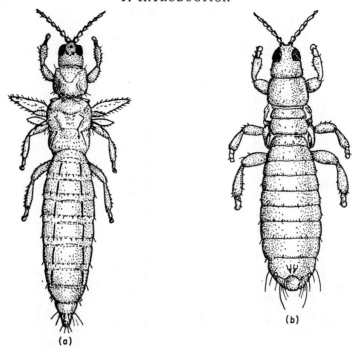

FIG. 4. Examples of (a) short-winged (brachypterous) female, *Thrips angusticeps* (original); (b) wingless (apterous) male, *Limothrips cerealium* (after Melis, 1960).

population. Sometimes one or both sexes, but usually males, have no wings at all (apterous) (Fig. 4).

Ten of the eleven segments comprising the abdomen are visible. The terminal segments of Terebrantia usually taper to a cone in females and are bluntly rounded in males (Fig. 4b), whereas in both sexes of Tubulifera the tenth segment forms a tube bearing a terminal whorl of setae. Other setae on the back of Tubulifera hold the wings in a resting position. Only the Terebrantia have an ovipositor, with four curved, saw-like valves, carried beneath the 8th and 9th abdominal segments (Fig. 3). Tubuliferan females have their genital opening between the 9th and 10th segments.

There are usually two pairs of thoracic and two pairs of abdominal spiracles.

Further morphological and anatomical details, including descriptions of immature stages, are placed in the text where their relevance to the biology and ecology of thrips can be more easily appreciated.

Many aspects of the systematic classification of the Order remain controversial among specialist Thysanopterists but a summary of Priesner's (1949, 1964b) widely accepted scheme is adequate for the purposes of this book (Fig. 5). The Aeolothripidae occur mostly in the temperate regions of Northern and Southern Hemispheres and many are facultative predators

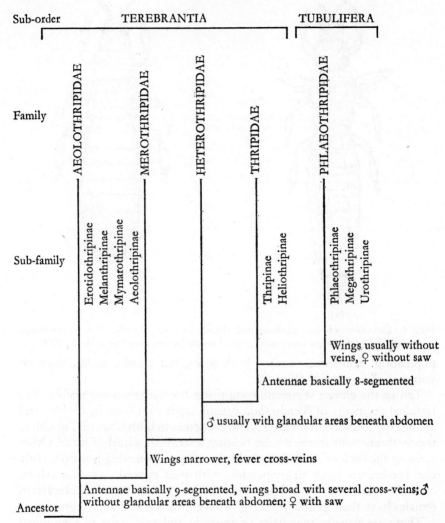

FIG. 5. Summarized classification of the Order Thysanoptera based on Priesner's (1964b) scheme, arranged to show the main characteristics of families and affinities of sub-families (Stannard, 1968).

on other small arthropods. The Merothripidae are minute, often wingless insects living in leaf litter and bark in the tropics and sub-tropics, and the Heterothripidae are a flower-dwelling family mainly from America. The great majority of thrips including almost all species of economic importance, belong to the families Thripidae and Phlaeothripidae, which are distributed throughout the world.

BIOLOGY

2 Reproduction, development and life histories

Sexual Differences

The sex of a few thrips can be distinguished by the naked eye but examination with a magnifying glass or microscope is usually necessary. Sexual differences are more obvious in the Terebrantia than in the Tubulifera. Males of Terebrantia are always smaller and usually paler in colour than females, and have an elongated, almost parallel-sided abdomen, bluntly rounded at the tip. Under magnification, claspers are visible on the 9th abdominal segment of many Aeolothripid males, and some other genera bear either abdominal spines (*Odontothrips*), horns (*Kakothrips*) or forked or feather-like appendages (*Scirtothrips*) which perhaps assist copulation. All Terebrantian females have a conspicuous saw-like ovipositor which usually lies retracted beneath a gradually tapered abdomen.

In the Tubulifera the difference in size is less distinct, and males are often stouter than females, with enlarged forelegs. Females have a longer tube than males and a short, chitinous rod, the fustis, is visible under magnification lying internally just anterior to the base of the tube. Details of the structure of external genitalia are described by Melis (1935), Doeksen (1941), Jones (1954) and Priesner (1964a).

Just occasionally gynandromorphs, in which male and female parts develop in the same individual due to aberrant distribution of sex chromosomes, are found, e.g. in *Oxythrips ajugae* (Morison, 1949; Mound, 1970d).

Internally the female reproductive organs consist of a pair of ovaries each composed of four ovarioles grouped at either side of the abdomen. The ovarioles are long tubes in which the oocytes lie in a single chain, becoming progressively more mature towards the posterior end. The ovarioles in each group open into an oviduct and these join to form a common median oviduct leading to the vagina. A narrow tube leads from the vagina to the *receptaculum seminis*. The ovaries are similar in both suborders except that in Tubulifera the terminal filaments of the ovarioles are connected to the long salivary glands. When mature, the ovaries occupy most of the abdominal cavity; their size indicates the approximate age of individuals (Fig. 6; Lewis, 1959a).

In males the testes lie in abdominal segments 6 to 8. From each testis a tube leads backwards into a slight dilation, the *vas deferens*, before joining

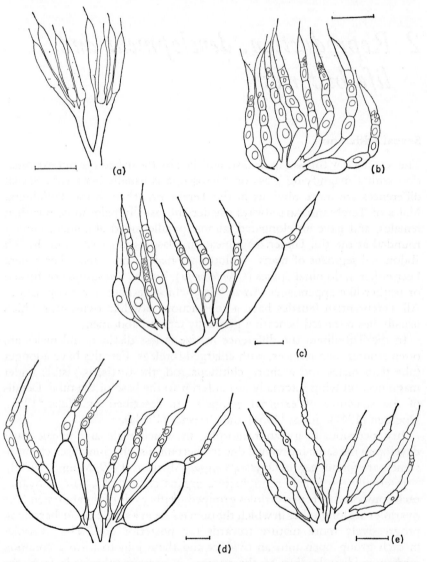

FIG. 6. Ovaries of *Limothrips cerealium* showing stages of development (a) undifferentiated; (b) early maturing; (c) late maturing; (d) laying; (e) senile. The horizontal scales each represent 100 μ (after Lewis, 1959a).

the opposite tube to form a common ejaculatory duct opening into the phallus. In Terebrantia one pair, and in Tubulifera two pairs, of accessory glands open into the upper end of the ejaculatory duct which may also expand into a spermatheca.

Anatomical details of the reproductive organs are given by Sharga (1933b), Melis (1935) and Bonnemaison and Bournier (1964); the post-embryonic development of the female reproductive system is described by Davies (1961) and of the male reproductive system by Heming (1970).

Sex Ratios

Field populations of most species are bisexual but females often predominate. In some species males are rare or unknown, and reproduction is partly or wholly parthenogenetic. In many species where the sexes are produced in equal numbers the females apparently predominate because they often live longer than the males. Spurious ratios may also occur in species with flightless males, when plants other than the larval hosts are sampled, when one sex is more active than the other, or when sexes are attracted differentially to traps (Lewis, 1961). In species in which only the females hibernate, the sex ratio changes in spring as new males are gradually produced. Koppa (1969a) suggested that in species that have flightless males, the sex ratio can be used as a comparative index of suitability of different host plants; the greater the proportion of males on the plants, the more breeding has occurred there. For example, in Finland, the proportion of males of *Limothrips denticornis* measured throughout six growing seasons was greater on winter rye and barley, on which this species reproduces vigorously, than on wheat or oats, which are less suitable.

In a number of cosmopolitan species the sex ratio differs in different regions. This may be dependent on temperature with fewer males occurring where it is warmer. In England, males of *Thrips tabaci* may appear locally out of doors, but never in the artificial sub-tropical climate of glasshouses where reproduction is always parthenogenetic (Morison, 1957). A different explanation is suggested by O'Neill (1960), who believes that parthenogenesis, and the scarcity of males, is most common in introduced species, because parthenogenetic forms are more easily spread than sexual forms. In the eastern Mediterranean region and Iran, the sex ratio of *T. tabaci*, the onion thrips, is about 1:1 (Mound, *in litt.*), in contrast to most other parts of the world where males are rare, for example 1♂:1,000♀♀ in Hawaii (Sakimura, 1932) and 0♂:3,000♀♀ in the Sudan (MacGill, 1927). The wild progenitor of the cultivated onion probably originated in central Asia, and it was well known in Egypt in the times of the Pharaohs (Chittenden, 1951). It is therefore possible that this area is the true home of *T. tabaci* and that the striking differences in its sex ratio in other parts of the world occur because the species was introduced there. *Haplothrips gowdeyi* is another widely distributed species in the tropics, but males are only known from Africa, where it similarly may have originated.

Sex ratio sometimes changes within a few degrees of latitude. In Scotland (57°N), males and females of *Taeniothrips vulgatissimus* occur in about equal numbers but males become scarcer in northern England, and rare or absent in southern England (52°N) (Morison, 1957), the Frankish Alps (49°N) (Weitmeier, 1956) and the Rhein Main (50°N) regions of Germany (zur Strassen, 1967). Males of *Aptinothrips rufus* are generally rare, but in Scotland the sex ratio is 1♂:150♀♀ (Morison, 1957) compared with 1♂:3,000♀♀ in central France (45°N) (Pussard Radulesco, 1930).

Parthenogenesis

Female thrips are always diploid and males haploid (Whiting, 1945; Stannard, 1968) so males are derived only from unfertilized eggs. Such arrhenotoky occurs in a number of Thysanoptera and, as in other orders of insects, especially the Hymenoptera, it characteristically produces unequal sex ratios (Hamilton, 1967). In species whose reproduction is known to be predominantly of this type, e.g. *Taeniothrips simplex* (Bournier, 1956a) and *Liothrips oleae* (Bournier, 1956b), there are usually four times as many females as males. Reproduction in *Limothrips denticornis* is probably also arrhenotokous because the sex ratio in aggregations found in the breeding sites is about 1♂:4♀♀ (Pussard Radulesco, 1930). Virgin females of *Thrips linarius* (Zawirska, 1963), *Caliothrips fasciatus* (Bailey, 1933), *Haplothrips verbasci* (Shull, 1917) and *Scirtothrips citri* (Munger, 1942) produce only male offspring, whereas fertilized females produce mostly females with some males from non-inseminated eggs.

By contrast, reproduction in species without males can only be by female to female parthenogenesis (thelotoky). For example, populations of *Heliothrips haemorrhoidalis* consist entirely of virgin females producing female offspring; only two males have ever been found (Crawford, 1940; Bournier, 1956a). Among species living in glasshouses there are many with males unknown or very rare including *Hercinothrips bicinctus*, *Heliono-thrips errans*, *Scirtothrips longipennis*, *Leucothrips nigripennis* and *Chaeta-naphothrips orchidii* (Morison, 1957), but this may be because they have been spread by man from their countries of origin.

Some species are predominantly thelotokous, but males occasionally appear in considerable numbers. For example, males of *Haplothrips tritici* are normally rare (Priesner, 1928) but Bournier (1956a) found a population of this species in central France in July 1953 containing 20% males (i.e. 1♂:4♀♀), which suggests that arrhenotoky may occur sometimes

in this normally thelotokous species. In North America, female parthenogenesis is obligatory in *Taeniothrips inconsequens*, and in Europe males of
this species are rarely seen. Nevertheless, in southern France, Bournier
(1956a) also found that 30% of the females in a population on *Prunus* were
inseminated, and that 20% of the II stage emerging from eggs laid by these
females appeared destined to be males. This 1:4 sex ratio again suggests
arrhenotoky. Bournier thinks that males may be more common than
suspected in Europe, and that their scarcity on fruit trees is perhaps
because they die at the overwintering sites soon after mating in spring, so
cannot fly to trees with the females.

The erratic changes in the proportion of males in natural populations
may be caused by changes of temperature. In warm glasshouses males of
Parthenothrips dracaenae are rare with 6–7 per 100♀♀ but in cooler houses
they are more common (Jordan, 1888). Pussard Radulesco (1930) suggests
that a parthenogenetic and a sexual race of this species live together,
because males caged with females known to be parthenogenetic, would not
mate, but they readily mated with certain females from a "wild" population. If changes of temperature do indeed affect the sex ratio in natural
populations, it is arguable that species have cyclical changes in their type
of reproduction, as occurs in many species of aphids, with thelotoky
predominating in summer and arrhenotoky or sexual reproduction in
autumn.

Details of the cytology of ovogenesis and spermatogenesis in a number
of species are given by Bournier (1956a) and Risler and Kempter (1961).

Mating

In typical bisexual species adults usually mate within 2 or 3 days of the
last pupal moult. Males are promiscuous and can each fertilize a number
of females. The sexes find each other by means of sense cones on the antennae. When the pair are side by side the male grasps the female's ptero-
thorax with his forelegs and mounts her, twisting his abdomen beneath
hers, and moving his copulatory organs back and forth until they encounter the female's genitalia (Fig. 7). When males fail at the first attempt they
may dismount and try to mate again from the opposite side. Various appendages and setae on his abdomen help the male to grip while mating (see
p. 9), and in Terebrantia a secretion from glandular areas of the abdominal sternites may also assist (Klocke, 1926). However, Pelikan (1951)
believes that these glandular areas secrete a lipoid substance containing an
aromatic component that soothes the excited female and discourages her
from running around while the male attempts to copulate. Sometimes
males spread their wings, perhaps also to help them balance. Males of
some Aeolothripids and Phlaeothripids with toothed tarsi are often

carried by the female during mating. Mating generally lasts longer in Terebrantia than in Tubulifera; in Aeolothripids from 20–60 min (Buffa, 1907) compared with 3–10 min in *Caliothrips fasciatus* (Russell, 1912a) and a few seconds (16–18) in *Haplothrips verbasci* (Shull, 1914b). *Caliothrips fasciatus* mate much more quickly when the temperature exceeds 35°C. In this species also, females which have mated only once, lay unfertilized eggs towards the end of their lives either because the sperm is used up in fertilizing the early eggs, or because it becomes impotent (Bailey, 1933).

A most unusual form of mating occurs in *Limothrips denticornis* in which adult males mate with the female prepupae whose reproductive organs develop prematurely. In the breeding season colonies of about 25 prepupae and 3 or 4 males occur in the leaf sheaths of host grasses. The

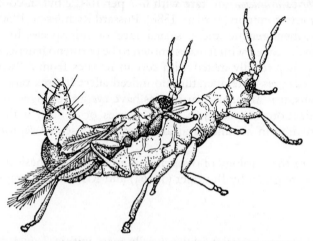

Fig. 7. Copulating *Aeolothrips fasciatus* (after Buffa, 1907).

prepupae have obviously mated because their spermathecae contain spermatozoa and they do not mate again after they have developed into adult females. Sperm is stored in the spermatheca during hibernation and used to fertilize the eggs laid next spring (Pussard Radulesco, 1930). Very occasionally, males and females of different species make unsuccessful attempts to copulate, but Koppa (1969a) claims to have found a male *Anaphothrips obscurus* actually copulating with a female *Frankliniella tenuicornis*.

Eggs and Oviposition

Mature eggs are large in relation to the thrips' abdomen so females usually contain only a few ready for laying at a time. The eggs of Terebrantia are cylindrical, slightly kidney-shaped with smooth, delicate, pale white or yellow shells. In the Melanthripinae and Aeolothripinae the bottom of the egg is rounded but the top is flattened and oblique to the axis; in Thripidae both ends are rounded. Tubuliferan eggs are oval, either symmetrical or constricted at the top, and often have a pink, yellow or darkly coloured

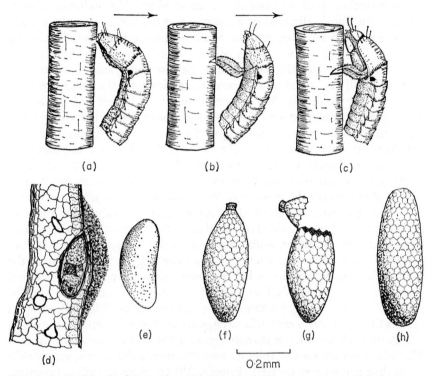

(a)   (b)   (c)

(e)   (f)   (g)   (h)

(d)

0·2mm

FIG. 8. Oviposition and examples of eggs. (a)–(c) Sequence of movements during oviposition typical of a thripid; original interpretation after descriptions by Russell (1912a), Cameron and Treherne (1918), Sharga (1933a), Bournier (1956a) (see text). (d) L.S. of egg of *Heliothrips haemorrhoidalis* embedded in plant tissue and covered with a blob of excrement (original). (e) Typical terebrantian egg (original). (f), (g) Unhatched egg and empty shell of the tubuliferan *Haplothrips leucanthemi* (after Loan and Holdaway, 1955). (h) Egg of *Hoplothrips pedicularius* (after Morison, 1947). The scale shows the approximate size of eggs (e)–(h).

shell sculptured with pentagonal or hexagonal reticulations. The micropyle at the top may be capped with a small knob through which passes the micropylar canal. Tubuliferan eggs measure 350–550 μ × 130–250 μ and Terebrantian eggs a little less (Figs 8e, f).

Most Terebrantia lay their eggs singly in an incision made in the plant tissue by the ovipositor. The part of the plant chosen differs between species; *Retithrips* spp. lay only on the underside of leaves, but others like *Limothrips* spp. and *Thrips tabaci* lay indiscriminately on leaves, cotyledons, petals, sepals or glumes. Usually the eggs are scattered but sometimes they are laid in short rows alongside or beneath veins. The following general description of oviposition is probably typical of most Thripidae. When the female is ready to lay, the tip of the abdomen is raised slightly and contractions push the egg down the oviduct. The abdomen is then arched with the tip directed downwards, and the thrips feels the surface of the plant with the long, sensory, terminal setae to find a suitable site for the egg. The ovipositor remains almost in its retracted position except that its tip protrudes slightly beyond the last abdominal segment, and while it is still supported in its groove the thrips stabs the point vertically into the plant tissue (Fig. 8a). It then pushes its abdomen backwards, and the ovipositor, pivoting on its articulation, comes out of the sheath completely and protrudes perpendicular to the axis of the abdomen (Fig. 8b). Another thrust with the full weight of the abdomen above it, pushes the ovipositor obliquely a little further into the plant (Fig. 8c). Working alternately the anterior and posterior blades saw the tissue and hollow out a cavity beneath the epidermis. When the ovipositor is almost buried the valves separate, and the abdomen contracts violently a few times to expel the egg, which is guided by the blades into the plant tissue. The small teeth that line the inner walls of the posterior blades in many species probably hook on to the shell and push the egg to the bottom of the cavity. The female withdraws her ovipositor slowly, and in some species, e.g. *Heliothrips haemorrhoidalis*, deposits a drop of excrement on the exposed tip of the egg, presumably to seal the cavity (Fig. 8d); in other species like *Scirtothrips citri*, whose eggs are embedded deeply in the plant, the incision closes almost completely after the ovipositor is withdrawn. The whole act lasts a few minutes and afterwards the female moves off a short distance to rest or feed (Russell, 1912a; Sharga, 1933a; Bournier, 1956a).

The eggs may be embedded in the tissue with their long axis inclined at a shallow angle to the surface (Figs 8d, 10b) or tilted more steeply at 30–45°. If thin leaves are illuminated from behind, eggs laid in them can usually be seen easily as translucent swellings, but in thick leaves, stems or fruit they are invisible. They can be made visible in thick leaves by dissolving the chlorophyll in a series of baths of alcohol, gradually increasing the strength from 50% to absolute, then clearing the leaves in clove oil. Once embedded in the plant, eggs may absorb water, and the eggs of *Scirtothrips aurantii* laid in citrus fruits quickly lose their characteristic kidney shape and swell into oval cylinders (Hall, 1930).

Citrus leaves react to the presence of eggs of *Heliothrips haemorrhoidalis* by developing cork cells in the surrounding tissue. This protects the egg from excessive water in the leaf, but in warm, moist weather, or in young leaves, the proliferation of these cells may be so great that the egg is crushed inside the cavity or even pushed out of the leaf. This may be why this thrips prefers to lay in old leaves, but if these wilt the thinner cork layer they produce is not resistant enough to water to protect the eggs and they too lose water and die (Rivnay, 1935).

Occasionally the ovipositor becomes immovably wedged in the plant and the laying female dies. In *Euchaetothrips kroli*, whose only known host plant is the semi-aquatic grass *Glyceria maxima* which often grows in exposed places, temporary anchoring by the ovipositor may help survival by ensuring that eggs can still be laid in strong winds, and that females are not washed away by floods. The cell walls of *Glyceria* are unusually strong, making withdrawal of the ovipositor difficult, and small, crescent-shaped feeding lesions can often be seen on the younger blades, produced by the anchored females as they feed on the only tissue they can reach. This pivoting on the ovipositor may eventually enlarge the incision and release them (Ahlberg, 1924; Lewis, 1955).

Among the Terebrantia, some species of *Chirothrips* are exceptional because their eggs are only partly embedded in plant tissue. *Chirothrips manicatus* lays its eggs in the tip of developing ovules of *Dactylis glomerata*, usually one to a seed, with only half its length in the tissue (Doull, 1956) and *Chirothrips crassus* deposits its eggs on the surface of leaves (Pesson, 1951). However, this is the usual method in all Tubulifera, which have no saw-like ovipositor. Their eggs are laid on flowers and leaves, under fungus, in galls and in bark crevices or galleries formed in wood by other insects, especially scolytid beetles. *Liothrips oleae* and *Aleurodothrips fasciapennis* even lay under the scales of certain coccids (Melis, 1935; Taylor, 1935). The eggs usually lie with their long axis horizontal to the supporting surface, and are attached to it with gelatinous substances. Eggs of gregarious species may be laid in hundreds in flat clusters, and some gall-forming species may lay several hundred eggs within the small internal cavity of a gall (Mound, 1971a).

The act of oviposition in Tubulifera lasts only a few seconds. The abdomen in the region of the 8th and 9th segments expands, the rounded end of the egg appears through the external opening and the female drops it quickly with a slight undulating movement of the abdomen (Loan and Holdaway, 1955).

Egg mortality is usually greater in Tubulifera than in Terebrantia perhaps because the exposed eggs are more vulnerable to desiccation and predators such as ants, rove beetles, bugs and predatory thrips (see pp. 65, 68). The West Indian tubuliferan, *Elaphrothrips brevicornis*, sits over her eggs after laying them on a leaf (Bagnall, 1915), perhaps to protect them from small predators.

At least two species of thrips living in temperate regions, *Megathrips lativentris* and *Caudothrips buffai*, are known to be ovoviviparous and a number of tropical species of *Diceratothrips* and *Anactinothrips* may also produce living young (Hood, 1950). In the region around Leningrad, U.S.S.R., eggs of *Megathrips lativentris* usually hatch after about 14 days, but occasionally, fully-formed larvae emerge within a few hours of laying, and caged females have produced larvae without empty egg shells being found (John, 1923). Ovoviviparity is much more common in *Caudothrips buffai* in southern France and may be the usual method of development in this species. The eggs are large when laid and stand erect on the substrate, stuck to it by the posterior pole. They hatch within a day and completely formed larvae emerge. Full-sized eggs dissected from the abdomen of gravid females also contain fully formed and chitinized larvae. *Caudothrips buffai* feeds on fungi that grow quickly in moist ephemeral environments and ovoviviparity may be an adaptation to this habitat, shortening the life-cycle and enabling the young to complete development before the food disappears (Bournier, 1957, 1966). This explanation of the phenomenon seems more plausible than that suggested by John (1923) for *Megathrips lativentris*, which was that cold slowed down metabolism, and presumably oviposition, so that the eggs were retained within the female whilst embryological development continued.

Fecundity and rate of oviposition

The total number of eggs laid by most female thrips ranges from about 30 to 300 depending on the species, the individual, and the amount and quality of food available. For example, *Thrips imaginis* lays practically no eggs if reared on a diet deficient in protein, an average of 20 per female when fed on *Antirrhinum* stamens with anthers removed, but an average of 209 when this diet is supplemented with pollen (Andrewartha, 1935). Although pollen is also the preferred diet of *Haplothrips leucanthemi*, females reared without it still produce fertile eggs (Loan and Holdaway, 1955). Some predatory species lay fewer eggs than average. Female *Scolothrips sexmaculatus* produce 4–5 offspring each (Bailey, 1939), *Trichinothrips breviceps* lays only 3–6 eggs (Seshadri, 1953), while females of *Leptothrips mali* in experimental cages produced one larva each, although 2 mature eggs were seen inside the abdomen of some (Bailey, 1940a).

In species having several generations per year, females of the later broods probably lay fewer eggs as food plants become scarce and less succulent. Temperature has probably little effect on total egg production once the threshold for laying is exceeded; *Thrips imaginis* lays no eggs at 8·5°C but the number laid at 12·5°C, 15°C, 20°C and 23°C is similar. Humidity affects the fecundity of *Haplothrips subtilissimus;* females reared in an almost saturated atmosphere each laid on average 129 eggs, but others in a drier atmosphere far fewer (Putman, 1942).

Although temperature has little effect on total egg production the duration of the oviposition period and the rate of oviposition depend on it. The preoviposition period, between the final pupal moult and the start of egg laying, rarely lasts longer than about two weeks, except in species that mate in late summer but delay oviposition until the following spring. The lower the temperature, the later females start to lay, and in South Carolina, U.S.A., where *Frankliniella tritici* breeds throughout the year, the preoviposition period increases from 1–4 days in summer to

TABLE 1

Examples of the approximate mean duration (in days) of stages in the reproductive life of female thrips

Species	Mean temp. °C	Pre- oviposition	Oviposition	Post- oviposition	Reference
Caliothrips fasciatus	—	—	67	—	Russell (1912b)
Frankliniella tritici	—	6	16	3	Watts (1936)
Haplothrips leucanthemi*	25	4	29	—	Loan and Holdaway (1955)
Haplothrips subtilissimus*	"warm"	3	35	2	Putman (1942)
Limothrips cerealium	23	16	31	5	Lewis (1958)
Liothrips vaneekii*	21	14	42	—	Hodson (1935)
Microcephalothrips abdominalis	28	4	—	—	Jagota (1961)
Odontothrips confusus	27	13	—	—	Bournier and Kochbav (1965)
Thrips imaginis	12·5	10	138	—	Andrewartha (1935)
Thrips imaginis	23	3	46	—	Andrewartha (1935)
Thrips tabaci	21	3	50	6	Sakimura (1937a)

A dash means the information is unrecorded. There is much variation in these periods between individuals reared at the same temperature, and the coefficient of variation for these means (S.E./mean × 100) is 30–60%.

* Tubuliferan.

10–35 days in winter (Watts, 1936). Once they have started, females continue to lay until a few days before death, unless it becomes too cold (Tables 1 and 2).

The rate of oviposition is generally similar in both sub-orders and in bisexual, parthenogenetic and even ovoviviparous species. Female *Thrips imaginis* and *Heliothrips haemorrhoidalis* lay steadily throughout their lives although the mean rate of oviposition increases roughly proportionally with increasing temperature (Fig. 9). A logistic curve fits the data

TABLE 2

Examples of the mean egg production and approximate rate of oviposition

Species	Type of Reproduction[†]	Temp. °C	Total eggs	Rate (eggs/day)	Reference
Caliothrips fasciatus	A	—	66	5·5	Russell (1912a)
Caudothrips buffai*	O	—	—	1–4	Bournier (1957a)
Haplothrips leucanthemi*	T	25	31	1·2	Loan and Holdaway (1955)
Haplothrips subtilissimus*	T	"warm"	129	3–5	Putman (1942)
Heliothrips haemorrhoidalis	T	15·5–20	25	0·6	Rivnay (1935)
Heliothrips haemorrhoidalis	T	25·5–28	47	1·4	Rivnay (1935)
Limothrips cerealium	B	23	43	1·4	Lewis (1958)
*Liothrips oleae	A	—	250	4–12	Melis (1935)
Scirtothrips dorsalis	B	25	—	2–3	Dev (1964)
Taeniothrips dianthi	B	23	55	3·0	Pelikan (1951)
Thrips imaginis	T	12·5	192	1·4	Andrewartha (1935)
Thrips imaginis	T	23	252	5·6	Andrewartha (1935)
Thrips tabaci	T	18	80	1·8	Sakimura (1937a)

The total number of eggs laid per female varies greatly between individuals. For example, the coefficient of variation for total eggs laid by *L. cerealium* is ±78%.

* Tubuliferan; † B = normal bisexual, A = arrhenotokous, T = thelotokous, O = ovoviviparous.

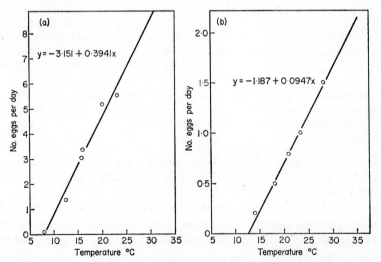

Fig. 9. The relationship between rate of egg production and temperature. Linear regressions fitted to data for (a) *Thrips imaginis* (Andrewartha, 1935) and (b) *Heliothrips haemorrhoidalis* (Rivnay, 1935).

for *T. imaginis* slightly better than a straight line, but this does not apply to the data for *H. haemorrhoidalis*. Extrapolation of the regression lines back to the abcissa on each graph gives the approximate threshold temperature for laying; for *T. imaginis* about 8·5°C and for *H. haemorrhoidalis* about 12·5°C. When *T. imaginis* was kept at four different temperatures the preoviposition period was completed in each instance after about 39 day degrees (number of degrees above 8·5° × number of days) had elapsed (Andrewartha, 1935).

Thrips tabaci similarly lays more eggs at higher temperatures, but unlike *T. imaginis* young females lay faster than old ones. Almost three times as many eggs are laid in the first 10 days of the oviposition period than in the last 10 (Sakimura, 1937a). *Frankliniella tritici* also lays slightly faster in the early part of its oviposition period than later (Watts, 1936).

Hatching

During incubation the shape of the egg changes gradually as the embryo inside develops, and when it is nearly ready to hatch, red or black pigmented eyes can sometimes be seen through the shell (Fig. 10b). Eggs usually hatch after 2 to 20 days, the quicker the higher the temperature.

In Terebrantia, the larva pushes itself about two-thirds of the way out of the egg by gently swaying backwards and forwards. It looks like a minute, translucent worm, with legs closely adpressed to the ventral surface of the body and the antennae bent below the head. At this stage it rests, often poised at an angle to the plant tissue, while the antennae and legs are slowly released. More flexing and twisting movements withdraw the abdomen until only the tapering end is still in the shell. The larva finally frees itself by pressing on the substrate with its feet and swaying backwards (Fig. 10c) (Hinds, 1902; Hall, 1930; Sharga, 1933a; Vité, 1956). The emergence hole left in empty egg shells of Terebrantia is roughly circular; in Tubulifera the broken edges of the hole follow the outline of the pentagonal or hexagonal reticulations on the shell.

The first-stage larva of *T. tabaci* actually emerges from the egg enclosed in a thin sheath, and remains inside this, with legs and mouthparts unable to function until it has wriggled to the leaf surface through the tunnel made by the female's ovipositor when the egg was laid. The sheath then splits along the mid-dorsal line and the active first-stage larva frees itself. In eggs dissected from leaves and hatched in petri dishes at 22°C this prelarval stage lasted 1–2 days, but it probably persists for a shorter time in eggs embedded in plant tissue (Gawaad and El Shazli, 1970). A similar stage may occur more widely among Terebrantia than hitherto recorded.

Hatching in Tubulifera is easier because the eggs are exposed. The larva simply breaks the shell, usually at the anterior, knobbed end of the

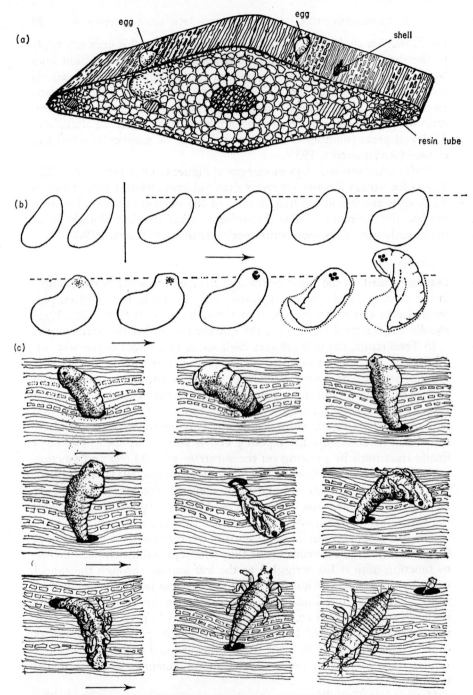

FIG. 10. Hatching in *Taeniothrips laricivorus* (after Vité, 1956). (a) Section of a larch needle showing the position of eggs buried in the plant, and an empty egg shell. (b) Diagram showing the sequence of changes in the appearance of an egg from before laying (top left) to hatching. The straight dotted line represents the surface of the larch needle; the solid vertical line, oviposition. The embryo's developing eyes are visible on the bottom row. (c) Successive stages in hatching.

egg, and wriggles free. It remains near the shell a short time, then slowly examines its surroundings, with repeated flexing of the antennae (Loan and Holdaway, 1955).

Many individuals probably die during hatching. In laboratory experiments, over half the larvae of *Scirtothrips aurantii* attempting to hatch from eggs in picked citrus fruits died before emergence was completed, and partly-emerged dead larvae are common on unpicked fruit. Most of

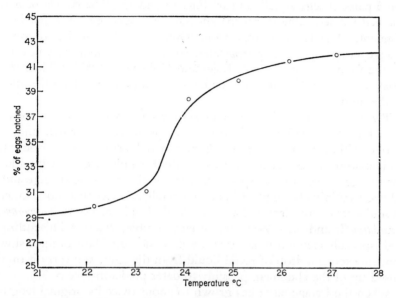

Fig. 11. The relationship between egg mortality and temperature for *Thrips tabaci* (after Ghabn, 1948).

the dead larvae were about two-thirds exposed, but had failed to free their antennae and legs (Hall, 1930). Batches of eggs of *T. tabaci* incubated at high temperatures not only hatched more quickly but more successfully than batches incubated at lower temperatures (Fig. 11) (Ghabn, 1948).

Development of Immature Stages

There are usually 4 or 5, in rare instances only 3, instars between the egg and adult. Customarily, the first two feeding instars are called larvae, and the later, non-feeding ones pupae. There are various objections to this terminology because some aspects of the unique development of thrips resemble the development of typical hemimetabolous insects whose young are called nymphs, more than holometabolous ones whose young are called larvae (Cott, 1956; Bailey, 1957). Nevertheless, these names are

retained here because they are established and understood in the literature
on thrips, and to change them now would be confusing.

Larvae

As in the adult, first-instar larvae have a head, 3 thoracic and 11 abdominal
segments. They have no ocelli, the eyes have only 3 to 4 facets, and the
antennae fewer segments than in the adult. There are no wing buds and
the 3 pairs of legs are all similar (Figs 12 and 13). The cuticle of newly
hatched larvae is almost transparent but soon develops a few patches of
pigment. Most larvae change from white or hyaline at first to yellow,
orange, crimson, or even purple later. Some colour changes may be caused
by the food (see pp. 62, 66) (Loan and Holdaway, 1955; Selhime *et al.*,
1963) and sap in the gut can often be seen as a dark line along the middle
of the body.

The two larval stages ingest all the food necessary for development to
the adult. They start to feed on the host plant soon after hatching, and
as oviposition is spread over a long period and larvae do not wander far,
all immature stages occur together on the host plant, often with adults.
First-stage larvae of many species that are predatory in the II stage and as
adults, feed initially on plant juices. Generally, species that move quickly
as adults have active larvae. Larvae of Aeolothripidae, *Thrips*, *Taeniothrips*,
Frankliniella and *Sericothrips* are usually nimble, whereas Phlaeothripid
and especially Heliothripine larvae are slow-moving. Many of these slower
movers excrete a drop of rectal liquid from the anus, and carry it on the
raised tip of the abdomen, possibly to deter predators (see p. 70).

When the I stage larva has grown to about twice its original length it
seeks a sheltered place on or near a plant, where it can moult undisturbed.
Just before moulting the eyes become detached from the old cornea and
the whole larva becomes a drab colour; yellow forms turn greyish and
red ones a dull brownish-red. The succeeding instar can be seen through
the old cuticle before it is shed. The cuticle splits along the mid-dorsal
line and the emerging II stage larvae are often smaller than fully fed I stage
ones, but during the instar they attain the size of the adult into which
they will eventually change. They differ from the first instar in colour,
cuticular structure, chaetotaxy and the shape of antennal segments (Speyer
and Parr, 1941; Priesner, 1964a), but both instars feed and walk in a
similar way. Larvae cannot jump, even in species whose adults leap readily.

When fully fed, the II stage larva is ready to change into a resting stage.
Most terebrantian larvae move into soil or litter to do this, usually con-
centrating beneath the food plants. For example, about three times as
many banana flower thrips, *Frankliniella parvula*, were found directly
beneath hanging fruit than in soil 2–4 metres from it (Harrison, 1963).
Larvae of many species, e.g. *Taeniothrips inconsequens*, have strong spines

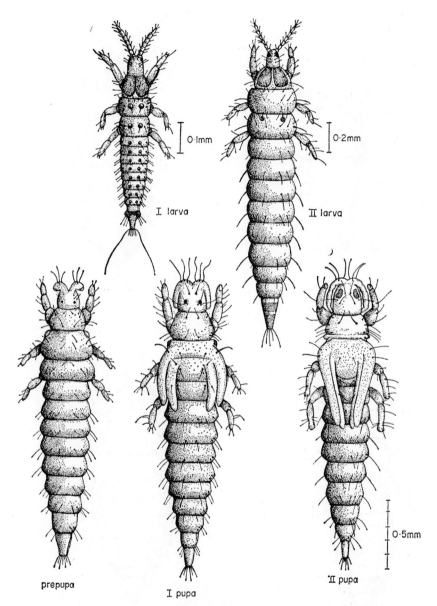

FIG. 12. Immature stages of *Haplothrips leucanthemi*, a typical tubuliferan (after Loan and Holdaway, 1955).

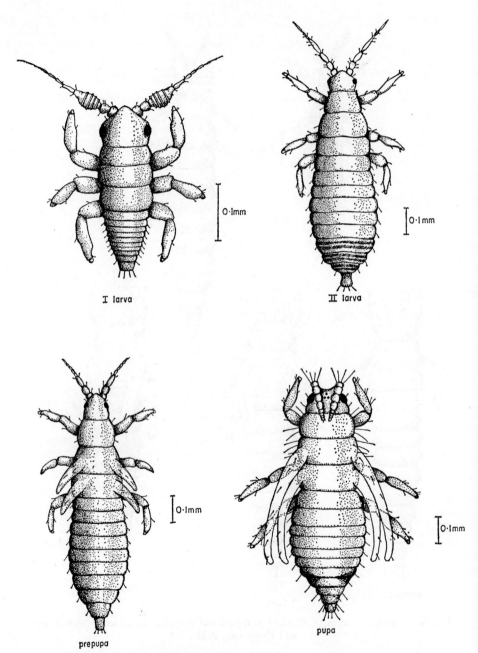

I larva   II larva

prepupa   pupa

FIG. 13. Immature stages of the bean thrips (*Caliothrips fasciatus*), a typical terebrantian. The I stage larva is newly hatched (after Bailey, 1933).

on the 9th and 10th abdominal segments which probably help them to penetrate the soil and mould a simple earthen cell in which to pupate (Bailey, 1934a). Larvae of *Odontothrips loti* make a more elaborate cell in which the soil particles are bound with silken threads. The construction of cells can be observed when larvae are placed on sandy soil in a narrow glass tube (Obrtel, 1963). Each larva levers itself forward through narrow crevices using the pronotum and hind margins of the 8th and 9th sternites; it can also back out of blind passages. When it finds a cavity with a diameter 1·5–2 times its length, the larva widens it by striking the walls violently with its abdomen and pushing the loosened particles to the circumference of the cell. The cell walls are then lined with a sparse web, woven with a thin transparent thread spun from the terminal abdominal glands. The cell takes about a day to construct and after another day's rest the larva moults and changes into the prepupa (Fig. 14a). Larvae of

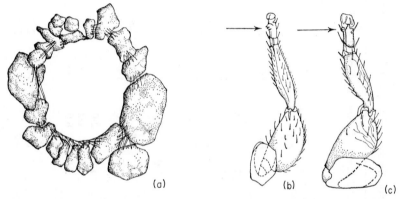

FIG. 14. (a) Diagram of pupal cell of *Odontothrips loti* showing how sand particles are fastened together by silk threads (after Obrtel, 1963). Cocoon-breaking hooks on the right forelegs of (b) *Aeolothrips melaleucus*; (c) *Heterothrips salicis* (after Stannard, 1968).

Aeolothrips and *Franklinothrips* spin a stronger but still loosely-woven cocoon either in soil or litter, or on leaves, using silk produced by one of the "salivary" glands (Priesner, 1964a) although Bailey (1940b) and Putman (1942) claim that larvae of many other species including *Ae. fasciatus*, *Ae. kuwanaii* and *Ae. melaleucus* spin their silk from the anus.

Larvae of *Thrips angusticeps* cannot make cells successfully in sand and require humus in the soil, presumably to help bind the particles together (Franssen and Huisman, 1958).

Many leaf-dwelling Terebrantia and most Tubulifera pupate on the host plant without making a cocoon, usually on the lower surfaces of leaves near the angles formed by protruding veins. Other species enter crevices on plants; the base of leaf stalks, crevices between sepals, leaf sheaths of grasses, or bark scales and hollow twigs, are favoured sites.

TABLE 3

Duration (in days) of stages of the life-cycle

A. AT CONSTANT TEMPERATURES

Species	Temp °C	Egg	I Larva	II Larva	Prepupa	Pupa I	Pupa II	Total	Adult longevity	Reference
Aleurodothrips fasciapennis†	27	10·0	6·0			7·0		23·0	—	Selhime et al. (1963)
Anaphothrips obscurus (macropterous)	24–26	5·1	6·6		1·2	2·4		15·3	—	Koppa (1970)
Caliothrips fasciatus	21	—	5·4	8·2	1·6	9·3		—	15·0	Bailey (1933)
Caliothrips fasciatus	38	—	3·0	2·0	1·0	1·8		—	7·0	Bailey (1933)
Dinurothrips bookeri	28	7·4		7·2	1·2	3·0		18·8	—	Callan (1951)
Frankliniella tenuicornis	23–24	4·2		6·4	1·1	2·0		13·7	—	Koppa (1970)
Haplothrips aculeatus*	24–25	5·4	14·3		1·0	1·0	2·8	24·5	—	Koppa (1970)
Haplothrips leucanthemi*†	25	4·6	6·0	7·0‡	1·5	1·4	3·7	24·2	29·2	Loan and Holdaway (1955)
Limothrips denticornis	25–26	3·6	6·3		1·0	1·8		12·7	—	Koppa (1970)
Microcephalothrips abdominalis	27–28	3·0	2–3	3·0	1·0	1–2		11–13	—	Jagota (1961)
Scirtothrips citri	25	9·0	1·9	2·0	1·0	2·1		16·0	—	Munger (1942)
Scirtothrips citri	31	8·0	1·5	1·2	0·7	1·5		12·9	26–30	Munger (1942)
Taeniothrips dianthi	23	7·0	5·5	8·0	1·0	5·0		26·5	—	Pelikan (1951)
Taeniothrips simplex	15	12·8	18·6			11·7		43·1	—	Herr (1934)
Taeniothrips simplex	30	2·9	3·9			3·5		10·3	—	Herr (1934)
Thrips tabaci	25	6·0	6·1		1·2	2·8		16·1	—	Harris et al. (1936)
Thrips tabaci	30	4·0	4·2		1·0	2·0		11·2	19·9	Harris et al. (1936)

B. AT FLUCTUATING TEMPERATURES

Species	Temperature									Reference
Frankliniella fusca	Summer (S. Carolina)	6·7	2·8	3·5	1·1	2·6	—	15·9	18·5	Watts (1936)
Frankliniella tritici	Summer (S. Carolina)	3·3	2·0	2·4	1·1	2·4	—	11·0	15·0	Watts (1936)
*Bagnalliella yuccae**	23–28	5–8	11–12	6–7	1–2	2–3	5–7	30–36	—	Derbeneva (1959)
Limothrips cerealium	Summer (U.K.)	10–13	5–7	8–10	2–3	6–7	—	30–35	—	Sharga (1933a)
Limothrips denticornis	Summer (Finland)	7·7	10·3		1·0	3·9	—	22·9	—	Koppa (1970)
Selenothrips rubrocinctus	Mean 20–21	8–16	8–16		1·4	4–7	—	28–43	—	Russell (1912b)
Taeniothrips laricivorus	Summer (Germany)	8–14	12–16		8–10		—	28–42	—	Vité (1956)
Thrips tabaci	Mean 30·8	4·8	5·9		1·4	2·4	—	13·9	20·2	Lall and Singh (1968)
Trichinothrips breviceps†	21–33	6–8	3–4	3–6	1·0	1·0	2–3	16–23	♀12–21 ♂3·6	Seshadri (1953)

* Tubuliferan; † Predator; ‡ There is usually a diapause of 8½–11 months at this stage.

Pupae

At the second moult a prepupae emerges from the cast cuticle; this is an intermediate stage between the larva and the true pupa. Wing buds are visible in terebrantian prepupae but not in tubuliferan prepupae; the rudimentary antennae appear as short sheaths with indistinct segmentation. Prepupae do not feed or excrete, and their rate of respiration is retarded (Priesner, 1964a), but those which do not occupy a cocoon can walk slowly if disturbed. Thrips in the genus *Franklinothrips* are exceptional because they have no prepupal stage.

The prepupa is followed by a single pupal stage in Terebrantia, and two stages in Tubulifera each separated by a moult. The pupal antennae of Terebrantia are longer than in the prepupa and turned back over the prothorax; wing sheaths are long in macropterus individuals and the body and appendages assume adult proportions. In Tubulifera the differences between I and II pupae are small; in both, the antennae lie at the side of the head and the wing sheaths are visible, but slightly larger in the II pupa (Figs 12 and 13) (Müller, 1927; Morison, 1947; Priesner, 1964a; Stannard, 1968).

The pupal stages of *Parthenothrips dracaenae* are less sensitive to light than larvae or adults (Müller, 1927) and pupae of *Haplothrips leucanthemi* are repelled by light in contrast to larvae and adults which are attracted to it (Loan and Holdaway, 1955).

At the final moult the adult emerges from the pupa. In species that spin cocoons, the adults have fore-tarsal hooks which probably help them to break the cocoon and escape (Figs 14b, c). Some species of *Odontothrips* and *Taeniothrips* also have claws on the fore-tibiae and fore-tarsi respectively which may be used to break the walls of their earthen pupal cells.

Speed of growth

There have been few studies on the duration of development of a single species over a wide range of temperatures, but many on the duration of the different stages and life-cycles of a number of species at one or two temperatures (Table 3). The shortest time required to complete the life-cycle in any species is about 10 days, some species require nearly a year and a few with prolonged adult diapause, about 20–22 months.

The rate of development of insects at different constant temperatures does not increase proportionately with rising temperatures throughout the range suitable for development, but is faster at temperatures in the median part of the range and slower at the cooler and warmer extremes. The relationship is best described by a logistic curve, with temperature plotted on the abcissa and the reciprocal of the duration of each stage on the ordinate. Commonly the reciprocal is multiplied by 100 so that it becomes per cent development per unit time (Davidson, 1944; Andrewartha and Birch, 1954). Few measurements have been made on thrips with the necessary precision, and at enough different temperatures, to illustrate the dependence of rate of development on temperature, but enough data have been collected for *Heliothrips haemorrhoidalis* (Rivnay, 1935) to show this relationship, for eggs, larvae and pupae, and for the complete life-cycle (egg to pre-oviposition of the adult) (Fig. 15). The right-hand ordinate $(100/y)$ represents the average per cent development per day and the logistic curves, which follow the observed data fairly closely, were fitted by Davidson's (1944) graphical method.

For each stage there is a peak temperature at which development is quickest, but above it the excessive heat retards development. The data for pupae do not show this effect so clearly, perhaps partly because their greater resistance to higher temperatures could help them to survive in the hot, dry soil. However, the discrepancy may also have arisen because the data are less precise than for the other stages; the pupal stage lasts for only about 3 days, and its duration was recorded approximately to the nearest half day, so small changes in rate of development near the peak temperature would probably have been overlooked. The data collected by Bailey (1933) for *Caliothrips fasciatus* and by Herr (1934) for *Taeniothrips simplex* follow a similar curve, but the logistic cannot be fitted satisfactorily because thrips were reared at too few different temperatures.

In nature the relationship between temperature and rate of development is less well defined because temperature and other components of weather fluctuate. Harmful high and low temperatures may occur (see pp. 176, 183) and impair subsequent development at favourable temperatures, or diapause may intervene. However, for species which do not diapause the rate of development at temperatures which fluctuate daily within a suitable range is probably similar to the rate at an equivalent constant temperature in this range. For example, *Thrips tabaci* develops from egg to adult in 11·2 days at a constant temperature of 30°C (Harris *et al.*, 1936), and in 13·9 days at a fluctuating temperature with a mean of 30·8°C (Lall and Singh, 1968). Adults lived an average 19·9 and 20·2 days respectively in these two temperature regimes (see Table 3).

There is evidence that male larvae of some species develop slightly more rapidly than females (Bournier 1956a). The average duration of the larval stage of six male *Limothrips denticornis* reared at 25·1–25·7°C was 5·7 ± 0·57 days, compared with 6·8 ± 0·66 days for females (Koppa, 1970).

Methods of relating speed of development of insects in the field to temperature are described by Andrewartha and Birch (1954) and Taylor (1957) for other orders of insects, but there is no data known for thrips suitable for analysis. Ewald and Burst (1959) claimed that *Taeniothrips laricivorus* required 11,210 ± 230 day degrees (C) to complete develop-

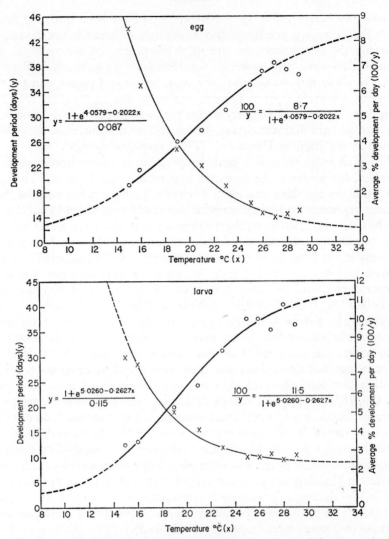

Fig. 15. The speed of development for three different stages and for the entire life history (egg-preoviposition) of *Heliothrips haemorrhoidalis* calculated from Rivnay's (1935) data. The days required to complete each stage (descending thin line), and the

ment in forests, and that its threshold temperature for development was 8°C.

Annual Cycles

Through its effect on rate of development, temperature largely determines the number of generations produced each year and each season. When the

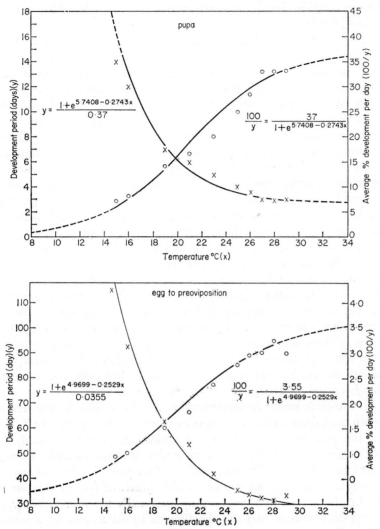

reciprocal of this time (ascending thick line), are plotted against temperature. The equation is given for each line.

life-cycle is uncomplicated by diapause, unfavourable weather, or shortage of food, breeding continues throughout the year, though the life-cycle takes longer in cool seasons. In South Carolina *Frankliniella tritici* on cotton has 12–15 generations a year, 10–11 developing in the six warmer months (April to September) and 4–5 in the cool season (October to March) (Watts, 1936). On citrus in Palestine, *Heliothrips haemorrhoidalis* has up to 7 overlapping generations a year (Rivnay, 1935), but in the warmer

c

and more equable climate in glasshouses it may complete 12 generations (Morison, 1957).

Out of doors in cool temperate regions most species have 1 or 2 generations per year and many diapause or are quiescent for a period during the life-cycle, either as fully grown II stage larvae or as adults. Diapausing larvae usually rest in soil and often remain there for 6–8 months. When there is only one generation a year, as with *Kakothrips robustus* in England, all the fully fed larvae usually enter the soil in mid-summer, pupate early in the following spring, and emerge as adults in late spring; nine to ten months of the year therefore, are often spent below ground (Williams, 1915). In other species such as *Haplothrips leucanthemi* in Ontario, eggs that are laid very early in the season complete their development and these young adults produce a second generation. Later larvae produced by the overwintered generation, and all second-generation larvae, diapause during the winter (Loan and Holdaway, 1955).

In Holland, *Thrips angusticeps* usually has 2 generations a year, but the life-cycle may be complicated by alternating generations of long-winged and short-winged adults. Brachypterous adults usually spend one winter in the soil and from the eggs they lay in spring a macropterous generation develops, whose eggs develop in late summer and autumn to produce the brachypterous adults which overwinter in the ground. A small proportion of these brachypterous females may remain in the ground for a second winter (Franssen and Huisman, 1958). In Finland, however, where summers are cooler and shorter, this species has only one generation which is entirely brachypterous (Hukkinen and Syrjanen, 1921).

This life-cycle might be interpreted as an illustration of how temperature can contribute to the wing-development of some species. In *T. angusticeps* the generations developing at cooler temperatures usually have shorter wings. *Haplothrips cottei* provides another example from the Tubulifera (Ghabn, in Priesner, 1964a). This possible effect of temperature may explain why macropterous individuals of species that are normally brachypterous or apterous are occasionally found in summer, but, as wing development may also be affected by daylength, light intensity and crowding, all of which increase in summer, and by the quality of food which changes as plants age, it is unlikely that temperature alone is responsible for wing polymorphism in all species. Indeed, studies on the parthenogenetic *Anaphothrips obscurus* in Finland (Koppa, 1970) and in Oregon, U.S.A. (Kamm, 1972), show that photoperiod is far more important than temperature, and seasonal differences in wing-length in other species, hitherto attributed to temperature, may instead have been caused partly by differing photoperiods. Both macropterous and brachypterous individuals of *A. obscurus* produce long-winged and short-winged progeny, and Koppa (1970) claimed that the ratio between the forms depended largely on photoperiod and nutrition, perhaps on temperature, but not on crowding. In an experimental lighting regime with 8 h light, 16 h dark, macropterous females produced only brachypterous offspring, but both forms appeared in longer natural (mid-summer) daylength. The wing-length depended on the photoperiod experienced by the eggs immediately before hatching or by the very young larvae. The ratio of the forms produced under natural day length also depended on the food available. Macropterous females reared on oats produced a larger proportion of macropterous offspring (66·9–81·3%) than females reared on timothy (11·9–54·2%). Samples taken from cereal crops over a period of four years contained fewer than 4·7% brachypterae, compared with 38·3% in samples taken from grass leys. Kamm (1972) similarly found that adults collected from grass in spring and exposed to a long photoperiod (16 h light, 8 h dark) produced a greater proportion of macropterous offspring than adults exposed to a short photoperiod (10 h, 14 h), and that temperature had no appreciable effect on the proportion of each form. But, in contrast to Koppa, he found that the ratio of macropterous to brachypterous forms *did* increase as a result of crowding during development and that the quality of the food had no effect. Perhaps the thrips in Koppa's experiments (max. density of 20 per grass plant) were simply not crowded enough to have any effect, whereas Kamm's (max. density of 45 per plant) were.

Some bark-dwelling phealothripids that feed on fungus or on the extra-cellular products of fungal decay, e.g. *Hoplothrips flumenellus* produce only brachypterae when the fungus is fresh and growing vigorously, but macropterae as it ages and deteriorates (Hood, 1940). Bournier (1961) noted similar differences in wing-development for species of *Megathrips* and *Cephalothrips* feeding on young and older fungi.

Whatever the cause, the development of complete wings in summer probably helps survival by enabling some individuals to migrate to fresh hosts when food plants dry, a phenomenon reminiscent of some aphids.

3 Interrelationships with host plants

Feeding Behaviour

Most thrips feed on either vascular plants, fungi or mosses; some are predatory on small arthropods and a few are omnivorous. Most Terebrantia suck juices from leaves, flowers, fruits and young shoots, and many flower-dwelling species swallow pollen grains or suck their contents. The Tubulifera usually suck juices from leaves but many feed on fungal hyphae or spores and some on bulbs, mosses and possibly algae. No root-feeders are known, although a few species, e.g. *Neurothrips indicus* and *Azaleothrips amabilis*, occur on the aerial roots of *Ficus* spp. and feed on the fungi growing there (Plate I).

Generally, species of Terebrantia have a wider host range than Tubulifera. For example, in Britain 47 species of Terebrantia have 1–3 species of larval food plant, 26 species 4–10 and 16 species 11 or more, whereas most species of Tubulifera are confined to a few species of fungi or flowering plants.

Structure of the mouthparts

The appearance and structure of the mouthparts differ between families (Borden, 1915; Peterson, 1915; Reyne, 1927; Ananthakrishnan, 1951; Jones, 1954; Risler, 1957; Davies, 1958; Mickoleit, 1963) but generally their action is similar when feeding. The mouthparts form a broad proboscis or cone attached to the underside of the head so that much of the cone lies beneath the first thoracic segment when the insect is resting (see Fig. 3). The apex of the cone is usually directed downwards but in some specialized groups backwards, and the tip may be blunt, as in most Thripidae, or taper to a sharp point as in many Phlaeothripidae. The mouthparts of all thrips are asymmetrical. The front of the cone is formed from the clypeus and labrum, separated by a membrane; the sides are formed by the galeal part of the maxillae, and the rear wall by the labium bearing a pair of palps at its tip. The labium often projects beyond the labrum and maxillae as a flexible flap. Small hooks on the

tips of the labium probably help the insect to grip the surface of the substrate when feeding (Figs. 16a, b).

Within this cone are sheathed the flexible, piercing organs, consisting

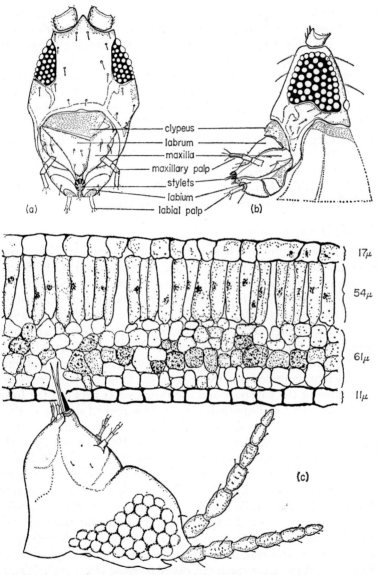

clypeus
labrum
maxilla
maxillary palp
stylets
labium
labial palp

(a)   (b)

17μ

54μ

61μ

11μ

(c)

FIG. 16. Head and mouthparts of *Chirothrips hamatus*. (a) Front view; (b) side view
(after Jones, 1954); (c) diagram of the head of *Thrips tabaci* feeding on the underside
of a cotton leaf, to show the depth to which the mandible and maxillae are inserted
(after Wardle and Simpson, 1927).

of the slender, paired maxillary stylets and the stouter, more strongly sclerotized left mandibular stylet. In some species the hypopharynx may serve as a fourth stylet (Borden, 1915; Wardle and Simpson, 1927). The mandible is hollow except for the solid pointed tip. Usually, the apical part of the maxillary stylets is extremely slender but in the Megathripinae the tips of these stylets are slightly broadened and asymmetrical, and fit together at the tip with a complicated array of overlapping projections to form a sub-apical opening (Plate II). The margins of these stylets also fit into each other lengthwise, by a tongue-and-groove system on each side, to form a long tube. Similar complex stylet apices are visible in some Aeolothripidae, and a tongue-and-groove structure in *Haplothrips statices* and *Thrips tabaci* (Mound, 1971b). In most Phlaeothripidae the maxillary stylets are longer and more deeply retracted into the head. An extreme example of this occurs in a group of thrips living on *Casuarina* trees in Australia; they have stylets so long that they are actually coiled in the head and are half as long as the body when extended (Mound, 1970b).

The larval mouthparts resemble those of adults.

Action of the mouthparts

The method of feeding of Thysanoptera is unique and still incompletely understood. On vascular plants, feeding thrips are variously described as puncturing tissues and draining the contents of cells causing their walls to collapse (Horton, 1918, for *Scirtothrips citri*), piercing the epidermis and rasping the leaf tissues within (Russell, 1912b, for *Selenothrips rubrocinctus*), or rasping leaf tissues and sucking the sap as its exudes (Cameron and Treherne, 1918, for *Taeniothrips inconsequens*; Bedford, 1921, for *Caliothrips indicus*). On green plants shallow and penetrating feeding are distinguishable. In the more common penetrating type the mouthcone is applied firmly to the surface of the substrate and the thrips sinks between the forelegs with the abdomen tilted upwards. The head is rocked slightly to shorten the flexible mouthcone and protrude the mandible which cannot be exserted by direct muscular action (Priesner, 1964a). Each upward and downward swing of the head alternately protrudes and withdraws the mandible, so that it gashes or tears a hole in the epidermal cells rather than pierces them. The fully protruded mandible probably rarely reaches the lower walls of the cells it enters and the length of each incision is about 50 μ depending on the species of thrips and the flexibility of the labium. After the outer surface of the leaf is broken, the longer maxillae are used as levers to break down the lateral walls of a number of mesophyll cells but they rarely reach the vascular tissue (Fig. 16c). When the walls are destroyed the head movements cease and the thrips sucks the juice that oozes from the ruptured cells by placing the tip of the mouthcone over the opening created (Wardle and Simpson, 1972) and perhaps extending the linked, tubular maxillary stylets into it (Mound, 1971b). In the shallow type of feeding the mouthparts hardly penetrate the leaves at all and only the epidermal cells and a few mesophyll cells are ruptured. The species with very long stylets living on *Casuarina* feed on the soft tissue at the bottom of deep furrows on the stem, and presumably can only extract food from these distant parts by sucking it through a tube formed from the maxillary stylets (Mound, 1970b).

Juices and chloroplasts are ingested by the pumping action of the muscles of the head and labrum, and the small circular muscles surrounding the two ends of the pharynx (Figs 17a, b). First the muscles contract to close the oesophagus and the pharyngeal pump dilates, sucking juices into the pharynx; then the mouth-opening closes, the oesophagus opens, and the large muscles relax to collapse the pharynx and squeeze the food down the oesophagus. The flow of saliva is controlled by a similar, smaller pump on the common salivary duct (Bonnemaison and Bournier, 1964). In experiments, adult female *Thrips tabaci* weighing about $2 \cdot 8 \times 10^{-2}$ mg and feeding intermittently on the leaves of chinese cabbage (*Brassica chinensis*) ingested about 17% of their weight of leaf tissue per hour at an average rate of $1 \cdot 0 \times 10^{-4}$ mg/min; larvae fed at about half this rate. Thrips fed through a plastic membrane on 5% sucrose solution ingested the liquid more quickly (Day and Irzykiewicz, 1954). Radio-labelling of adult *T. tabaci* has shown that they produce a glandular secretion, probably the saliva, that lubricates the mouthparts and is also released into the plant where it may partly predigest the food (Kloft and Ehrhardt, 1959). Klee (1958) claims that individual *Taeniothrips laricivorus* feeding on pine needles destroy about 30,000 cells in 8 h.

Many flower-dwelling Terebrantia can ingest pollen (Doeksen, 1941; Loan and Holdaway, 1955; Grinfel'd, 1959; Putman, 1965b) and it is essential for ovarian development in *Thrips imaginis* (Andrewartha, 1945). Grinfel'd lists 9 species in the genera *Aeolothrips, Odontothrips, Oxythrips, Taeniothrips, Thrips* and *Haplothrips* whose larvae and adults suck the liquid contents from pollen grains of many flowers including *Convolvulus, Anthemis, Sonchus* and *Centaurea*. These thrips are attracted to flowers with ripe anthers. They feed by scraping pollen grains towards the mouthcone with the fore-tarsi or tibiae and holding them, one at a time, with the maxillary and labial palps while the mandible tears the wall of the grain open (Fig. 17c). The contents are removed either by pressing the grain between the mouthcone and a solid surface to squeeze the contents through the puncture, or by sucking out the liquid, possibly with the maxillae functioning jointly as a tube. This takes 2–20 sec per grain and thrips usually empty 20–30 grains for a meal. Grinfel'd suggests that the asymmetry and asynchronous action of thysanopterous mouthparts have evolved as an adaptation to this specialized food. Pollen grains are so small, mostly 20–40 μ, that the use of more than a single stylet would squash, rather than puncture, the tough wall. Thus, to the ancestors of modern Thysanoptera trying to feed on pollen grains, two mandibles

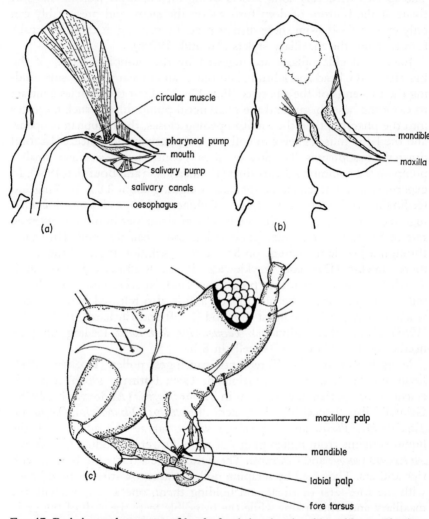

FIG. 17. Basic internal structure of head of a thrips showing (a) sucking mechanism; (b) stylets. Striations represent muscles (simplified after Peterson, 1915, and Reyne, 1927). (c) One method of feeding on a pollen grain (original).

were an evolutionary disadvantage, and atrophy of one has enabled thrips to exploit this rich source of food. The transference to feeding on plant juices is probably a later development, and fungus-feeding an even more modern adaptation (Stannard, 1957).

Some species also drink nectar from flowers (Hagerup, 1950) and, like some ants, one species of *Taeniothrips* induces mature larvae and pupae of lycaenid butterflies to exude, from their abdominal glands, secretions which the thrips drink (Downey, 1965).

Tubulifera are generally larger than Terebrantia, and many can ingest solid particles up to 10–30 μ diameter (Reyne, 1927). Sometimes they feed on fungal spores which they swallow complete and uncrushed. The gut of *Leptogastrothrips* sp. may contain many cross-septate four-celled spores of the Phaeophragmiae type, each about 6 × 20 μ (Stannard, 1968). Some Phlaeothripidae probably also feed on algae. All spore and algal feeders use the maxillary stylets to collect the solid particles, and the slightly asymmetrical and broadened stylets of the Megathripinae are especially adapted for brushing and scraping spores towards the mouth-opening.

Larvae and adults of carnivorous thrips pierce their prey and remove the contents in a similar way to sap feeders (see p. 65). Indeed, adults and larvae of some species of *Aeolothrips*, *Haplothrips* and *Scolothrips*, though usually predators, can feed on plant juices and pollen grains as well as on arthropods; *Scolothrips sexmaculatus* can survive on plant food alone for 3 days (Bailey, 1939).

FEEDING DAMAGE TO PLANTS

The surfaces of tissues that thrips have penetrated deeply develop a silvery sheen, due largely to air occupying the emptied cell cavities. The silvering is emphasized by the lens-like effect of the epidermis above the shattered cells. The damaged mesophyll cells often turn greenish-brown or yellow. Where many thrips have fed, the discoloured areas coalesce and the whole leaf may dry and fall prematurely. This silvering and yellowing of leaves is often termed "white blight", "white spot" or "yellow spot". Buds and leaves of some plants turn red when thrips have fed (Wood, 1956); leaves of *Ficus retusa* infested by *Gynaikothrips ficorum*, and the bulb-shaped leaf galls on *Acacia* produced by *Acaciothrips ebneri*, develop a reddish hue especially if exposed to sunlight (Priesner, 1964a). On flowers, the petals (Haq, 1961), stamens (Bournier and Koch-bav, 1963; Ananthakrishnan, 1955), or pistil may be attacked separately, or often the whole flower withers (Plate IIIa). Infested inflorescences of grasses and cereals wither and turn silvery-white producing "whiteheads" (Fig. 18) or "silvertop" (Plate IIIb) (Sharga, 1933; Rubtzov, 1935; Huk-kinen, 1936; Kamm, 1971). Stiffer organs like pine needles, stems and fruits, become distorted and scarred, and young shoots may die. Shallow feeding is usually confined to a small area and does not leave silver scars.

Figure 19 shows the histological appearance of degenerating cotton leaves after thrips have fed on their lower surfaces. In (a) the upper

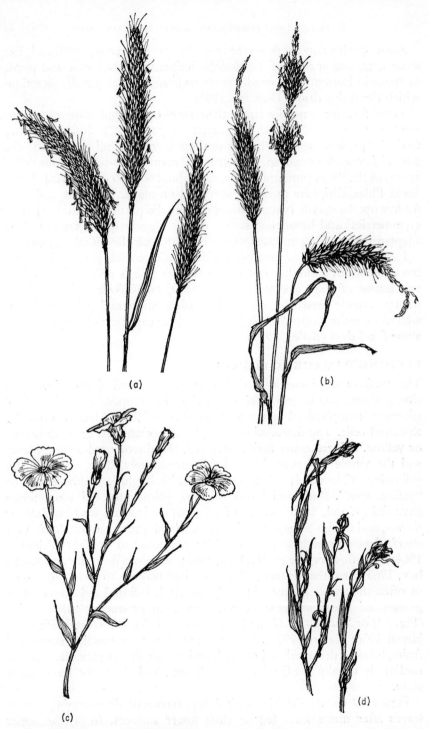

FIG. 18. Examples of feeding damage by thrips: (a) Healthy inflorescence of *Alopecurus pratensis*; (b) "whiteheads" caused by *Chirothrips hamatus* (original); (c) healthy flax flowers (*Linum usitatissimum*); and (d) withered flowers and leaves caused by *Thrips linarius* (original).

epidermis is intact; there are shrinkages in the lower epidermis beneath which some of the outer cells in the mesophyll layer are shattered and empty. In (b) the mesophyll is more distorted and by stage (c) the whole section is shrivelled, though the upper epidermis and most palisade cells remain intact; (d) shows the final stages of degeneration where only outlines of palisade cells remain and the epidermis is broken down. Such necrosis results almost entirely from mechanical damage (Wardle and

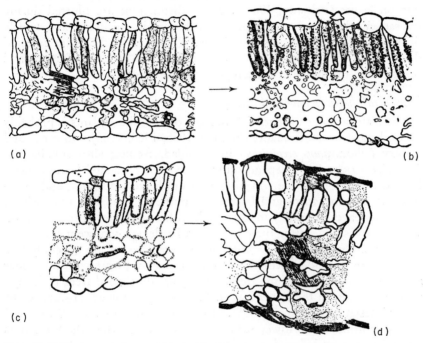

(a)

(b)

(c)

(d)

FIG. 19. T.S. of cotton leaves damaged by *Thrips tabaci* showing progressive histological degeneration (after Wardle and Simpson, 1927).

Simpson, 1927) but the contents of cells may be partly predigested by saliva (Kloft and Ehrhardt, 1959).

Faeces are deposited on or near the feeding site as a small speck of coloured material, surrounded by a larger volume of colourless, watery fluid. Brown fungi often grow on it as it dries and when thrips are abundant this produces an ugly discoloration on leaves and fruits (Plate IIIc). The larvae of *Heliothrips haemorrhoidalis*, *Selenothrips rubrocinctus* and closely related species often retain a large globule of faeces on the end of the abdomen between the anal hairs before depositing it on the plant (Fig. 27b). The hindgut of spore-feeding Megathripinae often contains a black bolus of undigested spore walls.

Feeding and Distribution on Individual Plants

Thrips feed on most parts of plants except the roots, but they usually concentrate on rapidly growing tissues such as young leaves, flowers and terminal buds. A few species spread over the whole plant, e.g. on Liliaceae, *Thrips tabaci* infests the bulbs, leaves and flowers, but most species concentrate on parts of the host. For example *Haplothrips reuteri, ochradeni* and *limoniastri* occur only on inflorescences, and most species of *Liothrips* live on leaves (Priesner, 1964a). These distinctions are blurred during migration when many individuals are blown in the wind and widely distributed over hosts and non-hosts. The part of the plant infested depends on the quality of the food and the protection afforded, and occasionally on interspecific competition (see p. 46).

Superficial-dwellers and food quality

The distribution of superficial-dwellers on their host plants is probably affected most by food quality (Fennah, 1955, 1963, 1965). On cacao and cashew (*Anacardium occidentale*) in Trinidad, feeding sites suitable for *Selenothrips rubrocinctus* depend on complex environmental factors affecting the physiological age and metabolism of leaves and pods, and therefore the amount, form and balance of nutrients available. On cashew, immature leaves are rarely infested and the heaviest infestations occur on mature leaves soon after their tissue has hardened; senescent leaves are unsuitable. On cacao also, young and senescent leaves are generally free of thrips; in unshaded plantations infestations are greatest on the upper and outermost leaves of the trees, but on lightly shaded trees thrips are most abundant on the lower leaves. Mature pods only are susceptible to heavy infestations (Fennah, 1955). In natural conditions cacao thrips feed on the morphological lower surface of leaves, even when the leaf hangs vertically, and on infested leaves a number of different but characteristic patterns of distribution develop (Figs 20, 21). On maturing leaves of cacao and cashew the commonest pattern is formed as thrips accumulate along the sides of the basal part of the midrib (Fig. 20a); pattern b develops from this. On physiologically old leaves thrips are distributed in small groups on the lamina, between the distal areas of the main veins (Fig. 20c), and on leaves with necrotic or damaged areas, larvae and adults usually occupy a zone up to 10 mm. wide bordering the affected tissue (Fig. 20d) or feed on surfaces above or below the injured area (Fig. 20e).

The susceptibilities of different trees to infestation and the different distribution patterns on leaves are induced by environmental stress on the plants, produced for example by drought, inadequate nutrition, flooding or excessively alkaline soil. Such conditions retard protein synthesis and increase the proportion of nitrogenous compounds in the

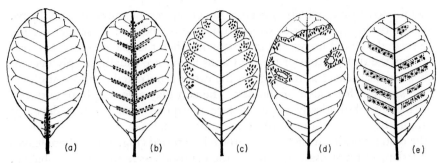

FIG. 20. Diagram of five distribution patterns formed by cacao thrips when feeding on cashew leaf. The insects are represented by the short-heavy lines. (a) Basal primary subvascular; (b) distal primary subvascular; (c) submarginal intervenal; (d) peritraumatic; (e) epitraumatic (after Fennah, 1963).

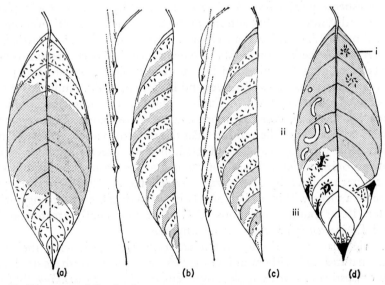

FIG. 21. Diagrams of distribution patterns adopted by cacao thrips feeding on a cacao leaf. The stippled areas represent normal green leaf tissue, light areas pale green or yellowish tissue and black areas dead tissue. Insects are represented by short, heavy lines. (a) Epitraumatic distribution on the isolated areas of a leaf that has been obliquely shaded; (b) epitraumatic distribution on a vertical cacao leaf of which the upper surface has regularly been raked by sunlight, as shown in profile on left; (c) epitraumatic distribution on a vertical cacao leaf of which the lower surface has regularly been raked by sunlight, as shown in profile on left; (d), (i) epitraumatic distribution of cacao thrips on areas damaged in the bud during the previous dry season, (d), (ii) lesions made by beetles in the early part of the wet season, (d), (iii) peritraumatic distribution of cacao thrips near the margins of dead tissue (after Fennah, 1965).

debilitated tissue, which is then a richer source of nutrients for thrips. When the trees are under stress from one or more of these factors, the insolated parts of leaves age more quickly than the shaded parts, and this affects the distribution of thrips. Cacao leaves are usually steeply decumbent so that leaves at the edge of the canopy are raked at a low angle by the sunlight for much of the day. The sun may shine on both surfaces; if the morphologically upper surface is exposed, water stress develops along the basal margins of the convex intervenal area (Fig. 21b) and along the distal margins if the lower surface is exposed (Fig. 21c); where leaves overlap, water stress is greater in the exposed tissue than in the shaded. Thus tissue exposed to sunlight, or around injured cells, is predisposed to infestation by cacao thrips which feed and breed where food is richest, thus producing these characteristic distribution patterns.

Generally, therefore, any environmental stress that weakens plants makes them more susceptible to infestation by thrips. In Puerto Rico, wilting, or tall, spindly *Cinchona* plants supported heavier infestations of thrips, especially *Scirtothrips longipennis*, than healthy or short, stocky plants (Plank and Winters, 1949), and in Queensland, Australia, unthrifty onions carried more *Thrips tabaci* than vigorous ones (Passlow, 1957). *Heliothrips haemorrhoidalis* feeding on plants of New Zealand spinach prefers stunted plants containing less nitrogen but more vitamin C, oxalate and phosphorus than normal ones (Wittwer and Haseman, 1945). Similarly, of two varieties of tobacco grown in Turkey, Malatya, the more susceptible to attack by *T. tabaci*, contains more oxalate, phosphate and water in its leaves and has less acidic cell sap than the resistant variety, Bursa (Tutel, 1963). Slow-growing varieties of gladioli have sweeter sap than rapid-growing ones, which perhaps contributes to the greater susceptibility of the former to attack by *Taeniothrips simplex* (Bailey, 1941). Even the presence of a virus within a plant affects its attractiveness to feeding thrips; Carter (1939) found populations of *T. tabaci* consistently greater on diseased *Emilia sonchifolia* than on healthy plants.

Species that can feed only on tender, young tissue, die when their host plant matures and withers unless they can migrate to fresh hosts, hibernate or aestivate. For example, adult female *Limothrips cerealium* fly from ripening wheat when the water content of the ears falls below about 45% but the males and larvae cannot fly, so die in the ears (Lewis, 1959a).

Interstitial-dwellers and protection

Many species seek protection in narrow crevices on their hosts and remain hidden for most of the time. Flowers and grasses with dense inflorescences protect these interstitial-dwellers, particularly the larvae, from exposure and desiccation as well as providing food. The dense inflorescence of Compositae are favoured by many species and a single head of field thistle (*Cirsium arvense*) or yarrow (*Achillea millefolium*) may contain over 100 thrips. Umbelliferae too are infested by many species of *Thrips* and *Taeniothrips*. Most species of *Haplothrips*, a genus that contains most of the flower-feeding Phlaeothripidae, prefer Compositae, grasses and sedges to leguminous flowers like trefoil and woundwort (Priesner, 1964a). The wide range of plant families inhabited by this genus suggests that the quality of food is less important to these thrips than the protection provided by the dense inflorescences. A similar habitat preference is shown by *Frankliniella intonsa* which is more common on the dense heads of white and red clover (*Trifolium repens* and *T. pratense*) than on the sparse inflorescences of the closely related sweet clovers (*Melilotus officinalis* and *M. alba*) that offer less protection (Shull, 1914a). The Graminae provide an abundance of suitable crevices for interstitial-dwellers in inflorescences and leaf sheaths and partly for this reason are hosts to more species of thrips than any other single family of plants.

There is little wandering from the favoured site on a plant. For example, on Compositae, Ward (1966) found only three individuals of *Thrips physapus*, *validus* and *angusticeps* on leaves, though thousands were in the flowers; conversely *T. nigropilosus* rarely entered flowers though abundant on leaves. Thrips are usually only distributed all over a plant when small flowers and leaves are interspersed along the stem.

Warm sunny weather may stimulate adults to leave the crevices. Often this is a prelude to take-off, but sometimes they bask on exposed surfaces, then return to crevices on the same plant. Species of *Limothrips* and *Chirothrips* often crawl over cereal and grass inflorescences in sunshine, and species of *Haplothrips* that usually hide inside the flowers of *Anthemis* and *Chrysanthemum* collect on the outside of the green sepals on sunny spring days (Knechtel, 1956).

On green plants, most restricted spaces between, for example, leaves and stems, or overlying petals, are moist enough for thrips to live in, and it is the width of the crevices that largely determines occupation. Suitable sites are chosen mainly by touch, and responses to light and gravity are less important. Many species are strongly thigmotactic and when crevices of different sizes are available each species chooses one of a characteristic width. In artificial crevices, made from sheets of glass (see p. 95), most adult female *Limothrips cerealium* chose crevices 200–500 μ wide (Lewis and Navas, 1962; de Mallmann, 1959, 1964) and preferred those with walls opposed at a shallow angle of between 0°53′ and 1°42′ to ones with walls diverging more steeply. *Chirothrips manicatus*, a smaller species, preferred narrower crevices (see Table 4). Males of the species listed in Table 4 were less strongly attracted to crevices of particular widths, than females, perhaps because they are slow-moving and thus less likely to enter a preferred zone during short-term experiments. Except for *Thrips tabaci*, few individuals of either sex wedged themselves tightly between upper and lower walls, but after exploring the crevice with their antennae, settled where the distance between the walls was greater than their own depth. Individuals with antennae removed occupied narrower crevices than normal insects. By contrast, *T. tabaci* were less fastidious and some individuals pushed themselves between the crevice walls to obtain maximum dorso-ventral contact, while others settled in wider parts.

Most individuals also sought lateral contact with the crevice walls. In

TABLE 4

Response to dorso-ventral contact in artificial crevices with walls opposed at angles between 0°38′ and 1°42′ (after de Mallmann, 1964)

Species		Optimum crevice width (μ)	Intensity of response
Limothrips cerealium	♀♀	200–500	strong
Limothrips cerealium	♂♂	150–600	weak
Chirothrips manicatus	♀♀	150–300	strong
Chirothrips manicatus	♂♂	50–450	weak
Aeolothrips fasciatus	♀♀	400–600	strong
Aeolothrips fasciatus	♂♂	150–600	weak
Aptinothrips rufus	♀♀	200–450	moderate
Thrips tabaci		—	very weak

artificial chambers (see p. 96, Fig. 38) with lateral walls at an angle of 5° and 20°, 60% to 75% of individuals settled with one or both sides in contact with the wall of a crevice (Table 5). Given a choice of artificial

TABLE 5

Response to lateral contact with walls opposed at angles of 5° and 20°. Percentage distribution of insect/wall contacts (after de Mallmann, 1964)

No. of lateral contacts between insect and walls	Limothrips cerealium		Thrips tabaci		Aptinothrips rufus	
	5°	20°	5°	20°	5°	20°
2	24·3	13·9	11·8	7·5	24·2	10·4
1	44·9	48·8	52·8	50·2	50·0	50·4
0	30·8	37·3	35·4	42·3	25·8	39·2

substrates, these species chose surfaces with a fibrous texture such as blotting paper or sand paper. For most of the time their positive thigmotactic response dominated their positive phototaxis (Lewis and Navas, 1962; Holtmann, 1963) and negative geotaxis (Holtmann, 1963; Mallmann, 1964), and this explains why interstitial-dwellers remain hidden on their host plants except before flight, or sometimes in sunny weather. Ovipositing females of *Isoneurothrips australis* also have tactile preferences. They can lay eggs in the middle of a flat artificial membrane as easily as on the surface of the *Eucalyptus* flower torus, but they prefer to push the tip of the abdomen into the angle of the outer rim of the torus, or into crevices between the bases of stamens (Laughlin, 1970).

Such delicate sensitivity to small differences in plant structure partly determines the suitability of plants as hosts. For example, the leaf structure of different varieties of onions affects the numbers of *T. tabaci* they can support. In heavily infested varieties the leaf blades usually have one flat side which in young leaves is closely pressed to the flat surface of the opposite leaf, providing many crevices and good protection for the larvae. In contrast, the variety White Persian has leaves almost circular in cross section with a wide angle between the two youngest leaves creating fewer crevices and therefore supporting only small infestations. When leaf blades of this variety were tied together to increase the number of contiguous surfaces and crevices, the population per plant increased 4-fold in 15 days (Jones, *et al.*, 1934; Sleesman, 1943). There are similar differences between types of onions grown in India where White Spanish and Country White varieties, which have less foliage and mature earlier than red varieties, are least susceptible to attack (Verma, 1966). Very dense hairs on the lower surface of leaves probably prevent feeding, but more widely spaced hairs do not deter thrips, and in fact the hairy American and Indian varieties of cotton are more heavily attacked than the smoother-leaved Egyptian varieties (Wardle and Simpson, 1927). The hairy undersides of plum and apple leaves provide ample shelter for all stages of the predatory *Haplothrips faurei*, whereas on peach which has glabrous leaves, the eggs, prepupae and pupae usually inhabit crevices on older twigs, and the few that occur on the leaves shelter near the midrib or under bits of detritus or dead scale insects (MacPhee, 1953; Putman, 1965b).

The smell and taste of different parts of plants probably also affects the position occupied by feeding thrips. When offered a choice of scents from different parts of oat plants (see p. 97) *Limothrips cerealium* and *Haplothrips aculeatus* preferred the smell from fresh green spikelets to that from leaf sheaths or stems (Holtmann, 1963). However, before the spikelets appear they feed on sheaths, and part of the attractiveness of spikelets must arise from the easily available moisture in them compared with the drying leaves. The alligatorweed thrips *Amynothrips andersoni* is also attracted to its host by odours (Maddox *et al.*, 1971).

GALLS

A highly specialized group of interstitial-dwellers is that whose feeding
stimulates plants to produce galls in which the thrips live protected from
enemies, heavy rainfall and drought. Most gall-forming species are
tropical and are commonest in Indo-Malaysia and Australia. Others occur
on Pacific Islands (Bagnall, 1928) and North Africa with a few in Europe
(Wahlgren, 1945), North and South America (Wood, 1960) and the West
Indies (Wyniger, 1962).

Only a small proportion of the Thysanoptera, about 30 species of
Terebrantia and 110 species of Phlaeothripinae, inhabit galls (Sakimura,
1947), though more are being described each year, and in the Indo-
Malayan region they cause about a tenth of all galls of animal origin
occurring on Pteridophytes, Gymnosperms, mono- and dicotyledons
(Docters van Leeuwen, 1956).

Most thrips galls are produced by the rolling, folding or wrinkling of
leaves; sometimes more elaborate structures develop on buds and stems,
as well as on leaves, and a few simple galls develop on flowers. The
structure of different leaf galls ranges from a mere curling of the two halves
of the leaf blade on either side of the midrib, without the margins meeting,
through more pronounced rolling, folding and swelling extending over
the whole leaf, to bizarre bladder, pouch and horn galls protruding from
a swollen distorted blade (Karny and Docters van Leeuwen, 1914; Mani,
1964a; Ananthakrishnan, 1969a; Ananthakrishnan and Jagadish, 1969)
(Fig. 22 and Plate IV).

Leaf roll galls, produced by ten species of *Gynaikothrips* and two of
Smerinthothrips, are especially common on species of *Ficus*, and occur in
Samoa, Tonga, Australia, Indonesia, Philippines, Formosa, Malaya, India,
North and South Africa, Italy, Canary Islands, Florida, West Indies and
Mexico (Bagnall, 1928; Priesner, 1939). *Gynaikothrips ficorum* is a cosmo-
politan species that feeds on the widely distributed *F. retusa*; the other
thrips have a more restricted range and feed on different species of
Ficus.

One of the simplest leaf rolls, on *Bridelia laurina* from New Caledonia, is
a swollen, bag-shaped malformation caused by the failure of the leaf rudi-
ment to open from the bud. The halves of the leaf blade on either side
of the midrib remain folded to form a gall cavity with openings to the
outside along the leaf margin, apically and basally (Houard, 1924). The
leaf fold gall of *Gynaikothrips pallipes* on *Piper sarmentosum* from Indonesia
also develops when thrips attack the tender leaves before the buds have
opened. One side of the blade is usually more heavily infested than the
other, and this half rolls upwards while the other half curves over the
rolled margin to form a sheath. The gall of *G. kuwanai* on *P. futokadsura*
is merely a narrow, convoluted infolding along the edges of the blade

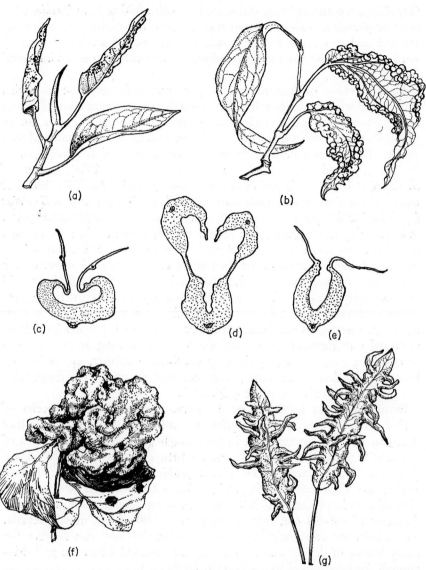

FIG. 22. Galls caused by thrips. (a) Simple leaf roll of *Gynaikothrips uzeli* on *Ficus retusa*. (b) Convoluted leaf roll of *G. kuwanai* on *Piper futokadsura* (after Takahashi, 1934). (c) T.S. gall of *G. crassipes* on leaf of *Piper*. (d) T.S. gall of *Eugynothrips intorquens* on leaf of *Smilax*. (e) T.S. gall of *G. chavicae* on *Piper* (after Mani, 1964). (f) Bud gall of *Austrothrips cochinchinensis* on *Calycopteryx floribunda* (original after Ramchandra Rao, 1924). (g) Horns galls of *G. heptapleuri* on *Schefflera elliptica* (after Docters van Leeuwen, 1956).

(Fig. 22b), in contrast to the elaborate leaf roll or fold of *Eothrips coimbator-ensis* on *Jasminum pubescens* from India, where the entire leaf is spirally twisted and rolled to produce a warty, cigar-shaped swelling about 75 cm long and 25–40 mm thick, containing a labyrinth of crevices inhabited by thrips (Mani, 1964).

Galls develop on infested plants presumably in response to toxins introduced with the thrips saliva. These inhibit the differentiation of tissues and stimulate cell proliferation and hypertrophy mainly in the meso-phyll, so that as the gall grows the thrips usually become enclosed in a cavity. The development of horn galls on leaflets of *Schefflera* spp. in Indonesia provides an example (Docters van Leeuwen, 1956). Leaflets of *S. elliptica* are infested, just after they have expanded from the bud but before the leaf tissues are fully differentiated, by females of *Gynaikothrips heptapleuri* which puncture the underside of leaves with their mouthparts. Within a day transparent spots appear on the upper surface above each puncture. The spots turn red and grow into a tubular gall 6–8 mm long in about 6 days. Female thrips then enter each hollow chamber to lay eggs that hatch and mature within the gall in about a week. Eggs are laid intermittently so each gall may contain adults, eggs and larvae in different stages of development. One female can induce a number of galls on each leaflet. No new galls are found on leaves from which females have been removed but immature galls continue to grow as long as they contain larvae. The galls eventually develop a tubular chamber, up to 30 mm long, surrounded by a spongy, succulent wall about 20 cells deep, with the phloem towards the inside (Fig. 22g).

Similar galls are produced by the same or closely related species of thrips on other species of *Schefflera* in Indonesia and the Philippines (Uichanco, 1919; Docters van Leeuwen, 1956). In Australia, species of *Kladothrips* produce hollow, spherical pouch galls on species of *Acacia* (Froggat, 1906; Karny, 1911; Hardy, 1916; Moulton, 1927; Mound, 1971a) and *Thaumatothrips froggatti* produces knob-shaped twig galls on *Casuarina stricta* (Karny, 1922). These galls on *Acacia*, with an internal diameter of 0·8–0·9 cm, may contain up to 600 adults, mostly female. The adults are extremely resistent to desiccation and can survive for many weeks in drying galls, until they are released when the gall dries enough for its aperture to re-open (Mound, 1971a).

The galls of Terebrantia are generally less elaborate than many pro-duced by Phlaeothripids. For example, in Europe *Thrips fulvipes* causes a small leaf gall on *Mercurialis perennis* and *T. tabaci* a flower gall on *Thlaspi arvense* (Wahlgren, 1945).

The species responsible for a gall usually live inside it, but they are sometimes accompanied by other species able to produce galls on the same or on different plants (Wood, 1960) and by non-gall-producing

inquilines. Sometimes as many as six species occupy a gall and it is difficult to determine the original gall-maker. The complex interrelationships
between gall-formers and the harmless and predatory inquilines that may
accompany them are illustrated in Fig. 23. As well as inquilinous thrips,
coccids, aphids, aleyrodids, anthocorids, psocids and even lepidopterous
larvae (Vuillet, 1914) may occur within thrips galls; in Brazil, 7 species of
parasites and predators, including a mite and a syrphid larva, attack
Gynaikothrips ficorum (Bennett, 1965). Thrips too, enter galls produced by
insects of other orders. In India *Dolicholepta inquilinus* occurs in psyllid
galls on *Zizyphus* (Ananthakrishnan, 1956) and in Europe, Priesner (1926–
1928) recorded *Liothrips setinodis* in galls of the mite *Eriophyes tristriatus*

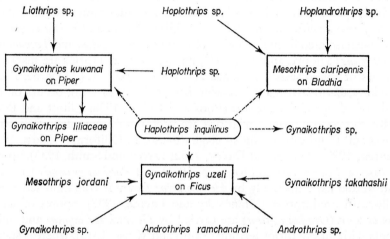

FIG. 23. Gall-forming (in rectangles), predatory (in oval) and harmless inquilinous
thrips inhabiting four genera of plants in Formosa. Solid arrows show interrelationships between inquilines and gall-formers, and dotted arrows point to species attacked
by the predators (after Takahashi, 1934).

Nal. and a number of species in cynipid galls. Most of these thrips enter
for protection, thereby satisfying their strong positive thigmotactic
response, but predatory species such as *Haplothrips subtilissimus* may be
searching for food and attack the larvae of the original gall-maker.

TRANSMISSION OF PLANT DISEASES

Thrips may also affect their hosts unfavourably by transmitting diseases,
including toxaemias caused by toxin in saliva, bacterial and fungal
diseases spread by mechanical contact, and viruses transmitted during
feeding (Sakimura, 1947).

Toxaemias are distinguishable from galls, which are similarly induced
by salival toxins, only by the absence of elaborate distortions and cavities

in the plant tissue in which thrips can live and shelter. Proven toxaemias are few but a toxin from *Taeniothrips laricivorus* probably causes the malformations in young larch shoots and needles that so closely resemble the early stages of some bud galls (Kratochvil and Farsky, 1942), and peach plume disease in Italy may also be a toxaemia (Curzi, 1932).

Pathogenic bacteria and fungi already on the epidermis of healthy plants, or carried on the bodies of thrips, can infect plant tissues through feeding punctures. A few economically important diseases may be spread in this way, but as thrips largely assist the spread of organisms normally carried by wind or rain in far greater numbers, their contribution to epidemics of most bacterial and fungal diseases is probably small. In glasshouses, the lesions of bean bacteriosis (*Pseudomonas medicaginis* var. *phaseolicola*) are associated with the feeding punctures of *Hercinothrips femoralis* (Buchanan, 1932). Fire-blight bacteria (*Erwinia amylovora*) have been isolated from the bodies of *Frankliniella moultoni* and *Taeniothrips inconsequens* (Thomas and Ark, 1934; Bailey, 1935) and thrips are suspected vectors of this disease in Canada and Oregon, U.S.A. (Treherne, 1923). Similarly, *Bacterium stewarti*, causing bacterial wilt of corn, occurs on the body of *Anaphothrips obscurus* (Poos and Elliott, 1936; Elliott and Poos, 1940) and several species of thrips are vectors of a complex of bacteria, yeasts and fungi causing fig-spoilage diseases in California (Caldis, 1927; Hansen, 1929; Hansen and Davey, 1932; Davey and Smith, 1933). Spores of rusts are often carried (Johnson, 1911) and *Puccinia graminis* has been transmitted experimentally by *Hercinothrips femoralis* contaminated externally with uredospores (Granovsky and Levine, 1932). Spores of azalea flower spot (*Ovulina azaleae*) are carried by *Heterothrips azaleae* and infect plants without the thrips feeding (Smith and Weiss, 1942), whereas spores of *Pestalozzia* sp. infect camphor trees only around the feeding lesions of *Liothrips floridensis* (Howard, 1923). At least three species of thrips can feed on pathogenic fungi: in India, *Euphysothrips minozii* feeds on wheat rusts (Ramakrishna Ayyar, 1928) and *Megaphysothrips subrumanii* on coffee leaf rust (Ananthakrishnan, 1969a), and in North America *Thrips tabaci* feeds on spores of *Aspergillus niger* (Bailey, 1935) and on several species of powdery mildews infecting many plants including grape vine, rose, strawberry and cantaloupe (Yarwood, 1943). These thrips throve better on mildewed leaves than on normal ones, probably because of the abnormal nutritional qualities of the diseased tissue (see p. 46).

Tomato spotted wilt virus is the only plant virus known to be transmitted by thrips, although they are suspected of being vectors of manihot mosaic virus in Brazil (Bondar, 1924), pistachio rosette virus in southern Russia (Kreutzberg, 1940) and sunflower mosaic virus in Argentina (Traversi, 1949). However, these records and viruses are of doubtful authenticity. Bondar specifically stated that damage to cassava was entirely mechanical and caused by the feeding of *Scirtothrips manihoti* on the young leaves, and Kreutzberg's pistachio rosette virus was probably also merely a mechanical effect caused by a species of *Liothrips*. Similarly, Traversi's claim is dubious because a white fly and an aphid are also recorded as vectors, and the virus was probably tobacco mosaic virus (Mound, 1973). Strains of tomato spotted wilt virus, under many different names (Smith, 1957), are distributed throughout the world and sometimes damage crops including tomatoes, tobacco, pineapples, lettuce and potatoes and ornamental plants (see p. 238). Symptoms differ on different plants but include stunting, distortion and mosaic mottling of leaves, clearing of leaf veins and concentric rings or wavy lines on leaves and fruit (Smith, 1932, 1957) (Plate V). The known vectors are *T. tabaci*, a cosmopolitan thrips common in every region where the virus is present, *Frankliniella schultzei*, widely distributed in the Southern Hemisphere, and *F. occidentalis* and *F. fusca* in North America (Sakimura, 1960). Apparently all species transmit the different strains of the virus with similar efficiency except the yellow form of *F. schultzei*, sometimes called *sulphurea*, which cannot transmit (Sakimura, 1969). Occasionally, plants may be infected by two species of thrips, or the same plant species may be infected by different vectors in different regions.

Thrips are exceptional among virus vectors because to become infective they must feed on infected plants during their larval stages (Bald and Samuel, 1931), but once infective, both larvae and adults can transmit the virus. The reason for this is unknown. Conditions in the midgut suspected of affecting the thrips' ability to become infective, such as the oxidation–reduction potential and pH of the gut contents, are similar in larvae and adults of the vector, *T. tabaci*, and the non-vector, *T. imaginis*, so are unlikely to be responsible. Both tracheation of the midgut and rate of ingestion are less in larvae than in adults (Day and Irzykiewicz, 1954), so likewise these factors are unlikely to account for the exclusive ability of the larvae to acquire the virus. One possibility is that some tissue, perhaps the wall of the gut, loses its permeability to the virus as the insect becomes adult (Bawden, 1964). However, when the gut wall of adult *T. tabaci* was punctured after the thrips had fed on infected plants, the virus was still not transmitted by them to healthy plants (Day and Irzykiewicz, 1954). Perhaps the virus simply multiplies in larval, but not in adult, tissues.

The virus seems to be concentrated in the epidermal cells of plants (Yarwood, 1957). Thrips probably need to feed on an infected plant for at least 30 min to acquire it (Razvyazkina, 1953). Following acquisition there is a latent period of about 10 (3–18) days, during which the larvae usually pupate and emerge as adults. The thrips may remain infective for a few days or for life and transmit continuously or sporadically, possibly depending on the amount of virus originally acquired. About 15 min of shallow feeding is sufficient to infect a healthy host plant (Sakimura, 1960, 1963).

Pollination of Flowers

Thrips do not always affect their host adversely. Many species pollinate flowers, though where larger insects such as bees and large flies are common, their contribution is probably small. However, where large insects are scarce and flowers are usually self-pollinated or cross-pollinated by wind, thrips may be more important and sometimes essential. Their small size and the relatively few pollen grains each individual carries (Table 6)

TABLE 6

Ability of Thysanoptera to carry pollen (after Shaw, 1914; Annand, 1926)

Flower	Species of thrips	Number examined	Largest number grains per thrips	Average number grains per thrips
Sugar beet	*Thrips tabaci*	2	140	137
Alfalfa	*Frankliniella tritici*	130	16	2·3
Acacia	*Frankliniella tritici*	8	1	1
Plum	*Frankliniella tritici*	16	3	0·2
Plum	*Taeniothrips inconsequens*	16	4	1·2
Daisy	*Frankliniella minuta*	20	4	2
California poppy	*Anaphothrips secticornis*	10	3	3
California poppy	*Frankliniella tritici*	96	20	3·3
California poppy	*Frankliniella minuta*	3	3	2
Filaree	*Frankliniella tritici*	6	0	0
English laurel	*Taeniothrips inconsequens*	1	19	19
English laurel	*Frankliniella tritici*	12	6	4
Spirea	*Frankliniella tritici*	18	0	0
Lupin	*Frankliniella tritici*	19	76	26·5
Lilac	*Frankliniella tritici*	3	21	14
Lilac	*Thrips madronii*	2	7	6
Lilac	*Aeolothrips kuwanii*	1	13	13
Clarkia elegans	*Frankliniella tritici*	67	11	1·5

is compensated for by the great numbers that may be present on flowers. In temperate regions, hundreds of *Haplothrips*, *Taeniothrips* and *Frankliniella* with pollen grains on their bodies and appendages infest compound inflorescences and many occur on simpler flowers such as apple, pear and strawberry. Individuals of *Frankliniella tritici* caught flying over lucerne fields carried as many pollen grains as those on lucerne flowers

(Annand, 1926) and individuals of various species flying over sugar beet carried up to 44 grains each (Shaw, 1914), showing that the pollen acquired in flowers sticks long enough to be carried to other flowers and cross-pollinate them. The thrips' ability to carry pollen depends on the number and structure of their setae and on the size and viscosity of grains (Fig. 24c).

On caged onion inflorescences in California, introduced *Frankliniella occidentalis* increased pollination, though seed set was even better when coccinellids and houseflies were introduced as well as thrips. The introduced thrips multiplied rapidly, and up to 3,600 adults and immature

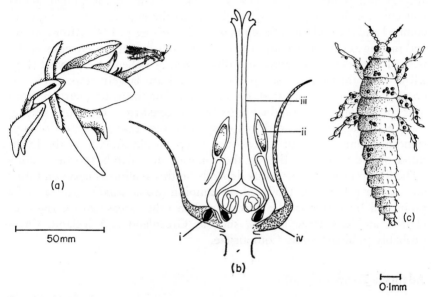

FIG. 24. (a) *Taeniothrips ericae* pollinating a flower of *Calluna*. (b) L.S. *Calluna* flower with (i) a thrips egg in the base. To reach and leave the oviposition site the thrips squeeze between (ii) stamens and (iii) style, collecting pollen on the way; (iv) nectaries (after Hagerup, 1950). (c) Larva of *Frankliniella tritici* carrying pollen grains of sugar beet (after Shaw, 1914).

stages occurred per inflorescence without causing serious damage (see p. 234; Carlson, 1964a). This thrips may also cross-pollinate beans (*Phaseolus vulgaris*). Several individuals may occur in each flower, and, in contrast to bees, they can penetrate a flower bud before it opens and therefore pollinate it earlier (Mackie and Smith, 1935; Vieira, 1960). Other species may help to cross-pollinate flax (Henry and Tu, 1928), and pyrethrum (Bullock, 1962). In Trinidad *F. parvula* perhaps contributes to the pollination of cacao; Billes (1941) found a mass of pollen containing a thrips' seta on the stigma of a thrips-inhabited flower. The presence of thrips on mesquite blooms in Texas supposedly deters honey bees from working the flowers (Parks, in Stannard, 1968) but no other records of this negative reaction are known.

In the Faroes where large pollinating insects are rare and long periods of heavy rain make wind pollination uncertain *Taeniothrips ericae* is the most important pollinator of *Calluna* in many summers. Often 4–6 young adults live inside each flower, feeding on the surface of the nectaries at the base of the petals, where even the mouthparts of bees cannot penetrate. When ready to fly the female thrips crawl to the tip of the protruding style which provides a vantage point for take-off, squeezing past the anthers and pollen on the way (Fig. 24a). Some of the pollen grains collected on the wings and body fall on to the stigma as the wings are combed and the abdomen flexed before take-off (see p. 134). After landing and mating on another plant, the females enter flowers to lay eggs at the base of the petals, collecting and disseminating pollen at each visit (Fig. 24b). Thus thrips assist self-pollination and are largely responsible for cross-pollination and set seed in most years. The thrips too, survive from year to year only because their eggs are protected from drowning during the wet winters in the dead, bell-shaped flowers that persist on the upstanding stems. Without *Calluna* the sheep would starve, so the livelihood of the crofters and even the name of the islands (Faroes means "The Sheep Islands") probably depends on this symbiosis between thrips and flower (Hagerup, 1950). Moss campion (*Silene acaulis*) may also be pollinated by thrips in the Faroes, and in other areas with a rigorous climate and few large insects, such as Greenland and Iceland, thrips probably pollinate some Compositae.

Adaptive Form and Colour

Many species with specialized habits or environments have evolved morphological features that facilitate movement and concealment in their habitat. Bark-dwellers especially have flattened bodies with enlarged muscular forelegs and prothorax, so that they can squeeze into crevices to search for fungal spores (Fig. 25e). Grass-dwellers living in the long, narrow cavities between culm and sheath, are usually either flattened, e.g. *Limothrips*, *Podothrips* (Fig. 25a), or slender-bodied with antennal segments almost parallel-sided, e.g. *Stenothrips*, *Sorghothrips*. The hind margins of their sternites, and to a lesser extent their tergites, are flanged, possibly to facilitate bending of the body.

Exposed foliage-feeders may have normal or flattened bodies but their cuticle is often sculptured with raised, net-like reticulations resembling the pattern of leaf surfaces (Plate VIa) (Fig. 25b); some tropical thrips living under fallen leaves have similar sculpturing (Priesner, 1964a). These

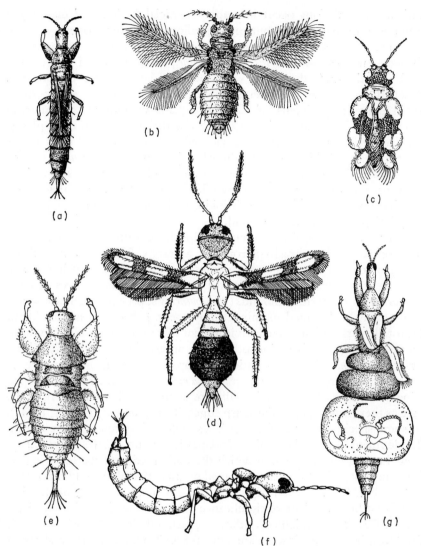

FIG. 25. Adaptive form in thrips. Drawings are to different scales; the approximate lengths are given below: (a) *Podothrips graminum* (2·6 mm) an elongated grass-dweller (original). (b) *Selenothrips rubrocinctus* (1·1 mm), a foliage-feeder, with reticulated cuticle (after Russell, 1912b). (c) *Arachisothrips millsi* (0·75 mm), the peanut-winged thrips (after Stannard, 1952). (d) *Franklinothrips myrmicaeformis* (2·8 mm), a fast-running predator (after Zanon, 1924). (e) *Hoplothrips pedicularius* (2·3 mm), a flat, powerfully built crevice-dweller (original). (f) *Leptogastrothrips* sp. (3 mm), an ant mimic (after Stannard, 1968). (g) *Kladothrips augonsaxxos* (2·7 mm), a bubble leaf-gall thrips, showing greatly distended, egg-filled abdomen (after Froggat, 1927; Moulton, 1927).

cuticular reticulations occur in many superficial feeders and may lessen water loss by partially stabilizing a shallow "boundary layer" of air a few microns thick around the animal. The hind tibiae of some superficial dwellers such as *Dendrothrips* and *Caliothrips* are elongated and usually held folded under the body with the tarsi beneath the mesothorax. When the hind legs are extended quickly the thrips leap and perhaps avoid danger; specimens of *Caliothrips fasciatus* with wings removed can still hop 25–30 cm (Bailey, 1933). Some thrips living on hairy plants, such as *Verbascum*, are themselves clothed with long, curved hairs and as these have evolved in different sub-orders (e.g. in *Parafrankliniella verbasci* and *Haplothrips verbasci*) they seem to be an adaptation for survival on hairy vegetation, possibly for camouflage. The very long anal setae on the first-instar larvae of Aeolothripidae and Phlaeothripidae may detect enemies approaching from behind.

Nearly all true leaf and flower thrips have fully developed wings (macropterous) but brachypterous or apterous forms, often without ocelli, are common in one or both sexes of many species that live where movement is restricted, such as in turf, bark, dark mouldy places or even in rats' nests and termite mounds (*Xylapothrips* spp.) where delicate wings might easily be damaged and hinder movement. It is surprising that life in the totally dark and protected environment of termite mounds has induced no morphological adaptation other than aptery (Hartwig, 1967). *Chirothrips ruptipennis* provides a curious example of apparent brachyptery. It lives in the inflorescences of the grass *Poa nemoralis* but the fully developed wings of the newly emerged adults are soon cut off unevenly near the base as they brush against the sharp edges of the glumes (Priesner, 1964a). In some species the size of wings is also affected by temperature during development (see p. 34). A bizarre adaptation occurs in *Arachisothrips* sp. from Jamaica and Mexico, in which the leading edge of the forewings is ballooned into a hollow, peanut-shaped outgrowth with heavily reticulated cuticle (Fig. 25c). These thrips live in the ground cover of neotropical rain forest but the adaptive significance of their wings is unknown (Stannard, 1952). Their indented outline suggests mimicry of small wasps or flat bugs, but as the insects are only 0·5–0·75 mm long their shape is unlikely to be conspicuous enough to deter vertebrate predators that might hunt them.

The males in some populations of thrips, mainly of tropical fungus-feeding Tubulifera, are polymorphic and include individuals with suppressed structural characters (gynaecoid) ranging through normal and intermediate forms to giants (oedymerous), which often have bizarre and grotesque modifications of their fore limbs, head and thorax (Ananthakrishnan, 1967, 1968, 1969b) (Plate VII). All forms copulate freely with females. Their role in the population is unknown but as they occur most often in species living in clusters on fungus-infested twigs and leaves, the larger forms may protect the eggs and early instars from small predators.

Another unusual adaptation occurs in at least two species of thrips which form galls on *Acacia* in Australia (Mound, 1970c, 1971a). The abdominal intersegmental membranes of ovipositing females become distended to twice their normal diameter so the abdomen looks like a series of beads, the membranous areas packed with eggs, alternating with the constricting segmental sclerites (Fig. 25g). Free-living thrips with such distended abdomens would be vulnerable to predators, but protected from enemies in the sealed gall, these females can utilize most of their energy for egg production, and their unusual shape is no encumbrance.

The taro thrips, *Organothrips bianchi*, lives in a most exacting habitat for Thysanoptera, and shows a remarkable ability to adapt to a completely aquatic environment. In the Hawaiian, Samoan and Palau Islands it infests cultivated and wild taro plants (*Colocasia esculenta*) growing in standing water, and lives in the slime that collects in the crevices between the bases of the thick, fleshy, overlapping stalks, a few centimetres above water level (Sakimura and Krauss, 1944, 1945). The only marked morphological adaptation to this semi-aquatic habitat is a pair of spoon-shaped scrapers, each with about 8 teeth, on the fore-tibiae, used by the thrips to clean themselves of algae and plant and snail slime (Hood, 1939). Specimens introduced to Germany on plants grown in warm-water aquaria developed more extreme behaviour and completed their life-cycle fully submerged on *Cryptocoryne* plants. No special organ for under-water respiration has been evolved by this unique aquatic thrips, but the subepidermal ends of the tracheae are extensively tufted, and sausage-shaped vesicles have developed in the legs and antennal segments to facilitate gaseous exchange through the cuticle (Titschack, 1969).

There are fewer known examples of adaptive colouring in thrips than adaptive form. The disruptive outline produced by the black and white stripes on the folded wings of many *Aeolothrips* spp. and sometimes on the abdomen of species such as *Ae. albicinctus*, *Ae. bicolor* and *Desmothrips reedi* may distract larger predatory insects and vertebrates long enough to increase the thrips' chance of escape. These three species, from Europe, North America and Australia respectively, all live in tussocks of grass, but the advantages of this pattern of marking, common to many species living in this habitat in different parts of the world, is unknown. The contrasting shades of yellow, dark brown or red on the body or wings of some grass- and leaf-dwellers living on exposed surfaces may be aposematic, although it is uncertain how useful warning coloration would be to such small creatures. For example, adults of *Trichromothrips bellus* have a vivid yellow body with grey-brown margins and a crimson proctodaem shining through the abdomen (Priesner, 1964a);

larvae of Phlaeothripidae such as *Haplothrips cahirensis* in Egypt and *H. subtilissimus* in Europe, often have alternate crimson and yellow segments, or a brilliant purple-red tip to the abdomen as in the North American *H. distalis*. By contrast, larvae of other species of *Haplothrips* and many other genera that live in concealed habitats where warning patterns could not be seen are uniformly red, orange or yellow; the tropical *Apelaunothrips* are exceptional because their larvae are striped even though they hide under fallen leaves (Priesner, 1964a). In hot climates the larvae of fungus-feeding Phlaeothripidae often live externally and are brightly coloured, but in cool climates they usually live in concealed habitats and are pale. Larvae of the predatory *Aleurodothrips fasciapennis* are peculiar because their colour is affected by food. They hide under old scales, exuviae and moulds on citrus trees and are uniformly light-lemon yellow except when reared on chaff scale (*Parlatoria pergandii* Comst.) when they develop orange-red to purplish-red spots on an off-white background (Selhime *et al.*, 1963).

4 Interrelationships with other animals and pathogens

The relationships between thrips and other animals are more varied and complex than between thrips and plants, but have been studied less. Species of potential economic importance have attracted most interest and studies of them embrace almost all knowledge of interrelationships between thrips and other arthropods and vertebrates. Appendix 3 lists predatory thrips and their hosts, and the natural enemies of thrips of economic importance.

Predatory Thrips

Prey

Predatory species usually feed on small, soft-bodied insects including other thrips, aphid nymphs, psocids, scales and especially on the eggs of mites and Lepidoptera. A few species may eat their own young.

Mites are common prey throughout temperate and tropical regions. For example, in Nova Scotian fruit orchards, *Haplothrips faurei* and *Aeolothrips melaleucus* often feed on the eggs of the tetranychids *Panonychus ulmi* (Koch) and *Bryobia rubrioculus* (Scheuten). When eggs are scarce *H. faurei* also attacks the active stages of the mites, but the thrips larvae then grow more slowly and adults are less fecund than when mite eggs are abundant (Putman, 1942, 1965b; MacPhee, 1953). In warmer parts of North America, and in Panama, Honduras, Barbados and Brazil, the black hunter, *Leptothrips mali*, also feeds on mites including common red spider [*Tetranychus telarius* (L.)], brown almond mite (*Bryobia praetiosa* Koch) and especially the eriophyid silver, or russet mites. It is the main predator of *Caliptrimerus baileyi* Keifer, a silver mite attacking apple leaves in parts of California, and the presence of this thrips often signifies infestation by eriophyids even when these are too few to be noticed (Bailey, 1940a). In central and southern U.S.A. and in parts of Africa, Australia, India, South America and the Hawaiian Islands, *Scolothrips sexmaculatus* attacks eggs, nymphs and adults of a variety of mites, many of them pests of orchard and field crops (Bailey, 1939), and *S. acariphagus* feeds on mites infesting cotton in much of central Asia (Yakhontov, 1935). Other thrips attack tetranychids during the growing season on grape vines in France (Rambier, 1958), on apples in Iraq (Mound, 1967a) and on beans in Germany (Fritzsche, 1958), where *Cryptothrips nigripes* is also recorded as feeding on dormant female mites late in the year. It is surprising that this member of the Megathripinae, all of which have been assumed to feed on fungal spores (Stannard, 1968), should be recorded as a predator, particularly as its stylets are not pointed.

In Nova Scotia, *L. mali* and particularly *H. faurei* eat many eggs of the eye-spotted bud moth [*Spilonota ocellana* (Schiff.)] (Stultz, 1955) and the codling moth [*Cydia pomonella* (L.)] (MacPhee, 1953), and in Ontario they destroy eggs of the oriental fruit moth [*Cydia molesta* (Busk.)] (Putman, 1942). Many scale insects are eaten by thrips in warmer climates. In citrus groves in Florida, U.S.A., *Aleurodothrips fasciapennis* prefers to feed on three species of armoured scale (Diaspididae) as well as on white fly and the mite, *Eotetranychus sexmaculatus* (Ril.) (Selhime *et al.*, 1953); in Fiji it feeds on coconut scale (*Aspidiotus destructor* Sign.), and on other armoured scales in Ceylon (Taylor, 1935). *Karnyothrips flavipes* commonly feeds on scales in the genera *Asterolecanium*, *Parlatoria* and *Saissetia* in the Mediterranean region and *Podothrips aegyptiacus* preys on mealybugs living under the leaf sheaths of grasses (Priesner, 1964a). In Central and South America, *Franklinothrips vespiformis* feeds on mites, leafhoppers and white fly in citrus and avocado groves (Moulton, 1932).

Thrips are sometimes killed by other thrips. *Aeolothrips fasciatus* attacks larvae and adults of the oats, onion, gladiolus and bean thrips (Kolobova, 1926; Maddock, 1949, Stathopoulos, 1964) and *Ae. intermedius* often attacks *T. tabaci* in southern France (Stradling, 1968). In experimental cages, *L. mali* eliminated colonies of *Frankliniella moultoni* and *Drepanothrips reuteri*, but free on grape vines or fruit trees, their diet includes mites and aphid nymphs as well (Bailey, 1940), so in natural conditions they probably have less drastic effects on wild populations. In Trinidad, *Franklinothrips tenuicornis* is a common predator on cacao and glasshouse thrips, and *F. vespiformis* occasionally attacks these pests as well as *Dinurothrips hookeri* on ornamental flowers and *Caliothrips insularis* on Sudan grass (Callan, 1943); in California it eats great numbers of *Heliothrips haemorrhoidalis* in avocado groves (McMurtry, 1961). The wide range of thrips species eaten by *Haplothrips inquilinus* in galls is shown in Fig. 23 (see p. 53).

When crowded, adults and larvae of *Scolothrips sexmaculatus* fight among themselves and become cannibalistic. Larvae and pupae deter attackers by flicking the abdomen rapidly from side to side, but weak or injured individuals quickly succumb; they do not attack other species of thrips (Bailey, 1939). Larvae of *Haplothrips leucanthemi* also feed on one another when food is scarce (Loan and Holdaway, 1955), and *H. faurei* occasionally feeds on its own eggs and pupae when confined in cages (MacPhee, 1953).

Most predatory thrips are therefore usually polyphagous predators and many can also survive temporarily on plant food. Probably only the leaf-dwelling aeolothripids such as *Aeolothrips versicolor* and larvae of *Franklinothrips* are entirely carnivorous; the others supplement their diet with plant juices or pollen. This is a great advantage for maintaining populations when prey is scarce. For example, *Haplothrips faurei* can complete development on fresh pollen of *Chenopodium album*, though the adults are undersized. In orchards where there is little prey, small populations of this thrips may still survive, probably by eating pollen blown on to leaves as well as the few arthropods available. Adults can survive for long periods on leaf juices alone, and those feeding on winter eggs of tetranychids die unless they also have access to peach leaves (Putman, 1965b). Similarly, larvae and adults of *L. mali* can survive up to five days on juice sucked from apple leaves (Bailey, 1940a). On rare occasions species that are normally entirely phytophagous, such as *T. tabaci*, may feed on small arthropods.

Searching and feeding behaviour

Predatory thrips closely resemble plant-feeders and have few obvious morphological adaptations to their way of life. Their mouthparts are structurally and functionally similar. The Aeolothripidae, however, rely on rapid movement to hunt and seize their prey. Thus their legs are generally slender and support the body well above the surface of the vegetation. A small hook and tooth on the second tarsal joint of aeolothripids, probably evolved to help the adult emerge from the pupal cocoon (Fig. 14b), may also help in grasping prey. The predatory species *Franklinothrips myrmicaeformis* and *F. vespiformis* have the base of the abdomen constricted and ant-like; this perhaps enables these thrips to run especially fast, or might even deter attack by vertebrate predators (Fig. 25d). By contrast, most predatory Tubulifera, e.g. *Haplothrips faurei*, walk slowly and do not use their forelegs to catch prey, so are less successful predators on the active stages of arthropods than are aeolothripids (Putman, 1942, 1965b).

Thrips probably find their prey by random searching and recognize it only from very near or upon contact; individuals overlap their tracks many times seeking food. Even a starved thrips may pass close to suitable prey and fail to find it.

Egg-feeders pierce the egg shell with their mandible, aided by a slight

D

rotation of the head. The maxillary stylets are then inserted into the egg with a jabbing, whipping motion, and the mouthcone pressed against the surface of the egg shell while feeding (MacPhee, 1953). The puncture is seldom visible, and unless the embryo is about to hatch the contents are almost entirely removed apart from a small remnant that imparts a faint pink or orange colour to the empty shell. These minute, neat punctures distinguish eggs sucked by thrips from those attacked by predators with coarser sucking, or biting, mouthparts.

Active stages of the prey usually resist the predator. *Scolothrips sexmaculatus* uses its forelegs to hold and turn mites before piercing their ventral surface (Bailey, 1939). *Trichinothrips breviceps* subdues psocids up to twice its own size by grasping each victim with its forelegs and leaping on to its back. The thrips thrusts its stylets into the psocid and when most of the viscera have been removed rolls it round with its forelegs and punctures the body from different sides to empty it completely (Seshadri, 1953). By contrast, *Aleurodothrips fasciapennis* moves rapidly between coconut scales (*Aspidiotus destructor*), usually stopping to suck from each for about half a minute and then running on without removing all the body contents. Occasionally it feeds on a scale long enough to empty it, or it may drag it from beneath its protective covering, and feed while grasping it with the forelegs as a mantis holds a moth (Taylor, 1935).

The body contents of the prey often colour the intestine of the predator. Yellow aeolothripid larvae appear pinkish after feeding on red *Haplothrips* larvae, and larvae of *Aleurodothrips fasciapennis* vary in colour from light lemon to dark yellow after feeding on Florida red scale (*Chrysomphalus ficus* Ashm.) or white fly, to off-white with orange-red to purplish-red spots when they have fed on chaff scale (*Parlatoria pergandii*; Selhime et al., 1963).

The rate of feeding and total consumption differs between species and instars. *Scolothrips sexmaculatus* can empty the contents of a red spider egg in less than a minute, and of an adult in 5–20 min. Bailey (1939) observed one adult thrips kill three brown almond mites and destroy six eggs in half an hour, but such periods of voracious feeding were separated by long intervals. Ananthakrishnan (1969a) mentions a record of one II instar larva of *Scolothrips indicus* consuming 55 eggs, 34 larvae, 7 nymphs and 6 adults of the spider mite *Panonychus ulmi* in only three days. Females of *Haplothrips faurei* reared in vials with unrestricted food require about 44 eggs per day to maintain full egg production, and each larva eats an average of 143 summer eggs of *Panonychus ulmi* during the 8–11 days required for development at 24°C. They do not feed at temperatures cooler than about 11°C (Putman, 1965b). Adults and larvae eat about three of the larger eggs of codling moth per day (MacPhee, 1953). Adult *Trichinothrips breviceps* feed for about 90 min on each psocid and consume 2–3 mature, or 3–5 young specimens, per day (Seshadri, 1953).

The nutritive value of food eaten by larvae partly determines the duration of their development, and the size and fecundity of adults. Larvae of *H. faurei* reared at 24°C on mite eggs develop in 8–11 days compared with 14–22 days for those fed on young mites. Larvae do not grow when fed on young leaves only, but may survive on them when prey

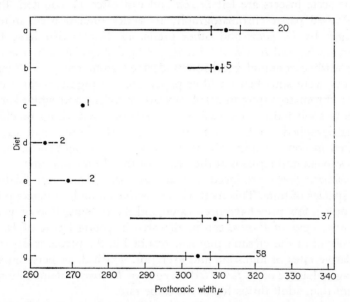

Fig. 26. The effect of diet on the size of adult *Haplothrips faurei*. Prothoracic width was used as an index of size; dots represent mean width, horizontal lines the range of measurements, and vertical lines the S.E. of means. The numbers show how many individuals were measured in each category. The diets were: (a) eggs of *C. molesta*; (b) unrestricted numbers of winter eggs of *P. ulmi*; (c) 10 eggs of *P. ulmi* per week; (d) 5 eggs of *P. ulmi* per week; (e) pollen; (f) natural diet—individuals collected from peach trees with moderate numbers of prey; (g) natural diet—individuals collected from peach trees with very few prey (after Putman, 1965b).

is scarce. Adults lay few eggs on a vegetarian diet but resume normal oviposition when prey becomes available.

The size of adult *H. faurei* collected from orchards and of others reared in vials on different foods is compared in Fig. 26. Individuals collected from different orchards were similar in size and did not differ significantly from individuals fed in vials on unlimited quantities of eggs of Lepidoptera and mites. However, none from orchards were as small as those fed on a restricted diet of mite eggs or pollen, though many larvae must experience food shortages through uneven distribution of prey, even in moderately mite-infested orchards. This suggests either that the larvae find alternative sources of food under natural conditions, or, more likely, that undernourished individuals are eliminated by natural hazards absent in the laboratory (Putman, 1965b).

Predation by many insects is difficult to detect in the field because the prey is often completely devoured; attack by thrips, however, is generally easier to detect because only the contents of victims are removed, and eggs or scale insects are left *in situ* and can often be counted. Putman (1965a) used paper chromatography to verify predation on mites and their eggs by *H. faurei* and other predators. Insects that have fed on *Panonychus ulmi* and *Bryobia rubrioculus* can be distinguished by the orange and reddish carotenoid biochromes derived from the prey. Suspected predators were squashed on filter paper and the pigments separated by circular chromatography in a 1:4 mixture of xylene and white kerosene. Pigments soon faded so squashes of predators were dried quickly and chromatographed on the same day. Pigments from small predators separated in about 5 min. The thrips' own orange pigment is insoluble in this solvent and remains at the origin of the chromatogram.

Haplothrips faurei produced similar chromatograms after feeding on either species of mite. This method is sensitive enough to detect pigment when only a few prey have been eaten and the consumption of as few as five winter eggs of *P. ulmi* can be detected in young larvae of *H. faurei*. The amount of the victim's pigment retained in the predator depends on the relative rates of ingestion and excretion which differ between species and stages. Larvae of *H. faurei* fed on eggs of *P. ulmi* contained more mite pigment than adult thrips fed at the same rate.

It is important to appreciate the limitations of this technique which can be used to assess the approximate proportions of predators in a population more accurately and much more simply than the precipitin test (Dempster, 1960), but as it is not quantitative the number of mites eaten by an individual predator cannot be ascertained. When insects are secondary predators on others that have already eaten mites, interpretation of chromatograms is difficult, but this probably rarely happens with thrips.

Studies on the effects of predatory thrips on the abundance of their prey have been almost entirely confined to species that attack crop pests. This topic is considered under the general heading of "Beneficial thrips" in Chapter 12.

Natural Enemies of Thrips

Predators

Thrips are eaten by many general predators including bugs, lacewings, dipterous larvae, other thrips (Appendix 3) and a few vertebrates. The soft-bodied and generally slow-moving thrips larvae are easy prey for small active predators, and in populations where all stages are present, they are usually attacked first.

Anthocorid bugs, especially in the genus *Orius*, are world-wide enemies (Russell, 1912a; Illingworth, 1931; Sakimura, 1937a; Tominic, 1950; Nolte, 1951; Priesner, 1964a; Bennett, 1965). Nymphs and adult bugs move among thrips on plants and pierce victims in either head, thorax or abdomen, often holding down the struggling prey with their forelegs. Feeding lasts 5–20 min and as the body fluids are removed, victims may shrink. After feeding the bug withdraws its stylets and usually cleans itself. When disturbed during a meal, adults of *Orius tristicolor* (White) may run or even fly with a struggling thrips larva impaled on the mouthparts (Bailey, 1933). In Spain, *Montandoniola moraguesi* Put. attacks the eggs of *Gynaikothrips ficorum* (Del Canizo, 1944), and in other parts of the world species of *Macrotracheliella*, *Ectemnus*, *Cardiastethus* and *Xylocoris* feed on larvae and adult thrips (Fig. 27a) (Melis, 1935; Bennett, 1965).

Mirid bugs attack thrips more often in warm than in cold regions. In India a species of *Psallus* feeds on *Taeniothrips distalis*. The young predators prefer thrips larvae and can eat 3–4 per day, but 4th and 5th instars and adults of the mirid attack all stages, consuming an average of two adult thrips or their equivalent daily. Young bugs usually pierce the immature thrips at the top of the abdomen but adults insert their proboscis anywhere (Rajasekhara *et al.*, 1964). Different species in the genus *Termatophylidea* attack cacao thrips in Jamaica (Myers, 1935), Surinam (van Doesburg, 1964) and Trinidad, where the lygaeid bug *Ninyas torvus* Dist. is also an enemy (Callan, 1943).

Some American sphecid wasps in the genera *Spilomena*, *Ammoplanus* and *Xysma* include immature thrips among the prey with which they provision their nests (Muesbeck *et al.*, 1951; Krombein, 1958) and in Europe *Spilomena troglodytes* Lind. have been seen storing thrips larvae, probably *Frankliniella* sp., in stems of *Rubus* to provide food for their brood. In Egypt, small *Ammoplanus* spp. may feed exclusively on thrips (Priesner, 1964a).

Larvae of chrysopid and coniopterygid lacewings probably consume more thrips per individual than other arthropod predators. For example, predatory *Franklinothrips* consumed one or two cacao thrips larvae daily, but in cages chrysopid larvae ate on average 14 larvae a day, and when hungry ate one or two thrips per minute over short periods (Callan, 1943). Other general predators recorded as destroying thrips in local situations include the larvae of the coccinellids *Hippodamia convergens* (Guér.) in California (Bailey, 1933), *Adalia bipunctata* (L.) in Europe (Priesner, 1964a) and *Coccinella undecimpunctata* L. in Egypt (Ghabn, 1948). Bark-dwellers are particularly susceptible to attack by staphylinid beetles, and flower-dwellers may be captured by ants (e.g. *Plagiolepis* sp.) that visit inflorescences. The ant *Wasmannia auropunctata* Roger occasionally carries larvae of cacao thrips in its jaws (Callan, 1943). Predatory syrphid larvae, including *Baccha norina* Curr. and *B. livida* Schiner in South America (Bennett, 1965), *Sphaerophoria quadrituberculata* Beazi in South Africa (Stuckenberg,

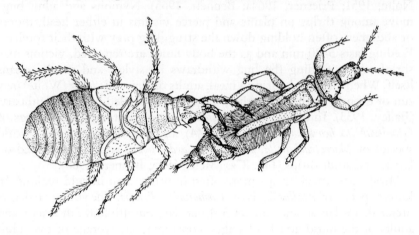

(a)

(b)

Fig. 27. (a) Immature cimicid bug, *Ectemnus reduwinus* (H.-S.), feeding on *Liothrips oleae* (after Melis, 1935). (b) Larva of *Heliothrips haemorrhoidalis* carrying a blob of excrement —perhaps a means of deterring predators (after Buffa, 1911).

1954) and *Syrphus corollae* F. in Europe and North Africa (Ghabn, 1948), attack immature and adult thrips, and predatory cecidomyiid larvae may destroy immature stages (Barnes, 1930; 1948; Melis, 1935; Bennett, 1965). Even a cricket, *Oecanthus turanicus* Uv., attacks *Thrips tabaci* in Egypt (Ghabn, 1948), and many flying thrips are caught in spiders' webs. In California, a tube-weaving spider in the genus *Dictynus* often spins its web across a citrus leaf, and up to 50 adults and larvae of the citrus thrips may be caught in a single web. The jumping spider, *Thiodina puerperis* Wtstn, also pounces on thrips and can eat up to 10 in succession, draining the juice from each victim and discarding the skeleton (Horton, 1918).

Many mesostigmatid and trombidiid mites are ectoparasites on thrips, especially Thripidae and Aeolothripidae. The immature stages of these mites usually attach themselves to their hosts at the intersegmental membrane between thorax and abdomen. For example, *Typhlodromus thripsi* MacGill seizes larval *T. tabaci* at this point and although the victims lash their abdomen from side to side in an attempt to dislodge the mite they rarely succeed in freeing themselves. Each mite may kill 5–6 thrips per day (MacGill, 1939). In Egypt, *Adactylidium* sp. is an efficient predator on eggs of *Gynaikothrips*, probably because it can complete its life-cycle in as little as four days at 30°C (Elbadry and Tawfik, 1966), and in Brazil a species of *Pediculoides* (Pyemotidae) attacks *G. ficorum* (Bennett, 1965). In Yugoslavia, a species of *Cheyletus* eats eggs of *Liothrips oleae* (Tominic, 1950). Although some species of mite kill their hosts quickly other thrips can walk and even fly with mites attached.

A Malaysian species of the genus *Psilogaster* (Eucharitidae) has a remarkable but puzzling association with the cacao thrips, *Selenothrips rubrocinctus*. All species of this family of tropical chalcids are believed to be parasitic on the larvae or pupae of ants, and lay their eggs remote from the ant host. This species of *Psilogaster* lays 50 to 100 eggs in a circle around a single egg of a cacao thrips. When the thrips hatches, it tries to escape from the surrounding palisade of *Psilogaster* eggs, thereby stimulating some of them to hatch, and the emerging planidia attach themselves to the young thrips larva (Clausen, in Kirkpatrick, 1957). It is not known how this behaviour enables them to reach their ant host, but some ants do occasionally carry larvae of cacao thrips in their jaws (Callan, 1943), and this might provide an opportunity for the planidia of *Psilogaster* to transfer.

Most species of thrips escape from predators by jumping or flying, but a few turf- and bark-dwellers retract their legs and antennae to feign death. When captured, many species raise and lower their abdomen in attempts to free themselves, and some heliothripine larvae keep the abdomen raised with a drop of intestinal fluid on the tip to deter attackers (Fig. 27b). The pungent odour emitted by the lily thrips, *Liothrips vaneekei*, may keep them free from infestation by bulb mite, *Rhizoglyphus echinopus* (Fum. and Rob.), which often injures other arthropods present on the bulbs (Hodson, 1935).

Thrips form part of the diet of some amphibians, reptiles and insectivorous birds, and are eaten accidentally with flowers, fruit and grasses by larger birds, poultry and grazing animals (Theobald, 1926; Morison,

1948). They were present in almost all the stomachs of a sample of 400 young American toads (*Bufo americanus*) collected at different sites and dates in the U.S.A., and constituted about 10% of the bulk of food consumed (Hamilton, 1930). *Frankliniella tritici* has been recovered from stomachs of the salamander *Ambystoma texanum* in Illinois (Moll, 1963) and several other species from the stomachs of the lizards *Anolis stratulus* and *A. pulchellus* in Puerto Rico (Morgan, 1925), *Uta s. stansburiana* and *Sceloperus g. graciosus* in U.S.A. (Knowlton, 1938), and *Cnemidophorus lemniscatus* in Trinidad (Lewis, unpublished). Some of the thrips were found in the anterior part of lizards' guts wrapped in mucilage and intact. Humming birds are known to eat thrips, but whether deliberately or incidentally as they sip nectar is unknown (Shull, 1911; Hood, 1914); in Europe, Buhl (1937) claimed that tits (*Parus* spp.), the chiffchaff (*Phylloscopus collybita*), willow warbler (*P. trochitus*) and whitethroats (*Sylvia* spp.), ate many larvae of *Kakothrips robustus*.

Internal parasites

Despite their small size thrips are attacked by parasitic insects and nematodes. Their insect parasites are all minute wasps in the families Eulophidae (Fig. 28a) which mostly attack larvae (Fig. 28b), and Trichogrammatidae and Mymaridae which are egg parasites (Appendix 3). Sakimura (1937b) describes the life-cycle and effect on the host of a fairly typical eulophid, *Thripoctenus brui* Vuil. a species which parasitizes thripids in widely separated parts of the world. In Japan, on *T. tabaci*, the wasp completes 4–5 generations per year, and hibernates as a pupa. Its life-cycle is completed in 24–39 days depending on temperature, and adults live about 20 days. The wasps prefer to lay in young larvae and the parasite pupates within the prepupa of the host and kills it. When wasps lay in a late II stage thrips larva the immature wasps cannot develop successfully, but they nevertheless sterilize the adult thrips and shorten its life. At first, attacked larvae appear and behave normally, and signs of the parasite only appear about a day before the change to prepupae when the shape of the parasitic larva inside shows faintly through the host's cuticle extending from the metathorax to the apical third of the abdomen. The head and abdominal tip of the parasitized thrips become paler as fluid is withdrawn from these parts, its abdomen lengthens and swells, and it walks with difficulty. It manages to moult into a prepupa, but the wing pads are short or absent. As the parasite grows it deforms the antennae and legs of the host and the thrips dies within one or two days as the parasite swells to occupy most of its body. Finally, the thrips prepupa splits behind the head to expose the pale parasitic pupa which quickly darkens (Figs 28c, d). The life-cycle of *Thripoctenus russelli* Crwf. on *Caliothrips fasciatus* in California is very similar (Russell, 1912c) though

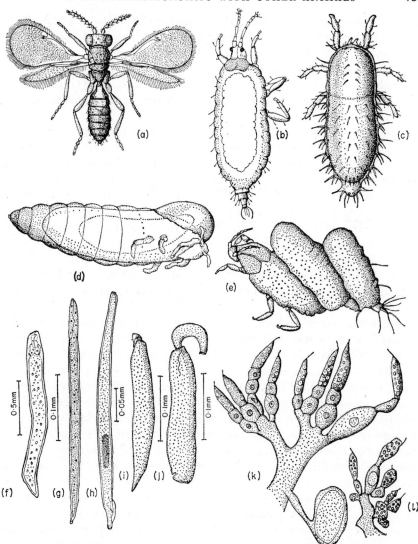

FIG. 28. Internal parasites of thrips. (a) Adult *Tetrastichus gentilei* (Eulophidae) (after Melis, 1935). (b) Larva of *Liothrips oleae* parasitized by a larva of *T. gentilei* (after Del Guercio, 1911). (c) Larva of *Thrips tabaci* bloated and distorted by larva of *Thripoctenus brui* (after Sakimura, 1937b). (d) Larva of *Kakothrips robustus* deformed and killed by pupating *T. brui* (after Kütter, 1936). (e) Larva of either *Megalothrips spinosus* or *Cryptothrips rectangularis* parasitized by three larvae of *Thripoctenus nubilipennis* Wms (after L. T. Williams, 1916). (f)–(j) Some forms of the nematode *Anguillulina aptini* found inside the body cavity of *Aptinothrips rufus*; (f) I stage larva; (g) fully developed female larva before free-living stage; (h) male larva before free-living stage; (i, j) sac-like females after free-living stage. (k) Normal ovary of *A. rufus*. (l) Degenerate ovary of parasitized insect (after Sharga, 1932).

details of the development of other parasites differ. For example, *Tetra-stichus gentilei* Del G. faces the hind end of its host larva, *Gynaikothrips ficorum*, so bursts from the terminal abdominal tergits (Plate VIII) (Bournier, 1967), and when the same parasite attacks *Liothrips oleae*, it diapauses from November to April in the hibernating thrips larva (Melis, 1935). *Dasyscapus parvipennis* Gah. kills its host in the larval stage (McMurtry, 1961).

When host larvae are abundant *Thripoctenus* females can lay eggs rapidly. In caged cultures Russell (1912c) observed *Thripoctenus russelli* laying 38 eggs in 36 larvae in an hour, but in the field when parasites must search for hosts, the rate of oviposition is probably much slower. In other species the act of oviposition may last several minutes. Usually only one egg is laid in each thrips' larva; if more are laid in terebrantian hosts, only one develops, though polyparasitism may occur in larger tubuliferan larvae (Fig. 28e) (L. T. Williams, 1916).

Egg parasites lay their eggs singly in thrips' eggs embedded in plant tissue, and these develop and kill the host egg before it can hatch. The parasites emerge from round exit holes cut in the egg blisters (McMurtry, 1961).

There are a few records, mostly from temperate regions, of thrips infested with nematodes probably of the family Allantonematidae (Uzel, 1895; Russell, 1912a). *Anguillulina aptini* (Sharga) commonly attacks larval, pupal and adult *Aptinothrips rufus* in Great Britain (Sharga, 1932; Lysaght, 1936, 1937). One or two gravid female nematodes occupy the abdominal or thoracic cavity of infected thrips, where they lay about 50 (10–110) eggs. The larval worms develop in the coelom of the host and eventually escape by boring through the wall of the midgut and passing out through the thrips' anus. After a short, free-living stage the infective vermiform females penetrate the cuticle of the larva or pupa, then swell into a sac-shape inside the new host (Figs 28f–j). Infected thrips may be found at any time of the year, but the number of eggs and larvae in each individual is greatest in spring and summer. For unknown reasons the proportion of infected individuals is less in populations living on coarse tufted grasses such as *Holcus* than on finer, more uniform swards. The nematodes do not affect the appearance or movement of their hosts, but cause their ovaries to degenerate (Figs 28k, l), so although the thrips are not killed immediately the rate of increase of infected populations is retarded as egg production ceases. In England, *Stenothrips graminum* is attacked by a similar, or possibly the same, allantonematid (Lewis, 1961), and in Russia, Kolobova (1926) also found *S. graminum* infested with an undetermined nematode which sterilized it.

It is appropriate to include fungal parasites with these animal parasites because both affect mortality in natural populations in a similar way. The fungi attacking thrips are probably not specific to them but are general and widespread entomogenous species (Steinhaus, 1949). The body cavity and appendages of II stage thripids, especially *Limothrips*, *Frankliniella*, *Melanthrips* and *Odontothrips*, sometimes contain red spheres about 40 μ diameter, each made up of fine filaments radiating from a central body, which may be compact fungal mycelia (Priesner, 1964a). Lysaght (1936) found specimens of *Aptinothrips rufus* with hundreds of colourless, elongated, but unidentified spores in the body cavity. As with infection by nematodes, the thrips appeared healthy but were sterile. *Beauvaria globulifera* attacks cacao thrips in the West Indies and Central America, especially in damp places and seasons, appearing as a white mould on the bodies of adults and larvae, and attaching them lightly to leaves (Nowell, 1916; Urich, 1928; Callan, 1943). In Surinam a species of *Cephalosporium* has also been found on living cacao thrips larvae (Reyne, 1921). *Entomophthora sphaerosperma* was recorded on *T. tabaci* in Massachusetts (Charles, 1941) and in Switzerland probably another species of *Entomophthora* (*Tarichium*) transforms II stage larvae of *T. tabaci* into blackened mummies packed with spores (Stradling, 1968). Thrips that overwinter or pupate in the soil are often killed by fungi, and deep ploughing of cereal stubble in autumn encourages the destruction of wheat and oats thrips by *Beauvaria bassiana* (Kurdjumov, 1913; Grivanov, 1939). In Bulgaria up to 20% of larvae of *Haplothrips tritici* may be infected (Lyubenov, 1961).

Incidence of attack by natural enemies

The effect of natural enemies, especially predators, on the size of field populations of thrips is rarely noticed because usually only a small proportion of each population is attacked. Occasionally, however, some parasites kill a large proportion of a field population. Examples of the amount of predation and parasitism recorded in the field are given here. The effect of attack by natural enemies on the dynamics of thrips populations is considered in Chapter 9 and their value as biological control agents in Chapter 12.

Generally, numbers of predators and prey fluctuate together, but the predators remain far less abundant. For example, Fig. 29 shows monthly changes in density of cacao thrips and two of its predators, larvae of *Franklinothrips* and chrysopids, in Trinidad. The prey was 30 to 1,400 times more abundant than *Franklinothrips* and 27 to 190 times more abundant than the chrysopids, but it is not certain whether changes in the numbers of prey and predator were directly related. The abundance of polyphagous predators such as these probably fluctuates as food supplies vary. There was no consistent time interval between peak populations of prey and predator, which suggests that predatory populations increased elsewhere, and were ready to exploit the large populations of cacao thrips when they developed.

By contrast, in Japan the relationship between *Thrips tabaci* and its parasite, *Thripoctenus brui* is density-dependent. Sakimura (1937c) sampled

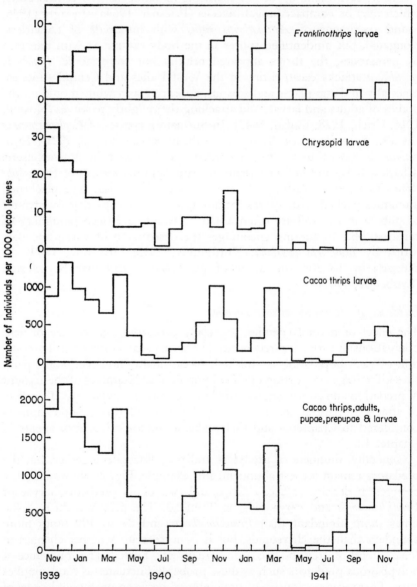

FIG. 29. Changes in the size of populations of cacao thrips and its more important predators in Trinidad (after Callan, 1943).

populations of this thrips in 25 different onion fields, on the same day by a standard method, and found generally that the greater the density, the greater the percentage of parasitism in the larvae. The regression fitted to his data (Fig. 30) shows this positive density-dependent relationship, but it does not show whether the increased parasitism at greater host densities was due to a greater number of thrips larvae attacked by each adult wasp, or to an increase in the number of wasps. Other populations of eulophid parasites and their thrips hosts are probably also density-dependent.

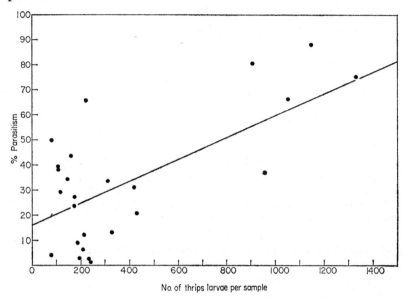

No. of thrips larvae per sample

FIG. 30. The relationship between the density of populations of *Thrips tabaci* and the percentage parasitism by *Thripoctenus brui* in Japan. $y = 15\cdot63 + 0\cdot047(x)$ (data from Sakimura, 1937c).

Several instances of 20–80% parasitism by eulophids have been observed. In Germany, Buhl (1937) found 36·6 and 23·2% of pea thrips larvae parasitized by *Thripoctenus brui* in consecutive years; in Switzerland, 68% of pea thrips were parasitized by *T. kutteri* Ferr. (Kütter, 1936), and in California 70% of bean thrips by *T. russelli* (Russell, 1912a). In Finland, Hukkinen (1936) found 58·2% of *Chirothrips hamatus* larvae and pupae destroyed by an unidentified hymenopterous parasite. In cacao fields in the Gold Coast (Ghana) up to 80% of cacao thrips were parasitized by *Dasyscapus parvipennis* in 1925 (Cotterell, 1927), which led to the introduction of this parasite to the West Indies between 1933 and 1937 where it established in Trinidad, Jamaica and Puerto Rico (Adamson, 1936; Callan, 1943).

Parasitism by some species is often very localized within the broader region they inhabit; for example, near Bangalore, India, up to 10% of *T. tabaci* were attacked by *Thripoctenus* sp. in some fields, but none in others nearby (Narayanan, 1970).

In populations of thrips infected with nematodes, up to 40% infection is common, and Lysaght (1937) found 64% of *Aptinothrips rufus* attacked in late summer, except in tufted grass where infection was much less.

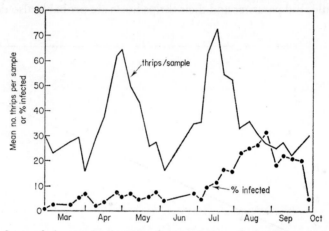

Fig. 31. Seasonal changes in the size of a population of *Aptinothrips rufus* and the percentage of thrips infected with *Anguillulina aptini* in England (after Lysaght, 1937).

Fig. 31 suggests that the percentage of thrips attacked was independent of their density.

Compatible Relationships

A few relationships between thrips and other animals appear mutually beneficial, or at least harmless. For example, inquilines living inside galls produced by other species benefit from the relationship and the original gall-former is probably unharmed (see p. 52). Occasionally *Aeolothrips* species are found in nests of house martins (*Delichon urbica*), and although probably blown there by chance, they may feed on the small scavenging dipterous larvae living in the bottom of the nest. In winter, old birds' nests in hedges often contain hibernating thrips, and *Xylaplothrips* is even tolerated by termites in their mounds (Hartwig, 1967). A strange and unexplained relationship exists between the Argentinian mantid, *Parastagmatoptera unipunctata* (Burm.) and *Symphyothrips concordiensis*, which lays its eggs on the mantid's oothecae (De Santis, 1971).

Direct Relationship with Man

As well as their widespread and sometimes serious infestations on crops (see Chapter 11), thrips occasionally irritate people out of doors and are a minor nuisance inside buildings. Trouble out of doors usually occurs in hot, sultry weather when large numbers of migrating thrips alight on bare skin. The itching and pricking they produce is probably caused by their attempts to obtain water from the moist surface, or perhaps they are attracted to volatile compounds in sweat such as capronic and lactic acid which Holtmann (1963) found were attractive to *Limothrips cerealium* and *Haplothrips aculeatus* in laboratory tests (see p. 98). These attempts to feed may provoke rashes on the skin, even though it is not completely pierced, or inflammation in the ears and nose. They can also produce a painful prick in the eye. These irritating habits have been reported for *Gynaikothrips uzeli* in Algeria (Senevet, 1922), *Caliothrips indicus* in the Sudan (Johnston, 1925), *Thrips imaginis* in Australia (Evans, 1932) and *Limothrips* in Europe (Körting, 1930). Körting, and Bailey (1936) also recorded swarms of cereal and flower thrips so dense that bathers abandoned beaches, but such nuisances are only likely to occur near to large breeding areas, usually crops, for a few days when the plants are dying and thrips need fresh food.

A few species can penetrate human skin and suck blood. *Karnyothrips flavipes* will feed on the wrist, raising small lumps and blotchy patches around each puncture; the ingested blood imparts a pale reddish colour to the thrips' body (Williams, 1921; Hood, 1927). Second-instar larvae of the phytophagous *Thrips tabaci* and *Frankliniella moultoni* can also pierce the skin, suck blood, and raise pinkish dots but they die after about 30 h on this diet, whereas the normally predatory *Scolothrips sexmaculatus*, *Leptothrips mali*, *Aeolothrips fasciatus* and *Ae. kuwanaii*, feed less willingly on man and cause little inflammation (Bailey, 1936).

Indoors, thrips may be a nuisance in summer, crawling over windows and washing, and in autumn and winter they crawl into crevices to hibernate often spoiling framed pictures by creeping between glass and mount.

SECTION II
TECHNIQUES

5 *Laboratory methods*

Rearing

It is often necessary to rear thrips in controlled environments in the laboratory to study life histories, confirm observations made on their behaviour in the field, and for virus-transmission experiments.

Examining and handling

Specimens caught in the field must be identified carefully before cultures are started. They can be inactivated for examination by an anaesthetic, by cooling, or by trapping them in a small space. The least harmful anaesthetic is carbon dioxide, which can be administered to thrips in a small, cloth-covered glass tube which is placed in a Buchner funnel connected to a CO_2 supply. The relationship between duration of exposure to the gas and duration of inactivity is shown for *Thrips tabaci* in Fig. 32. Cooling them for a few minutes at temperatures near freezing will inactivate them long enough to permit quick sorting of catches from a net; for a brief examination individuals can be trapped at the bottom of a glass tube by gently pressing them with a plug of cotton wool.

At room temperatures adults and larvae of many phytophagus species walk slowly and can be picked up with the tip of a fine camel-hair brush (size 00 or 000) slightly moistened with saliva. Species that run quickly or leap are more easily collected by shaking infested vegetation over a cloth or paper, and picking thrips from it with a pooter (aspirator) fitted into a small tube. Thrips can be transferred from this to a cage by removing the cap from the pooter and tapping the inverted tube to dislodge them (Sakimura, 1961). A watch-maker's eye glass, or low-power binocular microscope fastened on the forehead, make it easier to see the thrips. Eggs cannot be transferred without damage because they are either embedded in, or stuck to, plant tissue.

Cages

Many species are difficult to keep in cages because they are small and easily damaged, and because they have a remarkable ability to squeeze through almost imperceptible crevices. Thus cages must be thoroughly sealed with no gaps larger than 0·065–0·13 mm (200–400 mesh/in) depending on the size of the species. This restricts air movement and may encourage condensation on the cage walls to which the thrips often stick and die. To prevent this, large cloth-covered windows in the walls of cages, a small amount of food plant per cage, sub-surface watering of growing plants, and constant or slowly fluctuating temperatures, are recommended. Broadcloth, raw silk or sheer dacron are suitable coverings for cages, but batiste, organdie and sheer nylon are too coarse (Sakimura, 1961).

Species that complete their life-cycle on vegetation can be reared *en*

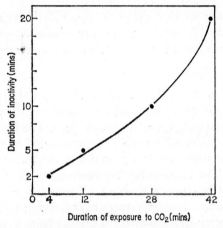

Fig. 32. Relationship between exposure to CO_2 and duration of inactivity for *T. tabac* (data from Sakimura, 1961).

masse on growing plants, in metal, wire or wooden-framed cages covered partly with plastic sheeting and partly with cloth. The base of the cage must be embedded in the soil to prevent thrips escaping. Similar wire-framed cages can be used to sleeve large, growing fruit, e.g. citrus, on which thrips are living. In tropical areas and glasshouses with small changes in temperature between day and night, food plants can be caged more successfully than in temperate regions, where it is difficult to prevent condensation forming at night inside plastic or glass covers. On small individual plants thrips can be confined in lantern chimneys, glass or plastic cylinders with their ends covered by cloth.

For restricting movement on plants, or for rearing individuals separately, different models of sandwich cages are useful. On plants with large, tough leaves cages can be made from rings of felt or foam polystyrene placed above and below the leaf surface and covered with transparent plastic, all the layers being held in position with a paper clip (Fig. 33b).

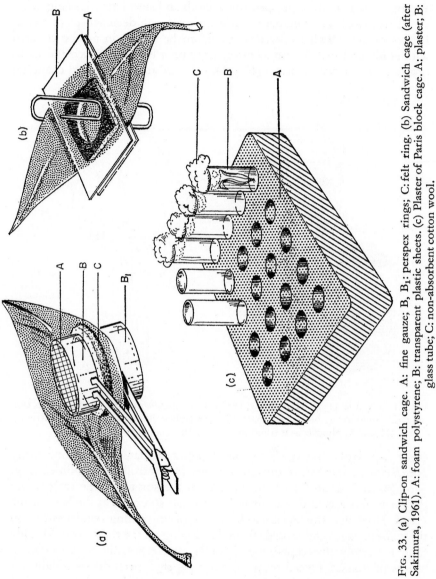

Fig. 33. (a) Clip-on sandwich cage. A: fine gauze; B, B₁: perspex rings; C: felt ring. (b) Sandwich cage (after Sakimura, 1961). A: foam polystyrene; B: transparent plastic sheets. (c) Plaster of Paris block cage. A: plaster; B: glass tube; C: non-absorbent cotton wool.

It is essential to bend the paper clip so that it applies even pressure over the whole cage, or the felt will tilt and the thrips escape. Cages on delicate leaves may need a wire support. A convenient clip-on sandwich cage can also be made from felt, perspex rings, cloth and steel hair clips (Fig. 33a).

Various cages are suitable for rearing thrips on detached plant tissue. Leaves or small twigs offered as food can be placed in a tube covered with cloth, and sealed with a cotton wool plug through which the stalks protrude into a dish of water (Fig. 34a). Twigs and leaves need changing

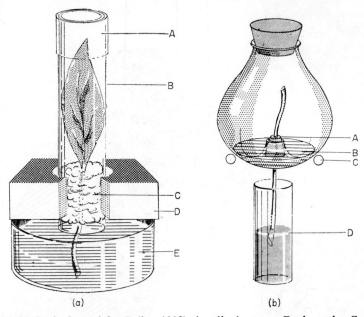

(a)   (b)

FIG. 34. (a) Cut-leaf cage (after Bailey, 1932). A: cellophane cap; B: glass tube; C: cotton plug; D: supporting block; E: water. (b) Controlled humidity cage (after Rivnay, 1935). A: plug; B: salt solution (see p. 93); C: support ring; D: water.

every 2–3 days. Rivnay (1935) developed a similar method for rearing *Heliothrips haemorroidalis* on citrus twigs in controlled humidities (Fig. 34b), but it is doubtful whether the air at the surface of the twig where the thrips fed was as dry as air nearer the salt solutions at the base of the vessel. However, this is an excellent way of preventing condensation on the walls of cages and moulds from developing on the food plant. Mouldy food is usually unacceptable and leaves, and especially whole inflorescences of grasses, placed on damp filter paper in petri dishes usually rot too quickly; also thrips often escape between the dish and lid.

These difficulties can be overcome by keeping the water in a small tube to reduce condensation, cutting the food with a sterilized blade to prevent contamination by moulds, and sealing the gap between the dish and lid with vaseline to prevent escape. Small cages can also be made from glass tubing with pieces of dacron fastened over each end with rubber bands; the food should be placed on blotting paper to absorb any liquid oozing from it. Immature stages of many leaf-feeding species can be reared uncaged on portions of leaves floating on water or a nutrient solution. This method has been used for *Selenothrips rubrocinctus, Dinurothrips hookeri, Caliothrips insularis* and *Heliothrips haemorrhoidalis* (Callan, 1947) and for *Limothrips, Chirothrips* and *Aptinothrips*. Cages need covering to confine winged species. Many grass-feeders can be reared individually on pieces of leaf or spikelets kept in small tubes having one end embedded in plaster of Paris and the other plugged with cotton wool; the cotton should be wrapped in gauze to prevent the thrips becoming entangled in the fibres. The moist plaster keeps food fresh for a few days and at the same time prevents condensation on the walls (Fig. 33c).

Species that pupate in the soil, and especially those with a long pupal stage are more difficult to rear than those whose life-cycle is completed on plants. Moist earth or sand at least 3 cm deep, and fine enough to pass through a 2 mm sieve, should be provided beneath the host plant for a pupation site. If the soil is in a glass vessel its walls should be covered to darken the sides. Some larvae may then pupate near the walls and the pupae can be observed later without disturbing them (Obrtel, 1963).

Predatory species thrive in small tubes stoppered with moistened, gauze-covered cotton wool (Putman, 1942; MacPhee, 1953). They could probably also be reared in the tubes embedded in plaster of Paris, mentioned above (Fig. 33c). Moth or mite eggs can be supplied to egg-feeders on bits of leaf or waxed paper. For general predators, larvae of other species of thrips or small aphids may be offered as food, though this carnivorous diet should perhaps be supplemented with vegetation or pollen.

VIRUS-TRANSMISSION EXPERIMENTS

Techniques for rearing and caging thrips in transmission experiments with tomato spotted wilt virus have been developed by Sakimura (1961). Because thrips can only acquire this virus by feeding on infected plants as larvae (see p. 55) the plant species used in transmission experiments should provide adequate food for the larvae as well as providing them with adequate amounts of the virus; it should also be one which readily shows symptoms of virus infection after infected adults have fed on it. Often the preferred natural host plant is not the most suitable one for experimentation. Plants suitable for rearing the thrips should be tolerant of feeding damage and have foliage on which the thrips remain easily visible; those with crevices or with folded or densely hairy leaves are unsuitable. The plants should be ones which have no necrotic reaction to infection. Infection should be systemic and should produce a high concentration of virus in both inoculated and systemically infected leaves for a prolonged period. *Emilia sonchifolia* (Compositae) combines many of these requirements; it is a satisfactory food, source of virus, and test-plant for infecting *T. tabaci, T. nigropilosus, Frankliniella schultzei, F. fusca, Chaetanaphothrips orchidii, Hercinothrips femoralis, Kurtomothrips morrilli* and *Parthenothrips dracaenae.* In glasshouses it grows readily with little attention. When larvae of *Frankliniella occidentalis* are used in transmission tests, they are better fed on bean pods (*Phaseolus vulgaris*) after acquiring the virus when young from leaves of *Emilia*, because they prefer floral tissue as food. Some other flower-feeders can be reared on fruit after acquiring the virus.

Tomato spotted wilt virus is most concentrated in *Emilia* about two weeks after inoculation; most virus is in the inoculated leaves and the next two or three leaves above them, especially in areas where symptoms are conspicuous. To ensure that the thrips acquire the virus from growing plants the youngest two or three leaves of a plant, 10–12 cm tall, should be sap-inoculated and the older leaves below these removed. The plants can be caged in plastic cylinders covered with cloth. One or two days later two or three female thrips are put into the cage; more may inflict too severe feeding damage on the leaves. These females lay on the young inoculated leaves, and on the new leaves that grow above them, so the progeny hatching 4–5 days later feed for the next 10–15 days on leaves where the virus is most concentrated. Infective adults of the new generation appear in the cage four weeks after introducing their mothers.

A sandwich cage (Fig. 33b) or a sticky barrier can also be used to restrict wandering or to confine feeding to certain parts of the plant during the acquisition period. The sticky barrier is made from a narrow ring (2–3 cm in diameter) of tree-banding grease smeared on to a detached leaf. The leaf is anchored with pins to a piece of wood stuck to the bottom of an open petri dish filled with water or sucrose solution to a level just above that of the wood. Twenty to thirty young larvae can be put inside the ring with a paint-brush and left to feed for two days; if more larvae are used, few infective adults are obtained. When the cages are kept cool and in subdued light only a few larvae are trapped in the grease. Larvae should be captured by approaching them quickly from the head end with a small pooter; when chased from behind they usually run into the grease and are damaged.

Thrips can be tested for infectivity on whole plants or by confining them in sandwich cages on certain parts of a plant. For routine transmission tests, an *Emilia* plant with 5–6 leaves, covered with a cylindrical cage, is satisfactory but individual sandwich cages are more suitable for serial transfers. Two adults per plant normally transmit enough virus for rapid development of the symptoms. The first systemic symptoms, weak vein clearing with conspicuous twisting, may appear on the terminal leaf about a week after feeding starts. When more than two thrips are used, the direct feeding injuries become severe; even with two thrips per plant, test feeding should be stopped before feeding damage becomes serious by spraying the plants with a suitable insecticide (nicotine sulphate, Malathion or Phosdrin).

To decrease the chance of escapes, transfers should be made in the cooler part of the day or in isolated compartments. At the end of observations, all plants, cages, pots and soil should be fumigated or heated to prevent accidental infestation and infection in laboratories or glasshouses.

Artificial feeding

A method of rearing thrips on a controlled artificial diet might occasionally be required for work on virus transmission. Second-instar larvae and adults of *Thrips tabaci* can be reared on 3% sucrose solution contained in tubes capped with "fish skin" mesenteric membrane, but very young larvae may be unable to pierce it (Sakimura and Carter, 1934). Perhaps stretched parafilm membranes used for feeding aphids on artificial diets (Mittler and Dadd, 1962) would be more suitable but, unlike aphids, thrips probably do not rely on turgor pressure to force sap into the pharynx, so presenting the food under pressure (van Emden, 1967) is unlikely to help. A trace of methylene blue (0·01–0·1%) or haematoxylin (1%) in the food enables its passage to be traced through the gut without harming the thrips.

LONG-DISTANCE TRANSPORT OF PARASITES

Before the advent of air transport parasites of thrips used for biological control were transported most successfully within parasitized hosts on food plants. For example, *Thripoctenus brui* survived best on the 8-day sea voyage from Japan to Hawaii when sent as larvae parasitizing larval *T. tabaci*. The parasitized thrips were kept on onion plants growing in metal cans. These were packed in cages made from wooden frames covered with brass gauze and lined on top and sides with tightly woven silk cloth. Joints in the cages were sealed with putty and adhesive tape, and a drainage hole drilled in the metal floor. The cans were embedded in moist peat to keep the vegetation fresh and to provide sites for thrips pupating *en route*. Cages were kept cool during the journey. Few parasites emerged from thrips larvae sent by different methods (Sakimura, 1937d).

Air transport has greatly improved the chances of survival of parasites on long journeys, and these elaborate methods are now unnecessary. For example, the procedure followed by the Commonwealth Institute of Biological Control for sending *Dasyscapus parvipennis* long distances, is to dislodge pupae carefully from leaves with a fine brush and glue them with "Gloy" to a card cut to fit snugly into a 2·5 × 12·5 cm (1″ × 5″) glass tube. About 50 pupae are fixed to each card and fine honey droplets sprayed on to the sides of the tube as food for adults emerging during the journey. Tubes are wrapped in soft paper tissue and packed in strong cardboard boxes for shipment. In the 1960's two batches sent from Trinidad to California, totalling 105 pupae, were five days in transit, but enough survived to start a laboratory culture. Two batches sent to Brazil, totalling 445 pupae, arrived within two days and most adults emerged in good condition (F. D. Bennett, *in litt.*).

Measuring Responses to Physical Variables

The effects of temperature, humidity, touch, light, scent and wind on the development or behaviour of small populations of a few species have been studied separately in the laboratory. Usually the responses of insects to one variable only have been observed, often in specialized apparatus, so it is difficult to apply the information obtained to natural situations where many physical factors change simultaneously. Moreover, the continually changing physical conditions in the natural micro-environments they occupy are difficult to measure (see Lewis and Siddorn, 1972) and difficult to simulate in laboratory apparatus. Many laboratory observations have also been made on the responses of individual thrips to single physical variables, and while these may be a guide to behaviour and mortality in field populations, conclusions from such studies should be applied cautiously. The methods described here are therefore largely confined to measuring responses in populations of thrips kept in as natural conditions as possible. The list is not exhaustive, but chosen to show the type of laboratory experiments and analysis required to produce information useful for comparison with responses of insects in the field.

Temperature

Responses to heat and cold are fairly easy to detect; constant or fluctuating temperatures can be maintained in modern controlled-environment cabinets. In experiments at different temperatures, humidity should also be measured and controlled where possible because it is dependent on temperature, and both affect survival and activity.

Hibernating thrips can be kept at temperatures below freezing (see p. 177) in crevices formed in rolls of paper tissue, placed in large glass tubes (15 × 3 cm diameter) half-filled with moist plaster of Paris to saturate the air, and plugged with non-absorbent cotton wool. The symptoms of cold injury in thrips exposed to severe or prolonged chilling are not clearly visible until at least two days after removal from low temperatures. In *Limothrips cerealium* damage cannot be assessed with certainty until 1–3 weeks later. Slightly injured thrips walk crookedly and slowly, and wave their antennae vigorously, but may recover. Seriously injured ones cannot walk at all, their appendages tremor uncontrollably, and they eventually die (Lewis, 1962).

In studies on the effect of high temperatures on survival and growth, humidity is more difficult to control because the thrips need food which itself affects humidity, particularly when whole plants are used. The humidity of the mass of air in cages can be controlled by placing them in an enclosed chamber above saturated solutions of appropriate salts or different concentrations of sulphuric acid (Andrewartha, 1935; Rivnay, 1935), but the humidity at the surface of vegetation where the thrips feed remains unknown (Fig. 34b). To obtain uniform evaporation from vegetation it is preferable to keep saturation deficit stable rather than relative humidity. Comparisons of mortality in populations kept at different temperatures should be made from estimates obtained by probit analysis, and expressed as the duration of treatment required to kill half the test population rather than all the insects (Lewis, 1962). The resistance of individuals to heating, and upper thermal death-points measured on individual specimens heated in tubes (Cederholm, 1963) probably have little relevance to wild populations.

Temperature preferences are difficult to detect. Cederholm (1963) made two pieces of apparatus to study them in *Limothrips*. One was a shallow circular chamber whose two halves were heated independently by small thermostatically-controlled elements, to offer distinct alternative temperatures to thrips in the chamber. The thrips showed no preference when offered choices between 16 and 25°, and not until the warm side reached 30°C did they begin to move to the cooler side, some remaining in the warm side until it reached 43°C. His other apparatus consisted of a circular plastic chamber (62 mm diam. × 7 mm deep) with a copper rod inserted through the sides. The rod was heated at one end and cooled at the other to produce gradients inside the chamber which were measured with fine thermocouples. The response of the thrips to the gradients was erratic and most of them simply ran around the arena. Thus, even sophisticated apparatus in which only one variable is changed, may produce unnatural behaviour because other important physical factors such as humidity, rough surfaces, or crevices are absent.

The choice of method for studying the effect of temperature on spontaneous activity depends on the temperature range to be covered. The threshold temperature for walking can be measured by placing thrips on moist paper marked with a grid on the bottom of a glass petri dish, and noting changes in position at different low temperatures without disturbing them. At higher temperatures the threshold temperature for take-off can be measured by placing about 50 thrips on an isolated platform in a large bell jar or glass-fronted cage, slowly increasing the temperature of the room and cage, and counting the number that fly to the walls of the cage for each degree rise in temperature. The temperature can be measured with a mercury thermometer with a shielded bulb, or a thermocouple, suspended as near the platform as possible. Thrips launch themselves into the air most readily from an exposed projection, so a circle of stiff paper with toothed edges or a bunch of dead grass stems held on a wire stand make a suitable take-off site. The stand is placed in a dish of water or oil to prevent thrips crawling away. This arrangement ensures

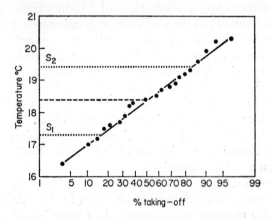

FIG. 35. The cumulative percentage of flight-fit *Limothrips cerealium* taking-off as the temperature increased beyond the threshold for take-off, plotted on arithmetical-probability paper. The mean threshold for take-off for this population was 18·4°C and the standard deviation $= \frac{1}{2}(S_2 - S_1) = 1\cdot05$°C (after Lewis, 1963).

that all insects seen on the walls of the cage have flown there. The intensity and perhaps quality of light required for take-off may differ between species and the most suitable illumination can only be assessed by experiment, but thrips are more likely to take-off in daylight than in artificial light. Specimens used for take-off experiments must be physiologically ready to fly (see pp. 132, 139). The threshold temperature for take-off for the population tested can be estimated by plotting the cumulative percentage of thrips taking-off at each temperature on arithmetical-probability graph paper and reading off the mean and standard deviation (Fig. 35) (Harding, 1949; Lewis and Taylor, 1967).

A sensitive activity recorder designed by Grobbelaar *et al.* (1967) to detect and record slight movements of insects, may be useful for measuring responses of thrips to heat or other stimuli, though results obtained with individuals should be applied cautiously to field populations. The instrument comprises a sensor, in which an insect is confined, an oscillator, frequency detector, amplifier and recorder. When the insect moves, the oscillator frequency is changed and this is converted to a visual record using a series of electronic circuits. The instrument is sensitive enough to detect small flickering movements of the abdomen of a single thrips while it is resting, and a walking thrips causes full-scale deflection of the recorder needle.

Humidity

Survival of thrips in hot, dry places (see p. 182) depends on their tolerance of low humidities which is closely linked with temperature (see p. 94). The effect of drying can be measured most easily for pupae and hibernating or aestivating adults that do not feed. The thrips should be caged in batches of ten in glass tubes loosely plugged with cotton wool, or in perforated gelatin capsules, and held in desiccators over saturated salt solutions at fixed temperatures (Stokes and Robinson, 1949; Solomon, 1951). At intervals, about 50 thrips are taken from each treatment, allowed to recover in a moist atmosphere, then the dead counted. Preliminary tests should be made to find out the period of exposure required to kill approximately 20% and 80% of the thrips; then the experiment should be made in full, choosing periods of exposure between these two limits. Percentage mortality at each humidity should be transformed to probits, a linear regression of probit mortality × saturation deficit fitted, and from it the L.D.$_{50}$ estimated (number of hours required to kill half the thrips). A series of L.D.$_{50}$'s measured at different temperatures can be plotted to show how long the insects can survive dry conditions at stated temperatures (Fig. 36) (Lewis, 1962). A similar method can be used to study the survival of tubuliferan eggs, except that they must be incubated in moist air after drying. Eggs of Terebrantia are embedded in plant tissue so it is impossible to control the humidity around them, and estimates of their survival in dry air in the laboratory have little relevance to field populations. Similarly, it is impractical to measure the resistance to dryness of thrips feeding on plants because their food affects the humidity of the air around them.

Useful measurements can possibly be made for predatory species since the amount of moist tissue offered as food is smaller and less likely to change the humidity within the cage. Putman (1965a) reared larvae of Haplothrips faurei on moth eggs in dry air by placing them over saturated salt solutions in sealed jars, but even so the humidity within the jars near the larvae was not known.

In experiments on the effect of soil moisture on the viability of pupal stages of Thrips imaginis, Andrewartha (1934) used dry soil that was fine enough to pass through a 2 mm sieve, made up to the required moisture content by mixing with water. A rose bud with 20 late II stage larvae was placed on the surface of the soil in a 100 ml beaker, which was covered with fine calico and the sides darkened with paper. When larvae were ready to pupate most entered the soil; those that did not were removed

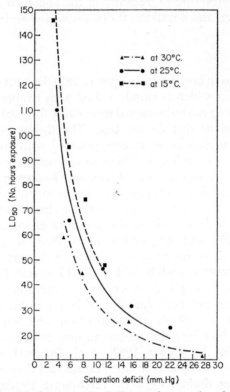

FIG. 36. The relationship between the L.D.$_{50}$ and saturation deficit for *Limothrips cerealium* at three different temperatures (after Lewis, 1962).

with the rose bud after two days. The beakers were kept in desiccators at 97% R.H. so that changes in the soil moisture were slow. At the beginning and end of the experiment the water content of the soil in each beaker was expressed as a percentage of the dry weight of the soil, and of its "field capacity", determined by immersing a box of similar soil in water and allowing it to drain until equilibrium was reached. Most pupae died when the soil was drier than 25%, or wetter than 85%, of field capacity (Fig. 37). Similar experiments on the survival of thrips during flooding could also be designed to be analysed by probit analysis and thus to give the percentage mortality for a given period of inundation.

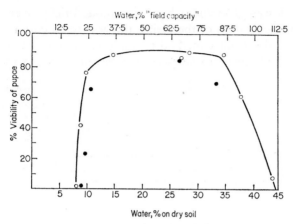

FIG. 37. The relationship between soil moisture and viability of the pupal stages of *Thrips imaginis*; ● at 14°C, ○ at 23°C (after Andrewartha, 1934).

Touch

In the laboratory, simple artificial crevices are suitable for observing the responses of thrips to touch, and their reactions to natural crevices in plants or bark seem similar (see p. 49). Reaction to dorso-ventral contact can be measured by offering a choice of six crevices of different width, made from seven thin glass cover slips (5 × 2 cm) stacked on top of each other but held apart by layers of paper inserted at one end of the stack to produce six adjacent crevices ranging from about 0·075 to 0·45 mm wide. The width required depends on the size of species tested. The stack of glass slips is bound with tape around the end with the inserts, and this end dipped in molten paraffin wax to seal the rough edges of the paper which might attract the insects. As many stacks (blocks) as there are different crevice-widths (treatments) should be made, with the sequence of crevices in each stack conforming to the arrangement of columns in a latin square so that a crevice of each width occupies each position in the experiment. About 100–150 thrips should be placed in smooth-sided plastic boxes, each containing one stack, and left in darkness for about 18 h at a temperature warm enough for them to walk into the crevices. The boxes should then be cooled to immobilize the thrips before opening and dismantling the crevices to count the thrips in them (Lewis and Navas, 1962; see p. 174).

A different method used by de Mallmann (1964) was to place thrips in a wedge-shaped cell made from two sheets of glass held apart with plasticine at shallow angles ranging from 0°38′ to 2°24′ to produce crevices ranging from 0 to 700 μ deep. Thrips were placed in the cell in the dark. At intervals of 20 min a piece of photographic paper was placed beneath the lower glass and a light was shone briefly over the cage to photograph the position of the thrips; this was repeated until movement ceased. He used a similar method to study lateral contact (Figs 38a, b). The reactions of thrips to surfaces of different texture were compared

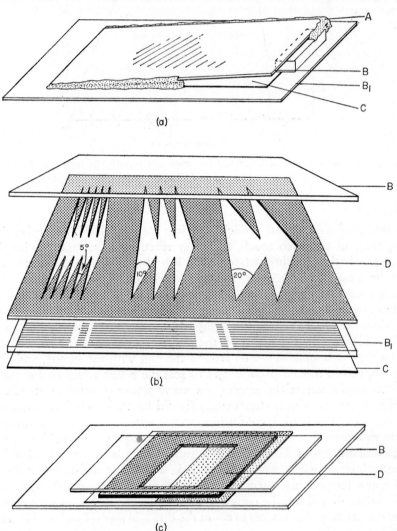

Fig. 38. (a) Cage for studying reaction to ventro-dorsal contacts. A: plasticine; B, B_1: glass; C: photographic paper beneath lower sheet of glass. (b) Exploded diagram of cage for studying reaction to lateral contacts in crevices with walls sloping at 5, 10 and 20° (lettering as above; D: cardboard sheet). (c) Cage for detecting preferences for surfaces of different texture (lettering as above); see rough (stippled) and smooth surfaces offered in the cell (after de Mallmann, 1964).

in a shallow chamber with a floor of contrasting papers ranging from smooth writing paper to sandpaper (Fig. 38c).

Light and gravity

The attractiveness of light can be tested by placing 10–50 insects in a long transparent box or tube, covering half of it with black paper, then placing it in a well-lit place to see where they collect. For experiments on *Limothrips cerealium* and *Haplothrips aculeatus*, Holtmann (1963) used a 100-watt lamp suspended 25 cm above a half-covered black box, and at 20°C allowed about 15 min for thrips to choose a position, which was almost always in the bright half of the box. He also tested their ability to orientate to a restricted source of light by tracing the track of individual insects released at the end of a sheet of blotting paper (70 × 50 cm). Two 40-watt bulbs were suspended 30 cm above the two sides of the paper, and by shining these alternately and measuring the angle through which the walking thrips turned in response to the changed light source he concluded that they tried to keep a constant angle to the light, and thus are orientated by menotaxis. Light of different colours [U.V. (365 mμ)–Red (650 mμ)] and intensities (2–890 lux) was produced by different filters and lamps. These were shone into a black box and the responses of thrips observed. These two species of cereal thrips walked towards even the dimmest light tested, and a comparison of the number of responses to coloured and white lights showed that they could distinguish between light of different wavelengths, and that the strongest response was to blue and violet tones.

Responses to gravity can be measured by placing thrips in a uniformly lit, tall cylinder with sealed ends and then counting the number that collect at the top or bottom. Cereal thrips are negatively geotactic so collect at the top (Holtmann, 1963). The interaction between the effects of touch, gravity and light can be tested in cylinders, or in cells like de Mallmann's (1964), by lining them with different surfaces, tilting, and partially covering them.

Scent

The smell of plants or certain parts of them may help thrips to choose where to live, and with this in mind, Holtmann (1963) compared the responses of *Limothrips cerealium* and *Haplothrips aculeatus* to a number of natural and artificial compounds. He used a choice-chamber to determine the concentrations of each compound that affected the animals, and a Y olfactometer to find which substances attracted or repelled the insects.

The choice-chamber was made from a 9 cm diameter petri dish placed in one of 15 cm. The floor of the larger dish was covered with damp blotting paper and the substance to be tested put into the smaller dish.

E

Both dishes were covered with a piece of fine perlon fabric, and the lid of the larger dish inverted over it to hold the fabric in place and cover the whole chamber. A black cylinder, 25 cm deep, was put around it and a 40-watt bulb was suspended 50 cm above it. Two or three thrips were placed on the fabric above the smaller dish, and the number of times they approached its edge and turned back towards its centre, or walked beyond the edge, was recorded.

The Y olfactometer was made from three 15 cm lengths of 5 mm diameter glass tube, with the arms of the Y separated at an angle of 60°. The substances tested were put in turn into an open tube and fixed to the end of one arm, while a similar tube containing distilled water was fixed to the other. The median tube was tilted upwards at 110° from the horizontal to encourage the negatively geotactic thrips to run against the wind when air was sucked down the arms and along the median tube.

TABLE 7

Attractiveness of organic compounds to *L. cerealium* and *H. aculeatus*
(after Holtmann, 1963)

Scent	Minimum dilution to produce a response	Dilution giving maximum response	Olfactometer: % of thrips attracted to scent
Total oat			22
Sheaths			13
Spikelets			82
Spikelet extract			77
Acetic acid	1:1,000	1:500	21
Butyric acid	1:500	1:200	9
Capronic acid	1:10,000,000	1:500,000	75
Capric acid	1:10,000	1:1,000	85
Acetic ether	1:100,000	1:1,000	83
Geranoil	1:10,000,000	1:1,000,000	79

The apparatus was first tested without scents to ensure that its structure did not encourage thrips to enter one side more than the other. For testing scents, five to eight thrips were put into the median tube to make their choice at the junction and about 50 thrips were used to test each compound. The scents tested included those from oat spikelets placed loose in the test tube, from sheaths and stems of oats cut into 5 cm lengths, and from two extracts of oat plants, one of "total oat" made from squashed oat stems and spikelets, and one made from 20 fresh, fully-developed, green oat spikelets boiled for 10 min in 120 ml of distilled water. Some of the other natural and artificial compounds tested, and the dilutions to which the thrips responded, are listed in Table 7.

Wind

Fast winds may blow resting thrips from vegetation. Their ability to cling to vegetation in strong winds can be demonstrated by putting individuals on a stalk placed about 10 cm inside a glass tube, 1 m long and 5 cm diameter. The stalk should be attached to the wall of the tube with plasticine and smeared with grease near the plasticine to prevent the thrips walking off it. Thrips are placed on the stalk, and air is sucked with a pump into the tube from the other end. The rate of flow can be measured with a flow meter and the wind speed blowing past the thrips calculated. *Limothrips* spp. cling in winds up to at least 13 k.p.h.

It is difficult to follow flying thrips in the field because once airborne they are blown away by the gentlest breeze and quickly disappear against a background of sky and vegetation. Their flight speed, and thus the speed of wind above which control over their track is minimal, can be assessed approximately in still air in a glasshouse by timing the interval between their take-off from the dimmer centre of the house and their landing on the brighter roof, and then measuring the distance flown. The insects' track is often slightly spiral, so the direct distance may be less than the actual distance. Records of flights shorter than about 2 m should be discarded because the timing errors are large. The speed of flight depends on size, but probably ranges from about 10 cm/s for small species to 50 cm/s for large ones.

Some species will fly in a flight chamber. A model in which individual insects fly upwards towards a light but are prevented from reaching it by a downward draught of wind (Kennedy and Booth, 1956) is moderately satisfactory. The downward speed of wind needed to keep the insect hovering at a chosen height in the chamber is equal to its vertical air-speed. With careful manipulation of the wind, thrips will fly untethered for a few minutes in this type of chamber, but it is less successful for thrips than for aphids because they are easily disorientated, fall out of the light beam and are lost.

6 Sampling

Thrips can be collected easily by sweeping vegetation of even height, by beating individual plants or bushes over a sheet, or by extracting them from leaf-litter and soil. Each collector has favourite methods of collecting and of preserving and mounting specimens for identification and morphological or taxonomic studies. Some methods of preparing dead thrips for detailed structural examination are described in Appendix 2.

The scope of this chapter extends beyond simple collecting, to describe methods of sampling that give qualitative and quantitative estimates of the size and distribution of living populations. In addition to many of the difficulties associated with sampling populations of insects (Southwood, 1966) thrips pose special problems because individuals are so small, many species are easily dried-out and killed, and different stages in the life-cycle often occupy different environments. They may inhabit parts of plants inaccessible to the observer and often enter crevices in sampling apparatus. Their behaviour often changes with changing weather, populations range from sparse to extremely dense, and spatial disposition differs between species. All these have discouraged ecological studies and especially the standardization of sampling methods. In this chapter the different sampling methods available are therefore described and compared, and those most suitable for each purpose recommended.

Apparatus for Sampling and Extraction

Populations of thrips in the three main environments they inhabit, soil, vegetation, and air, must be sampled by different methods.

POPULATIONS IN SOIL AND LITTER

Soil and litter containing thrips can be collected fairly easily but extraction of the animals from samples is difficult. Living surface vegetation should be removed before sampling. Soil samples are usually taken with a 10 cm diameter soil-corer (Bullock, 1964; Edwards and Fletcher, 1970). Where populations are dense, as beneath heavily infested host plants, a smaller core may suffice. Large corers compress the soil least, so thrips contained in them can probably emerge more easily. The depth of samples required depends on the species and season. Samples for pupae in the soil probably need not extend below 10–25 cm, but it may be necessary to penetrate to 80 cm for hibernating thrips in mid-winter (see p. 168). Individuals often burrow more deeply in loose, cultivated soil around crops than in undisturbed soil. For populations of pupae or larvae in soil around host plants, where the numbers of individuals decreases rapidly with increasing distance from the plant, a scissor-type grab sampler (Webley, 1957) with blades 10–15 cm deep may be more convenient than a corer. Cores should be bulked to obtain a mean of at least 1 thrips/sample, so for typical densities occurring in the field, sample size may range from 500 to 20,000 cm³ of soil (see p. 169).

Another method of measuring the depth to which thrips penetrate the soil to pupate is to use an emergence box with a transparent plastic bottom spread on the inside with sticky grease. After larvae have descended from plants the box is placed upside down over soil surfaces exposed at different depths and the emerging thrips collected on the grease as they ascend towards the light (Harrison, 1963). This method avoids soil washing, but is probably only satisfactory in dry weather and in hot climates where pupation is soon completed.

Litter can be sampled by area, weight or volume. The litter within a grid 1,000–10,000 cm² is scraped loose to soil level and collected (Faulkner, 1954; Lewis and Navas, 1962; Healey, 1964; Ward, 1966). A metal box without top or bottom but with a sharpened lower edge that can be pressed into the litter makes this easier. In studies on hibernating thrips, Wetzel (1963) collected samples each of 100 g of grass stalks and litter, but unless the plant material is dry the size of samples collected by weight may differ. An assessment of the distribution of host plants is also needed to obtain the density of thrips per unit area by this method. For sod-dwellers, Post and Thomasson (1966) found that turves 400 cm² × 7·5 cm deep were convenient units.

Thrips are easily trapped in condensation and quickly die, so samples should be stored in finely woven material bags, or if in plastic bags, kept cool (<5°C) to inactivate the thrips and prevent them walking into moist patches on the sides.

Thrips are usually extracted from samples of soil or litter by a dry funnel method based on those described by Berlesé (1905) or Tullgren (1918), in which the insects are stimulated to move from samples by heat applied from above by light bulbs or infra-red heaters (Faulkner, 1954; Lewis and Navas, 1962; Stannard, 1968). They are usually collected into 70% alcohol. Less often, flotation techniques have been used in which

TABLE 8

Comparison of methods of extracting Thysanoptera from clay-loam soil under three types of management
(after Edwards and Fletcher, 1970)

[Mean number per soil core (10 cm diam. × 5 cm deep) is expressed as log $(N + 1)$]

Method†	Dry funnel							Flotation				s.e.	l.s.d.	Variance ratio
	A	B	C	D	E	F	G	H	I	J	K			
Woodland	0·029	0·048	0·0	0·0	0·067	0·048	0·056	0·086	0·019	0·322	0·111	0·042	0·110	4·436**
Pasture	0·0	0·019	0·019	0·029	0·019	0·0	0·056	0·0	0·169	0·481	0·0	0·044	0·116	11·270**
Fallow	0·0	0·0	0·0	0·0	0·019	0·0	0·038	0·086	0·094	0·056	0·049	0·026	0·068	1·799

† A. Simple plastic funnels (Edwards and Fletcher, 1970); B. Rothamsted controlled gradient funnels without heat (Edwards and Fletcher, 1970); C. Rothamsted controlled gradient funnels with heat (Edwards and Fletcher, 1970); D. Split funnels (Murphy, 1962); E. High-gradient funnels (moist regime) (MacFadyen, 1962); F. High-gradient cylinder extractor (MacFadyen, 1962); G. Infra-red extractor (Kempson et al., 1963); H. Simple brine flotation (Edwards and Fletcher, 1970); I. Salt and Hollick flotation (Salt and Hollick, 1944); J. Mechanized flotation (Edwards and Heath, 1963); K. Grease film extractor (Aucamp and Ryke, 1964).
  **Significantly different at 0·01% level. s.e. = standard error of means. l.s.d. = least significant difference.

thrips and other invertebrates, with a specific gravity between 1·0 and 1·1 (Edwards, 1967), float to the surface of solutions with s.g. of 1·2 or more, whereas denser soil sinks (Bullock, 1964). Flotation methods collect dead and living animals indiscriminately and delicate specimens may be damaged during extraction. Funnel methods are simpler to operate and the thrips recovered are in good condition, but they are probably less efficient for extracting larvae than adults, and useless for pupae.

The absolute efficiency of these extraction methods is unknown, but a useful comparison between eleven methods of extracting soil animals has been made by Edwards and Fletcher (1970; Table 8). For thrips, flotation methods, especially the Salt and Hollick (1944) method (Fig. 39) and a

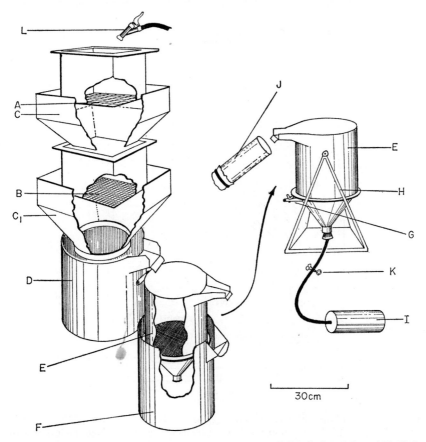

30 cm

FIG. 39. Exploded diagram of soil washing apparatus (modified after Salt and Hollick, 1944, and Ward, 1966). A: coarse sieve; B: medium sieve; C, C₁: splash guards; D: settling can; E: Ladell can; F: drainage tank; G: air outlet tap; H: pivot stand; I: air pump; J: collecting tube; K: pinch cock; L: adjustable spray jet nozzle. See text for procedure.

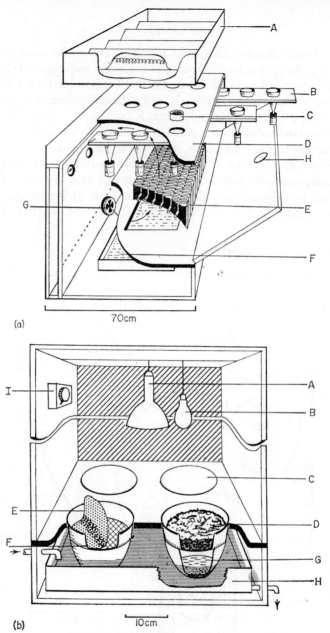

(a)

70cm

(b)

10cm

Fig. 40. (a) Exploded diagram of high-gradient cylinder extractor (redesigned after Macfadyen, 1961). A: tray housing heating elements (each 20 gauge nichrome wire, with resistance of 7·5 ohm); B: sliding racks holding aluminium funnels and collecting tubes containing alcohol; C: soil core (5 cm × 10 cm diam.) in metal cylinder with gauze bottom; D: insulated plate with holes for core containers; E: refrigerated cooling unit; F: water tray in sealed lower compartment; G: fan; H: hygrometer. Arrows show the path of moist, cool air when the insulated box is sealed during extraction. (b) Exploded diagram of infra-red bowl extractor (after Kempson *et al.*, 1963). A: 250 W infra-red lamp; B: 100 W lamp; C: screen; D: leaf litter; E: fine gauze; F: coarse grid; G: aq. picric acid solution; H: cold water bath; I: Sunvic simmerstat.

mechanized version of it (Edwards and Heath, 1963), were generally more efficient than the more widely used funnel methods. There were no significant differences between the dry methods tested, but the high-gradient funnel extractor (Macfadyen, 1962), in which the funnels are cooled while the upper surface of the soil core is heated and humidity is controlled, was probably most useful (Fig. 40a). Thrips are especially vulnerable to condensation on the inner walls of funnels, but a gap between the wall and the sieve containing the sample prevents this forming (Haarløv, 1947). Insects should be extracted in this apparatus as quickly as possible but without heating the upper surface of cores above 30°C on the first day and 45°C thereafter. The temperature and rate of drying can be regulated with dampers. Complete extraction of thrips from undisturbed cores may take about 7 days and from litter up to 3 days. Extraction was also satisfactory by the infra-red bowl method (Kempson et al., 1963), in which the funnels are replaced by a wide-mouthed bowl filled with a saturated aqueous solution of picric acid (Fig. 40b). However, the surface of the preservative is only a few centimetres below the sample, so a high humidity is produced by evaporation in the small volume of air between them. This may be a slight deterrent to thrips because a comparison of high-gradient funnels using moist (90% R.H.) and dry (70%) regimes suggested that they were extracted more efficiently from litter when the surrounding air was dry (Macfadyen, 1961).

When funnel extractors are used, recovery is greater from unbroken soil cores inverted on the retaining gauze, than from crumbled or upstanding samples, and cores should not touch the sides of the container or thrips may be trapped in condensed water droplets. For adults attracted to light, a spotlight shining below the collecting tube, shielded by a water jacket to prevent heating, may attract individuals and discourage wandering around the walls of the extractor (Dietrick et al., 1959).

Thus, because they are much more convenient, funnel extractors are useful for relative comparisons of faunas, but when reliable estimates of the absolute numbers in soil are required, for example, estimates of the density of overwintering larvae in a field, a flotation method of extraction is recommended. A brief account of the procedure based mainly on descriptions by Salt and Hollick (1944) and Raw (1955, 1962) is therefore included (see also Fig. 39).

(i) *Dispersion.* Soil particles, especially clay soils, can be dispersed by soaking cores in water, deep freezing them and thawing. Particles in light to medium soils can also be dispersed by soaking cores in sodium citrate or sodium oxalate. For heavy clay soils, a combination of physical and chemical methods may be necessary: cores are crumbled in a plastic container and covered with a solution of sodium hexametaphosphate (50 g) and sodium carbonate (20 g) (sold commercially as "Calgon") in a litre of

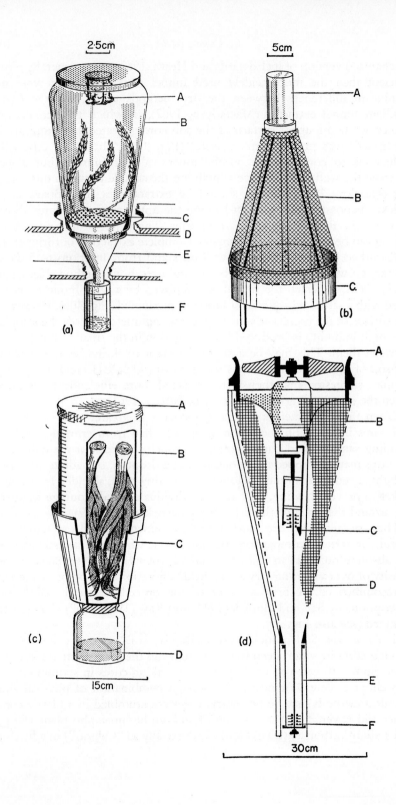

(a)   (b)   (c)   (d)

water and placed in a vacuum desiccator to extract the air. Atmospheric pressure is then restored and samples are frozen at 10°C for at least 48 h (Raw, 1955).

(ii) *Washing.* The thawed sample is placed in the upper, coarsest sieve of the washing apparatus and the soil washed through by gentle jets of water, whilst the vegetation and large lumps of soil are teased apart.

The mesh of the lower sieve should be coarse enough to allow thrips to fall into the settling can, which can be swivelled on its pivot to tip them into the Ladell can. The mesh size in the Ladell should be about 125 μ to retain all thrips, and the water level in the drainage tank is best maintained just above this sieve to prevent blockage. Air locks that might form under such a fine sieve can be prevented by fitting an outlet tube with a tap to the side of the Ladell just below the sieve (Ward, 1966). When most of the water in the Ladell has drained through the sieve, the Ladell is lifted from its stand and allowed to drain completely.

(iii) *Flotation.* The lower opening of the Ladell is closed with a bung and the flotation liquid, usually magnesium sulphate (s.g. = 1·2), poured in until the Ladell is about two-thirds full. Air is then bubbled from the bottom of the Ladell to free any animals trapped in the sieve, and more flotation liquid added from below until the floating debris flows over the lip to be collected in a tube with a piece of bolting-silk or fine mesh gauze fastened over the end. The sides of the Ladell near the level of the overflow should be waxed to prevent thrips sticking. The flotation liquid is retained for further flotations.

(iv) *Separation.* In arable soils containing little residual vegetation thrips can often be picked directly from the other insects at this stage, but in soil with much vegetation the insects and plant debris need separating. To do this the material collected from the flotation stage should be rinsed free of flotation liquid, placed in about 300 ml of water, and 10–15 ml of xylene added. The thrips and other insects will collect in the xylene layer; the xylene/water interface should be agitated to disperse bubbles and ensure that no thrips are missed. It is often possible to distinguish the different species and instars present at the interface without mounting the specimens.

FIG. 41. (a) Turpentine vapour extractor (after Lewis, 1960). A: cotton wool soaked in turpentine; B: lamp glass; C: perforated zinc gauze; D: wax seal; E: polythene funnel; F: water and detergent. (b) Emergence trap (after Wetzel, 1963). A: glass collecting tube; B: fine gauze tent on metal supports; C: metal cylinder. (c) Apparatus for extraction of thrips from vegetation by heat (after Shirck, 1948). The whole apparatus is placed in an incubator, usually at 49–52°C. A: cloth top; B: cardboard cylinder; C: plant pot; D: 0·5% formalin. (d) Suction trap with automatic segregating mechanism for sampling airborne populations (simplified after Taylor, 1951). A: fan inlet; B: fan motor; C: disc-dropping mechanism; D: gauze filter net inside iron framework; E: collecting tube; F: segregating discs.

The temperature and duration of storage affects the number of thrips extracted from soil cores. Samples should be stored at 5°C and extracted within a month of collection, otherwise recoveries decrease (Edwards and Fletcher, 1970).

Thrips can be expelled from dry litter with turpentine vapour (Lewis and Navas, 1962) or methyl isobutyl ketone (Post, 1947) (Fig. 41a). Wetzel (1963) placed samples of litter collected in winter in emergence cages (see Fig. 41b) and caught the thrips as they emerged naturally in spring. Populations that emerge from soil in spring can also be sampled with emergence cages placed on the ground. For collecting *Taeniothrips inconsequens* in pear orchards, Bailey (1944) used solid-sided pyramids with a base of 1 m² and a glass collecting bottle at the apex.

The number of species and individuals living in soil and litter differs markedly between wide geographical regions. For example, there are fewer soil and litter dwellers in the west of North America than in the east, and in Australia more thrips occur in dry litter in semi-arid areas than beneath rain forest (Mound, *in litt.*).

POPULATIONS ON VEGETATION

There are few species of plants on which thrips can be seen easily and on which they remain still for long enough to be counted reliably *in situ*, so estimates of population size are usually most precise when infested vegetation is removed for the thrips to be extracted and counted. Sometimes thrips can be dislodged from vegetation in the field, leaving the plants intact.

Estimates from collected vegetation. The methods of separating thrips from samples of vegetation differ, depending on the physical characteristics of the plant and the behaviour of the species of thrips. For species that hide inside crevices such as leaf sheaths or deep inside compact inflorescences, a dry, dynamic extraction method that stimulates thrips to crawl off the vegetation separates them most efficiently, whereas species living exposed on the plant are best killed and washed off. Heated funnels, similar in principle to those used for extraction from soil (see p. 101), are suitable for dry extraction, but they can be of simpler design because vegetation need not be heated for as long or as gently as soil cores, and temperature gradients within the funnels seem less important. Figure 41c shows a simple model which, with vegetation inside it, is placed completely in a warm incubator. Condensation on the inside is prevented by the cloth cover and the cardboard and porous pot walls, which absorb the moisture, though a few thrips may need picking from the bottom of the pot after extraction. For *Thrips tabaci* temperatures of 49–52°C stimulate the quickest emergence, and 92–96% of those extracted emerge within 8 h.

However, at 52°C half the larvae are killed before they leave the samples and extraction at 46°C for 24 h is more efficient (Fig. 42). Temperatures

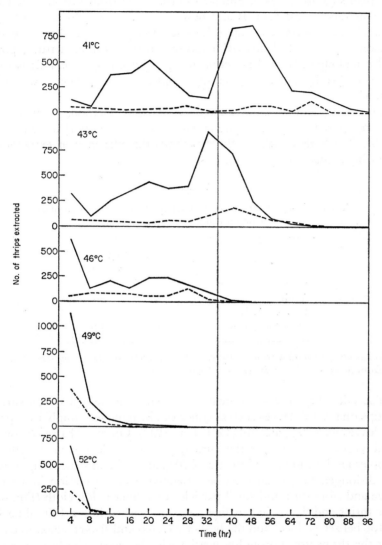

FIG. 42. Time required for the extraction of *Thrips tabaci* from samples of onion plants at different constant temperatures. Solid lines represent larvae, dotted lines adults (after Shirck, 1948).

below this are usually less satisfactory because eggs already on the vegetation hatch and young larvae, not present when the samples were collected, therefore emerge. However, temperatures as low as 43°C may be

needed for vegetation that wilts easily. The method is suitable for extracting thrips from onion plants and seed heads, lucerne, clover, bean and
pea plants (Shirck, 1948). Similar funnels made from cartons fitted into
long-stemmed glass funnels and heated in a cabinet (Hoerner, 1947) or
from above by infra-red lamps (Faulkner, 1954) have been used to extract
different species of thrips from lucerne, cotton, onion, lettuce, potato
and cantaloupe plants. Extraction using repellants (see p. 112) is much
quicker ($\frac{1}{2}$–1 h) and the apparatus simpler (Evans, 1933; Lewis, 1960;
Fig. 41a).

The most serious disadvantage of dynamic methods is that slow-
moving or sedentary stages are extracted less efficiently than active ones
(Table 9) whereas with washing methods the efficiency of extraction is
probably similar for all stages.

TABLE 9

The efficiency of extraction of different stages of thrips from dry
vegetation using turpentine vapour

| | % extracted | |
	(a)	(b)
Adults	85	86
Larvae I & II	67	50
Prepupae, pupae	19	17

(a) *Limothrips cerealium* from wheat ears (Lewis, 1960); (b) *Thrips physapus* and other
sp. from flowers of *Leontodon hispidus* (Ward, 1966).

The suitability of washing methods depends on the type of leaf surface.
Thrips on smooth leaves such as coffee can be removed simply by dipping
the leaves into 70% alcohol (Le Pelley, 1942). On more hairy or convoluted vegetation, such as the curled, pinnatisect leaves of pyrethrum,
samples are best immersed in jars of 70% alcohol and shaken vigorously;
the dislodged thrips can then be separated from vegetation in a coarse
sieve and from dust and small particles by a finer nylon cloth (Fig. 43a).
The thrips and debris in this fine sieve are washed into a 100 ml beaker,
a little benzene added, and the beaker covered and shaken again to ensure
that the thrips are thoroughly wetted with benzene. The beaker is then
placed in a 9 cm petri dish and the benzene and some water overflowed
into the dish by introducing a jet of water below the interface. Thrips
adhering to the outside of the beaker are brushed off and the beaker
removed. Before counting, the interface in the petri dish is completed by
adding benzene or water as appropriate, ensuring that the meniscus is
concave, otherwise thrips accumulate at the edges of the dish and are

difficult to count. Brushing the aggregations at the edges with absolute alcohol disperses the thrips (Bullock, 1963a).

To help count thrips washed from vegetation in this way, Bullock recommends that a light alloy grid be dipped into absolute alcohol and placed in the dish dividing it into areas each small enough to inspect with a traversing microscope (Fig. 43b). The disturbance at the interface

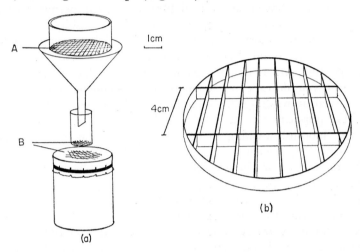

FIG. 43. (a) Simple apparatus for washing thrips from leaves, with 70% alcohol, A: coarse sieve; B: fine nylon (125 μ) mesh (see text for procedure). (b) Metal grid fitted into a petri dish to facilitate counting (after Bullock, 1963a).

caused by the alcohol on the grid makes the thrips lie near the middle of each grid compartment and facilitates counting. Samples containing up to 1,000 individuals can be extracted and counted in about 15 min using this method. It is also helpful when counting specimens collected in alcohol to add a few drops of borax carmine to stain the larvae and pupae, then to filter thrips from the alcohol and count them on the filter paper (Faulkner, 1954).

Bullock (1963a) also devised a quick, approximate method, convenient for use by farmers, for estimating populations on pyrethrum, which may also be suitable for other thrips living on leaves. Leaves are collected into wide-necked jars and shaken with about 100 ml of rectified or methylated spirit. The leaves are removed, about 50 ml of petrol added, then the whole mixture decanted into a dish through a coarse household sieve to retain the larger pieces of debris. Enough water is added to cause the petrol to float above the spirit layer, and enough methylene blue to dye the spirit bright blue so that thrips above the interface are clearly visible and can be counted by the naked eye.

A simple method of collecting *Frankliniella tritici* and *F. fusca* from cotton leaves and probably other simple, fairly smooth leaves, is to cool the plants in a sealed paper bag at 3–4°C for several hours to inactivate the thrips, then to dislodge them by striking the plants sharply against a cloth stretched over a frame (Hightower and Martin, 1956).

Taylor and Smith's (1955) comparison of three methods of extracting *F. tritici* from rose blooms, expulsion by turpentine, washing (10 min in detergent at a concentration 2 ml/litre water), and permitting natural development of larvae and pupae, showed that washing removed all thrips and was the most useful method for quick extraction, whereas about 20% were found dead in the flowers after overnight extraction with turpentine, and a similar percentage after ten days natural development. However, flowers may contain dead specimens and as these are also removed by washing this method may sometimes over-estimate the size of the population. Ota (1968) confirmed that for removing thrips from rose blooms, washing in either 70% ethyl alcohol or detergent solution (95–97% efficient) was superior to expulsion by ethyl acetate or methyl isobutyl ketone (36–39% efficient), or to simply shaking dissected blooms vigorously (38% efficient).

When young I stage larvae occur in tight leaf clusters or leaf sheaths it may be necessary to search for them under a binocular microscope. No extraction methods are suitable for eggs embedded in tissue. On thin leaves they can usually be seen easily as translucent swellings when held against the light and are visible in flowers pickled in alcoholic Bouin's fluid (Laughlin, 1970). Hall (1930) describes a method of making them visible in thicker tissue (see p. 16).

Bark from trees with girth greater than about 10 cm can be sampled by cutting squares of bark down to the phloem with a hammer and chisel. The thrips can be removed efficiently by breaking the bark into small pieces over a sieve and shaking them vigorously over a white tray. Sixty to 80% of the thrips fall on to the tray and most of the remainder can be removed in a turpentine extractor. The combined methods are 98–100% efficient for flaky, rough or smooth bark (Lewis and Navas, 1962).

Estimates from standing vegetation. The simplest and most widely used equipment for sampling thrips directly from vegetation of even height is a sweep net. The usual nets have a diameter of about 30 cm at the opening tapering to 10 cm at the bottom, with the net about 60 cm deep and a handle 80 cm long. A D-shaped net is more useful for sweeping vegetation shorter than about 15 cm. The fabric may be linen or strong cotton but should not be hairy or the thrips are difficult to remove from it. They stick to the slightest film of moisture in the net so synthetic fabrics are unsuitable and only dry vegetation can be swept satisfactorily. The efficiency of sweeping depends on the vigour of the stroke, the length and evenness of the vegetation, the vertical distribution of thrips present, and changes in their activity and position on the plant which depend on the weather and the time of day. Thus, there is often great variation between samples and although attempts have been made to convert swept samples into absolute estimates (e.g. Holtmann, 1962), sweeping is more useful for rapid qualitative assessment of faunas than for reliable absolute estimates of the size of populations, or even for detecting short-term changes in population density (see p. 125). It is however cheap and a quick way of collecting large samples, though thrips collected in a few minutes may take an hour to empty from the net.

Estimates of population density in short vegetation can also be obtained with suction samplers (e.g. Johnson *et al.*, 1957; Southwood and Pleasance, 1962) which permit the sampling area to be defined more precisely than with sweeping. Johnson *et al.* (1957) claimed a 97·5% extraction of Thysanoptera from rough grassland using this method. This was probably an optimistic estimate of its efficiency because it is unlikely that all thrips remaining in the grass were found when it was later searched by hand in the laboratory; the efficiency could only be evaluated properly by washing remaining thrips from the debris.

A comparison of sweep netting and suction methods by Heikinheimo and Raatikainen (1962) showed that on timothy leys, nets caught about 25 times more thrips per unit area than suction samplers, and on spring cereals about 18 times more, assuming that 50 sweeps covered 1 m². However, neither method is likely to sample satisfactorily larvae or species that live concealed on plants.

Dwellers on superficial parts can often be beaten from individual plants and collected on a tray for counting. For example, McGregor (1926) shook infested twigs of citrus over a sticky paper held in a frame and later removed it for counting. The paper was protected from leaves and coarse debris by a wire mesh shield above it. Numbers of *T. tabaci* on onion fields were estimated by Sakimura (1937a) by tapping all plants growing in a given 3 m of a row over a black card marked with two smaller squares equal to about 1/20 of its area, counting the thrips in the two sub-samples and multiplying the result by 20. This method is quick but not very accurate because thrips are likely to be concentrated near the middle of the collecting card and not scattered randomly over it. To sample *T. tabaci* on potato leaves, Powell and Landis (1965) placed a cotton lint cloth beneath plants and tapped them five times with a hand to dislodge the thrips, which were temporarily entangled in the fibres of the cloth and counted. This method was more efficient than sampling with a brushing machine (Henderson and McBurnie, 1943) and three times as quick.

Beating and shaking are more appropriate for sampling bushes or small trees (Kratochvil and Farsky, 1942; MacPhee, 1953; Putman, 1965a) but it is difficult to obtain absolute estimates by this method unless "units" of the tree, such as an "armful" of branches, are beaten so that the proportion of the whole tree included in the sample can be estimated, and from it the total population computed. As thrips are often distributed unevenly over trees, units representative of different heights and aspects should be taken. To estimate the numbers of *Taeniothrips laricivorous* on young spruce trees, Vité (1956) removed enough twigs to fill a 2-litre beaker from different heights of each tree. He then estimated what proportion of each section sampled was represented by the twigs in the beaker, and computed the numbers on the trees. MacPhee (1953) tried to collect the total population on a small apple tree by suspending a paper cone under it and applying a quick-action toxic dust (6% DDT, 10% derris and 30–40% pyrethrum) to foliage and branches with a small power-duster, but this method was tedious and many thrips remained in crevices. Jarring and brushing branches with a stiff brush and collecting the dislodged thrips on a white cloth held beneath is probably a more efficient method (Putman, 1965a). Weitmeier (1956) sampled twigs on bushes and trees using a hinged frame covered with white cloth. To sample, the jaws of the frame were snapped over twigs and the enclosed vegetation shaken to dislodge the thrips. A more sophisticated hinged box with leading edges faced with sponge rubber (Dempster, 1961) may be more useful. The insects on the twigs enclosed in this can be killed with ethyl acetate or CO_2 released into the closed box through a small hole. Larvae of *Isoneurothrips australis* falling from *Eucalyptus* trees were simply caught in a funnel that led into a bottle containing 50% alcohol (Laughlin, 1970).

It is not practicable to estimate the size of a population using marking and recapture methods because the number of thrips is usually so large that only a very small proportion can be marked. However, marked individuals may be used to trace movement on individual plants. Thrips should be removed from the plant, lightly dusted with zinc sulphide and returned to the plant. The marked thrips can be detected later under U.V. light when they fluoresce with an orange-yellow glow. Zinc sulphide will stick to thrips for up to 10 days without apparently harming them.

AERIAL POPULATIONS

In many ways it is easier to sample airborne populations than terrestrial ones because the air is homogeneous and the insects easily separated from it, but the results from aerial traps are far more difficult to interpret than counts from vegetation. False conclusions may be drawn from catches unless the way the different traps function is understood. It is therefore essential to define the object of trapping clearly before embarking on a programme because the aim determines the type of trap required. For example, the traps most suitable for determining the seasonal occurrence of a single species are quite different to those suitable for determining the absolute density of the aerial fauna or the age-structure of flying populations.

Samples from exposed sites

About two metres above the ground or above the tips of vegetation, the wind speed almost always exceeds the flight-speed of thrips, which are therefore blown at random on to impaction traps or into nets placed in such exposed situations. In general surveys of aerial fauna, thrips have been captured between 3 and 3,000 m in nets suspended from kites (Hardy and Milne, 1938), masts (Freeman, 1945), balloons (Johnson, 1950) and ships (Holzapfel and Harrell, 1968), and also on sticky screens attached to aircraft (Berland, 1935; Glick, 1939). Although such passive traps can be used to detect thrips, and with suitable corrections to catches, to estimate vertical profiles of relative densities (see p. 151), they cannot provide estimates of the mean absolute density of insects at a given position over a period. Suction traps which collect a known volume of air are required for this (Fig. 41d). At 3–15 m above the ground, models delivering about 3,000 m^3 of air per h are suitable; above 15 m larger traps may be necessary (Taylor and Palmer, 1972). The efficiency of suction traps changes with increasing wind speed, but corrections for this can be made from published tables (Taylor, 1962). When a vertical series of traps is used to define the density $\times$ height profile, total aerial populations can also be estimated (Johnson, 1957).

Samples from sheltered sites

Below about 2 m and especially among vegetation where wind is slower, flying thrips may often be able to control their track and alight on chosen sites. Thus thrips caught in these situations may have been blown to traps accidentally or have been attracted to and alighted on them deliberately. Quantitative and qualitative estimates of populations made with attractant and non-attractant traps in slow winds may therefore differ enormously, so in sheltered places the type of trap used must be chosen carefully. Because estimates of aerial populations are usually made in this partially sheltered zone, the models of traps available and the best ways of using them are discussed below.

Absolute estimates. Suction traps are most useful for measuring the absolute density of aerial populations, the daily periodicity of flight, the specific structure of faunas and for relating numbers taking-off to known weather conditions. Below 2 m a small, exposed-cone model with a 23 cm fan (Fig. 41d) (Plate IX) delivering about 460 m^3 of air per h (Taylor,

1951) usually catches samples of convenient size (10–100 thrips per h) with individuals clean and easily identifiable. When the orifice of the trap needs to be near the ground the cone should be lowered into a long slit trench to allow air to escape freely and to prevent recirculation of sampled air.

Wind is the only environmental variable affecting the efficiency of this type of suction trap, but its effect is small in wind speeds less than 15 k.p.h. and can be allowed for if necessary when counting (Taylor, 1955, 1962). There is no evidence that these traps attract thrips. The catch can be collected directly into 60% alcohol in a polythene bottle fastened beneath the gauze cone; the bottle should have a small piece of gauze welded into one side to act as an overflow in rain. Alternatively, the catch can be collected in a metal tube, and for studies on periodicity and flight thresholds, segregated automatically into hourly samples by discs that fall into it. The thrips are killed and lightly stuck to the discs by a film of pyrethrum and light machine oil mixed in 100 ml of petrol ether and applied to the discs before they are loaded into the trap.

In dry weather, individuals caught on discs may be slightly desiccated and unfit for internal dissection.

The numbers of insects caught in wind speeds of less than 7 k.p.h. can be converted directly into terms of aerial density using the formula:

$$\log \text{catch/h} + \text{conversion factor} = \log \text{density}/10^6 \text{ ft}^3 \text{ of air.} (\equiv 28{,}300 \text{ m}^3)$$

The appropriate conversion factor for thrips of different sizes is given in Table 10. At wind speeds above 7 k.p.h. a small correction to the figures is necessary to give precise estimates of aerial density because the

TABLE 10

Conversion factors for 23 cm Vent-Axia suction trap (after Taylor, 1962)

Insect	Wind speed (k.p.h.)				
	0–3	4–6	7–10	11–13	14–16
All Terebrantia and Tubulifera					
<4 mm long	1·89	2·00	2·12	2·23	2·34
Large Tubulifera					
>4 mm long	1·97	2·08	2·20	2·31	2·42

efficiency of the trap is lowered in strong winds (Taylor, 1962), but this is a refinement that need rarely be used for populations of thrips near vegetation level.

The main disadvantages of suction traps are that they are expensive compared with simpler traps and require a source of electric power. Nevertheless, it is cheaper in the end to use them to collect reliable data on the aspects of aerial populations mentioned, than to spend time collecting, by simpler methods, samples that are unreliable, but which still require as much effort and time to sort and count. Where populations are more dense than about 1 per m³, an order of density typical over heavily infested crops, catches in traps delivering 460 m³ of air per h may be too large. A smaller trap can then be used, or catches may be sub-sampled for counting and identification, by taking aliquots from an agitated suspension of thrips in alcohol.

Relative estimates. Most studies of aerial populations of thrips have been made with simple traps, regardless of the type of information required. Sticky traps and water traps, whose efficiency is unknown and depends on the size and shape of trap, wind speed, and probably the size of thrips, have been most popular.

Flat sticky surfaces in horizontal and vertical positions have been used, but cylindrical surfaces are more efficient than either these or the swivel-ling impaction traps used by Johansson (1946) because the airflow around them is less turbulent (Gregory, 1951), and because they also catch insects blown from all directions. They are useful for detecting the presence of flying thrips and especially to record times of emergence from hibernation and seasonal changes in activity. They are best placed within 2 m of crop level but can also be used to estimate vertical profiles of relative density at greater heights because the size of the catch on them depends largely on the population density and wind speed, for which a correction can be applied (see p. 151; Lewis, 1959b). Failure to correct the figures may result in misleading profiles with apparent maximum densities at different heights for different species, and lead to the false conclusion that different species concentrate or prefer to fly at a chosen height. Among field crops where there is little wind, sticky traps catch few thrips, and often become covered in debris.

The choice of the size of trap depends on the frequency with which it can be inspected and the probable density of the aerial population. Generally, the shorter the interval between inspections the larger the trap needed. Over cereals, where thrips are often abundant, the sticky surface need not exceed 100 cm² for daily inspection, but above vegetation supporting fewer thrips larger traps are needed to collect sufficiently large samples. Cylinders with a surface area of about 250 cm² have been used around onion fields (Wolfenbarger and Hibbs, 1958) and of 650 cm² around cotton fields (Hightower and Martin, 1956). Attempts have been made to use vertical, flat, sticky boards facing different directions to detect the source and direction of movement of populations (Körting, 1931; Harding, 1961) but insects as small as thrips accumulate on the leeward side of vertical barriers (Lewis and Stephenson, 1966) so numbers caught on opposite sides of vertical sheets give no clear indication of the direction of movement. Similarly, traps with vertical barriers designed to segregate insects entering from different directions (Coon and Rinicks, 1962) are likely to give suspect results with thrips.

Transparent or brown tree-banding grease is the best grease to use on sticky traps, but it is difficult to remove individuals from it without damage. When it is necessary to mount specimens for identification most greases can be dissolved from them by organic solvents. Turpentine or kerosene can be used to clean the traps. When sampling total populations or easily identifiable species, glass boiling tubes about 20 cm long × 3 cm diameter make convenient sticky traps. The grease is spread evenly over the outside and the tube supported in the field on an upright pipe or rod, preferably painted black (Plate IXa). The tube can then be taken to the laboratory for examination and replaced by a freshly greased one. It is easier to count thrips stuck in the grease if the exposed tube is supported over a white rod or illuminated strip-light. Sub-samples of any size can be counted by drawing a grid on the internal support, and counting only those thrips that occur between the chosen grid lines.

Water traps are useful for catching large numbers of thrips easily and for collecting flying individuals in a condition suitable for dissection. Thrips required for this purpose should be removed as soon as possible from the water, and never left in it for more than 24 h. Traps should be about 6 cm deep with a surface area of 250–500 cm², and preferably round, with the water level about 2 cm below the rim (Plate IXa). A few drops of detergent should be added to the water so that thrips sink and do not drift to the edges and escape; a drop of formaldehyde prevents algal and fungal growth. Thrips are easily removed with a pipette. Water traps cannot be used to estimate absolute densities and are unreliable for estimating relative densities at different heights because their efficiency is probably affected by different wind speeds at different heights and by the level of the water in the traps; the faster the wind, the greater the coefficient of variation of catches (see p. 127; Lewis, 1959b). They are best used at vegetation level but not where leaves and debris can fall into them, because these discolour the water. A coarse wire-netting guard may be necessary to prevent birds bathing in them.

Some species of thrips are attracted to pale colours, especially white. Black traps should therefore be used to avoid differential attraction when studying the specific structure of a fauna. The number of individuals, and the specific composition of the thrips community caught over a wheat field in water traps painted black, green and white, and in a black suction trap is shown in Fig. 44. The proportions of species in the suction trap, which was non-selective, may be taken as a standard with which to compare catches in the water traps. The specific composition of the catches from black and green traps closely resembled that from the suction trap, but *Taeniothrips* and *Thrips* spp. dominated the catch in white traps and suggested an entirely false community composition (Lewis, 1959b). Similar anomalies may occur between sexes in attractive and non-attractive traps: the ♀: ♂ ratio of *Stenothrips graminum* recorded by Lewis (1961) in white water traps was about 5:1 but in suction traps at the same site 40:1. Wilde (1962) also caught twice as many thrips, mainly *Taeniothrips* sp., on sticky white boards hung in pear and cherry orchards as on yellow

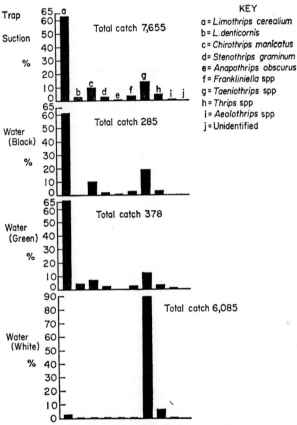

FIG. 44. The total catch and proportions (%) of different species caught in suction and coloured water traps over an English wheat field in June (after Lewis, 1959b).

ones, and Moffitt (1964) caught nine times as many *Frankliniella occidentalis* on white boards exposed among pear trees as on yellow ones. Mixed populations flying around cotton fields were much more strongly attracted to white, blue and, to a lesser extent, to golden aluminium foil than to silver, red, green, black or brown (Beckham, 1969).

It is a puzzling fact that while many Homoptera such as species of psyllids, aphids and jassids feed on young pale-green plant tissue similar to that chosen by thrips, no thrips are known to be markedly attracted to yellow whereas many species in these homopterous families are (Moericke, 1950; Kennedy *et al.*, 1961; Wilde, 1962). The difference may be because the homopterans have colour vision whereas Moller-Racke (1952) claims that Thysanoptera cannot see colours but only degrees of brightness. In her experiments she offered only orange and blue as a choice, and the thrips used were unnamed species from Compositae. Even if correct, this claim may not apply to all Thysanoptera because *L. cerealium* can distinguish between light of different wavelengths (Holtmann, 1963) which may be why it is not attracted to white. However, white, to which many other terebrantian species are attracted, would usually present the greatest contrast with the surroundings, and this might explain its attractiveness.

Attractive traps depend partly for their catch on stimulating thrips to settle, but the insects may respond to them differently than to vegetation. Coloured traps therefore cannot be used to estimate the number of thrips which have alighted on a unit area of vegetation, and should be avoided for quantitative studies. However, they are sometimes useful for collecting large numbers of certain species desired for other purposes, such as dissection. Similarly, aromatic substances such as cinnamic aldehyde, salicylaldehyde and anisic aldehyde added to water (Howlett, 1914; Morgan and Crumb, 1928) may attract species differentially so are undesirable for quantitative work.

Sometimes a few thrips are caught in light traps, but these are probably merely late-flying stragglers of large day-flying populations rather than true crepuscular or nocturnal fliers. Thus light traps are unlikely to be useful for ecological studies.

Sorting catches

A drawback common to all methods of trapping is sorting the thrips from the large numbers of other insects caught, and this usually takes far more effort and time than the trapping itself. Sorting is especially difficult when the thrips are mixed up with Lepidoptera in whose scales they lodge; with large numbers of small staphylinid beetles which they superficially resemble; or with delicate Diptera with whose appendages they become entangled. Catches collected in liquid can be sieved to remove the larger insects, but a mechanical vibrator is necessary to shake the sieves and to disentangle the thrips from them satisfactorily. A battery of mechanical sieves that separates typical suction trap catches into six groups of different-sized insects with about 92% of the thrips in the smallest category, helps to speed separation and counting (Arnold, 1971).

Most of the traps described and used for sampling aerial populations of thrips were originally designed to sample small insects generally (or aphids in particular), which are soft-bodied and easily squashed (Taylor and Palmer, 1972). Thrips differ structurally from many other insects commonly found in the air by being much smaller and tougher creatures, not easily squashed, and these features could perhaps be exploited in a trap designed specially for thrips and tough insects of similar size. One possibility might be to suck air through a coarse sieve to remove larger insects, vibrating the sieve to prevent clogging, and impacting the thrips and smaller insects on a lightly adhesive, moving, transparent plastic roll. They could then be examined directly or easily picked off for identification. The time of capture could be determined by the position of the thrips on the roll whose speed could also be regulated to space individuals conveniently for examination and counting.

Sampling Procedure and Precision

The principles of sampling, experimental design and analysis are obtainable from standard statistical textbooks, and the methods most appropriate for insects generally are summarized by Southwood (1966). Few of these methods have been used in field studies on thrips populations, so the accuracy of many published estimates of the size of natural populations and infestations on crops is unknown. Attention is therefore drawn in this section to aspects of sampling procedure especially relevant to populations of thrips, to encourage the collection of useful quantitative data.

TERRESTRIAL POPULATIONS

For populations in soil and litter the number of cores or other sampling units required, and the aggregate of units taken as a sample, can be calculated as follows. Given that 10 cm diameter cores are a convenient size to handle and that each usually contains at least one thrips, the number of units bulked to make a sample depends on the accuracy required and on the variance (s^2) of the sample units. Usually an error of 10% of the mean is satisfactory. The number of units needed to give this accuracy can be obtained from preliminary sampling which will provide a rough estimate of the mean and standard deviation (s) of units. These values can then be inserted in the formula:

$$n = \frac{100s^2}{m^2}$$

where n = number of sample units bulked in each sample

m = the sample average

s = the standard deviation

and an appropriate value for n chosen. For example when $m = 5$ and $s = 4$

$$n = \frac{1600}{25} = 64$$

When greater or less accuracy is required (an error of $c\%$) n can be obtained from the formula:

$$n = \left(\frac{100s}{cm}\right)^2$$

These formulae can also be used to determine the number of sample units of litter or vegetation needed to give a sample having the required accuracy.

The sites from which units are to be taken need choosing carefully. In fairly uniform habitats such as meadowland or in the soil beneath a cereal crop, random sampling is ideal; it is not very practicable, however, where host plants are clumped or on stony ground. Too few randomly chosen units may seriously bias a single sample, so random samples should contain at least five and preferably ten or more units. Where proper random sampling is not feasible, regular samples should be taken. In natural habitats procedures can be devised to suit the terrain. In artificially regular populations, such as those occurring on a row crop, the following method is useful. Count the rows (R) and their length (l) and compute the total length of row (R × l). Divide this by the required number of units (n) for the sample and so obtain the unit row length (lR/n). Beginning in the middle of the plot measure off a unit row and take a sample unit such as a leaf or plant. Repeat the process working along the row to the end and doubling back along the next row. Carry over any partly completed unit row length into the next row of plants each time. Upon reaching a corner of a plot, carry over the partly completed unit length to begin again at the diagonally opposite corner. The last sample unit should be taken near the starting point in the middle of the plot (Lewis and Taylor, 1967). Thrips are often more abundant near the edges of fields than in the centre, so when estimates of total populations are required sampling should be designed to include all parts of the field. When pupal populations in soil are sampled the method will need modifying to allow for the change in density of insects in the bare soil between rows.

Some of the difficulties of obtaining reliable estimates of populations living on vegetation and sampled by removing leaves from plants, were investigated by Bullock (1965) using *Thrips nigropilosus* on pyrethrum. Different vegetation and species of thrips present different problems but his work illustrates some general practical difficulties encountered when sampling and shows the preliminary approach necessary to define and overcome them. Pyrethrum plants are usually planted about 30 cm apart with 1 m between rows. When samples were taken by one of four methods: (i) ten leaves from each of 5 plants chosen at random; (ii) one leaf from 28 plants; (iii) one from 49, and (iv) five from 20 (the three latter being collected in different sequences of traverses across the rows), the sample from the five randomly chosen plants contained fewer thrips than the samples collected by other methods. This was for two reasons; (a) the thrips were aggregated in the plots, and (b) by limiting inspection to five plants some samples included no heavily infested ones. Thus with aggregated distributions, which are far more common in field populations than either regular or random ones, a sampling design that covers as many patches as possible and includes an effective proportion of the relatively few heavily infested plants must be used. Another factor affecting the accuracy of samples from populations of thrips on pyrethrum, and probably on most other row crops, is that the foliage of adjacent plants may touch, enabling larvae to crawl from plant to plant along the row but not between rows. Thus regular traverses across several rows, taking sample units from plants not in contact, will help to decrease the errors due to patchy distribution between and within rows. Sample units and the size of sample found convenient in different crops are listed in Table 11; unfortunately the accuracy of the estimates they provide is unknown.

Changes in the distribution of populations on growing plants, and the purpose for which estimates are required, also affect the suitability of sampling units. For example, mature pyrethrum leaves taken from the base of plants usually carry a greater population than an equal number of young leaves, but young leaves are smaller and the population on them is more dense. Thus leaves of different ages should be included in samples taken for detailed population studies. By contrast, young leaves are best excluded from samples taken for insecticidal assessments, except when trials are spread over a long period, because freshly opened leaves may have missed the spray application. Great differences between the numbers of thrips on leaves of different ages occur on many other species of plants and are especially noticeable where there are infestations of crevice-dwelling thrips attacking juicy tissues. For example, cereal thrips move to fresh tissue as it develops and are rarely found on the blades of mature leaves, so samples containing only old blades greatly underestimate the population on a crop. A problem with flower-dwelling species is that the number of flowers available to them is always changing, often rapidly, so that the periodic estimates of numbers per flower may not reflect changes in total population. The number of flowers, as well as thrips per flower, should be estimated simultaneously to derive a realistic estimate of the thrips population. Species that jump from foliage when disturbed are best sampled at cool times of the day when they are least active.

When a sampling procedure has been chosen it is useful to check the accuracy of the estimate it gives by taking two simultaneous samples and calculating their variance (Davidson and Andrewartha, 1948a). The accuracy will be greater for larger samples, but some precision must be

TABLE 11

Examples of convenient sample units and sample sizes when vegetation is removed from plants

Plant	Sample unit	Units per sample	References
Lucerne	Terminal 15 cm of shoot	10	Faulkner, 1954
Lettuce	1 leaf	6 (2 young, 2 medium, 2 old)	Faulkner, 1954
Cotton	1 whole plant (early season)	10	Faulkner, 1954
	1 fruit terminal (late season)	10	
Cotton	1 whole plant (1–6 weeks old)	20	Hightower and Martin, 1956
	1 terminal + 1 fruiting arm (>6 weeks old)	10	Hightower and Martin, 1956
Onions	All leaves cut to ground-level	10	Faulkner, 1954 Hoerner, 1947 Shirck, 1948
Cacao	2 leaves (of known age) or 1 flush (about 6 leaves)	1	Fennah, 1955
Rose flowers	1 flower	20	Davidson and Andrewartha 1948a
	1 flower	5	Ota, 1968
Leontodon (Compositae)	1 flower	10	Ward, 1966
Compound inflorescences (e.g. Umbelliferae, Labiatae)	No. of flowers loosely packed in 5 × 2·5 cm tube	10	Ward, 1966
Wheat	1 tiller (before flowering)	10	Lewis, 1959a
	1 ear (after flowering)	10	
Wheat	1 tiller (for dissection)	50	Holtmann, 1962
All cereals	1 tiller, ear or panicle	25 or 50	Koppa, 1967
Pyrethrum	1 leaf (see p. 122)	20 to 49	Bullock, 1965
Pine bark	6·5 cm^2 (1·0 in^2)	36 (from 2 trees)	Lewis and Navas, 1962

sacrificed to make the sampling practicable. For example, the mean coefficient of variation for Davidson and Andrewartha's samples of *Thrips imaginis* on rose flowers was about $\pm 12\%$ using 20 blooms per sample. Samples of 40 blooms gave a more accurate estimate but were too laborious to sort and count. The mean coefficient of variation for Bullock's (1965) sampling method of thrips on pyrethrum was about $\pm 28\%$, and ± 11 to 20% for Lewis's (1959a) method for thrips on wheat. The accuracy of these estimates from vegetation is comparable with that for estimates of aerial populations made with suction traps. Hosny (1964) found that the accuracy of estimates of the number of *T. tabaci* on cotton obtained by removing whole plants and washing off the thrips, and of estimates obtained by shaking growing plants over a cloth, were similar as long as at least 20–30 plants were examined. When the size of the populations is changing or fluctuating rapidly, the geometric mean gives a better description of the "typical" numbers than the arithmetic mean (Williams, 1964).

Thrips hibernating in tree bark are aggregated in crevices (see p. 221) so the accuracy of samples is affected by the distribution of the sample units over the trunks. With a total sample of 232 cm² (36 in²) of Austrian pine bark, consisting of 4×58 cm² units, from two trees, the coefficient of variation was $\pm 43\%$; but when a total sample of the same size consisted of $36 \times 6\cdot5$ cm² randomly sited sample units, the coefficient of variation decreased to $\pm 14\%$. The accuracy of samples of thrips in chestnut and poplar bark similarly improves when many small sample units are collected rather than a few large ones, providing that most units contain one or more individuals. When units become so small that many contain no individuals, accuracy again deteriorates. The position of sample units can be determined by reference to two series of random numbers representing the height and girth of tree trunks in inches (2·5 cm). These points can be located using a rule for vertical measurements and a tape for girth, and marking the point with a drawing pin. The sample unit is centred on the pin. On conifers, resin may exude from wounded trees, and run down the trunk embalming thrips in crevices beneath, so when sites for sample units determined by this method happen to fall on a resinous patch, the sample should be taken to one side of the resin as near to the marker as possible (Lewis and Navas, 1962).

Sweeping has been the most popular method of sampling populations on vegetation. Usually 25–50 sweeps (units) constitute a sample, taken while the collector walks through the habitat (von Oettingen, 1942; Holtmann, 1962; Cederholm, 1963; Koppa, 1967). There is often much variation between samples, and the method is unreliable for making quantitative comparisons between habitats or occasions unless there is a very great difference in the size of the populations sampled. This was stressed by Gray and Treloar (1933), who swept *Anaphothrips obscurus* from lucerne, taking 25 sweeps per sample. The mean coefficient of variation for 40 samples was ±74·9%; 50 sweeps decreased the value to ±47·8%, 100 to 36·9% and 200 to 24·1%. To restrict the error range to 50% and 10% of the grand mean, 36 and 897 samples respectively, each of 25 sweeps, would have been necessary. Thus, sweeping provides a much less precise estimate for a similar amount of effort, than sampling methods in which vegetation is removed from the field. Moreover, because the numbers in successive swept samples are often not correlated (for successive samples of *A. obscurus* on lucerne $r = 0·086$, $P = 0·55$) small changes in the size of populations cannot be detected. When used to detect qualitative differences between habitats, it should be remembered that the taller the vegetation, the smaller is the proportion of it sampled by the net, so leaf-dwellers are likely to be missed in tall, flowering stands. Samples should always be taken at the same time each day, preferably in the warmest part of the day between 12.00 and 15.00 h in temperate areas when most thrips are exposed on the surface of plants; in hot arid areas or if the population contains species that leap, larger catches may be obtained during the cooler times of the day. Holtmann (1962) attempted to quantify sweeping by covering a swath in cereals 2 m wide and 20 cm deep, so that each sweep covered approximately 0·6 m² and each sample of 50 sweeps 30 m², but a more precise way than this of estimating populations is to remove plants and extract the thrips from them.

AERIAL POPULATIONS

Sampling positions for aerial populations at high levels need not be randomized because the airborne insects are already randomized by the wind. For measuring vertical profiles the vertical distance between traps should be chosen so that when the data are plotted on a logarithmic scale the trapping heights are at approximately equal intervals along the axis representing height (usually the abscissa). Nearer the ground or vegetation, the vertical disposition of traps is critical because the behaviour of individuals and age structure of the population change rapidly with height in the first few metres, or even decimetres (see p. 151). Above the thin envelope of slow-moving air, or boundary layer (Taylor, 1960), whose depth varies from a few centimetres in windy weather to a few metres in almost still air, most of the thrips will be migrants, being blown from one site to another, whereas populations at or below vegetation level will include thrips in transitory flight from plant to plant, as well as migrants leaving and arriving on the vegetation. No traps distinguish between these different populations, although the proportion of immature females, which are probably migrants, can be detected by ovarian dissections from catches in water traps (see p. 10), and catches at vegetation level probably contain a higher proportion of new migrants early in the day than later on.

The accuracy of catches differs between types of trap and between the same type at different heights (Table 12). At vegetation level catches in

TABLE 12

The coefficient of variation obtained from catches of thrips in black (23 cm) suction traps at crop (wheat) level and in black sticky and white water traps at crop level and 2 m (Lewis, 1959b)

| Trap | Species | Coefficient of variation ($\pm\%$) | |
		Crop level	2 m
Suction	All thrips	7·8	—
Sticky	All thrips	21·4	20·2
Water	*Limothrips cerealium*	27·8	42·8
	Chirothrips manicatus	28·6	56·4
	Taeniothrips spp.	26·8	27·8
	Thrips spp.	29·4	36·7

a 23 cm suction trap are about three times as precise as in sticky traps and four times as precise as in water traps. Above vegetation where wind is faster unattractive sticky traps retain their precision, but unattractive water traps do not. When a vertical series of suction traps or sticky traps is used within the first few metres above the ground, it is useful to place an integrating anemometer by each trap to permit corrections to catches if necessary (see p. 116). At greater altitudes relative wind speeds at the height of each trap can be obtained from mean profiles calculated from the formula:

$$Ur = \frac{1}{k} \log_e (Zr - d) - \log_e Zo$$

where Ur = relative velocity at height Zr
K = von Karman's constant = 0·4;

d and Zo are constant for each type of vegetation and generally increase as the ground cover deepens. Typical values for various crops are given in Table 13 for substitution in this formula (Paeschke, 1927; Sutton, 1953; Jensen, 1954).

On a horizontal plane, sites in sheltered zones produced by hedges or trees should be avoided if mean densities over a field are required. The distance between the hedges and traps should be at least five times and preferably ten times the height of the sheltering vegetation. By

TABLE 13

Estimates of the height $(\bar{d})$ and roughness length (Zo) of different
crops (data from Paeschke, 1927; Jensen, 1954; and Tani *et al.*, 1955)

Crop	Crop height $(\bar{d})$ (cm)	Zo
Fallow land	—	2·1
Sugar beet	30	6·7
Short grass	10	3·2
Tall grass	13	4·0
Wheat	16	4·5
Rice	80	18·0
Tall potatoes	30	3·5
Short wheat	5	0·05
Tall wheat	80	20·0

contrast, when large samples of flying thrips are required it is helpful
to place traps in sheltered sites, though not beneath overhanging branches
where debris and larvae fall in.

SECTION III

ECOLOGY

7 Movement, migration and dispersal

Many thrips run rapidly for such small creatures; others crawl slowly and a few can jump. However, flight, for which the majority of species in the Order are so well adapted with their delicately fringed wings, is by far the most important natural form of dispersal. Therefore, this chapter concentrates on migration and dispersal by flight, after brief consideration of other forms of movement.

WALKING AND HOPPING

Most thrips walk with their bodies raised above the ground and often move with a slightly sinuous deflection of the abdomen suggestive of a stalking lizard. Males of many tubuliferan species raise and curl the tip of the abdomen forward when walking, and look like miniature scorpions. Wingless individuals, particularly phlaeothripids, usually have a more cumbersome walk, and appear to drag the abdomen in a serpentine manner. The legs move in the following sequence: left fore, right hind, left middle, right fore, left hind and right middle (McGuffin, 1933, in Stannard, 1968). Medium-sized thrips such as *Limothrips* and *Taeniothrips* can travel 12–20 cm/min on a flat surface (Klee, 1958), smaller species such as *Chirothrips* move more slowly but predatory *Aeolothrips* and *Scolothrips* faster (see p. 65). Many can walk backwards for a few steps as easily as forwards. Some species that live in exposed positions on plants, especially leaf-feeding Heliothripinae and Thripinae, can leap high into the air, a habit that probably helps them to escape from predators and also assists take-off (Plate VIb). For example, *Caliothrips fasciatus* can jump about 25 cm vertically even with its wings removed (Bailey, 1933).

Walking thrips are unlikely to travel far from their place of development. Distant dispersal in flightless species depends more on accidental movement, such as when they cling to drifting logs or vegetation, are carried in soil and on plants transported by Man, or are blown by winds in severe storms. Adults of the wingless *Aptinothrips rufus* have been trapped occasionally in the air in suction traps, and Glick (1960) even caught a thrips larva at 150 m. Some species may have been carried to tropical islands in stores or soil carried as ballast by early sailing ships (Mound, 1970a). However, the number of wingless or winged individuals spread by such sporadic accidents must be negligible compared with the vast numbers that are scattered, especially over land, during migratory flights from breeding, hibernation or aestivation sites on many days each year. Accidental introductions by Man from one part of the world to another are important economically, but are beyond the scope of this chapter and are considered on p. 226.

FLIGHT

Thrips are among the weakest flying insects, yet their finely fringed wings enable them to remain airborne long enough for the wind to blow them to great heights and for long distances. Indeed, migration by flight from breeding sites is a regular event in the life-cycle of many species. Although it produces widespread scattering, resulting in the loss of millions of individuals, it nevertheless ensures that fresh food is found for breeding populations, and sheltered sites for protection during unfavourable seasons.

Most species fall into two of the categories of migrants listed by Johnson (1969); namely Class Ib, in which populations of short-lived adults emigrate from their place of development never to return, spreading ubiquitously over the area where breeding is possible and even beyond it, and Class IIIa, in which this same type of migration is temporarily halted by a dry or cold season, when perhaps diapause intervenes, flight being renewed when the weather improves. Local flights are also made between plants within a habitat when thrips are searching for food or oviposition sites, but the airspeed of flying thrips is so slow that even on these appetitive flights the slightest breeze is often enough to displace an individual many metres before it can alight. Migratory flights consist of three basic stages; take-off, displacement and settling (Johnson, 1969). It is essential to appreciate the different characteristics of these stages and the stimuli to which individual thrips respond before and during flight, as a basis for understanding the processes involved in the collective migration and dispersal of field populations.

Individual Aspects of Migration

Flight Maturation

Thrips cannot fly immediately they emerge from the last pupal stage. Their wings and bodies must first harden and their flight muscles mature. This process of teneral development is probably temperature-dependent, as in aphids (Taylor, 1957), but its duration in thrips is unknown except for *Limothrips cerealium* in which it lasts about five hours at 20°C. This means that in cool temperate regions individuals emerging from pupae before about noon on a warm day can fly later the same day, whereas individuals emerging later in the day or during the night probably cannot fly until early the next day. Tropical species may have a much shorter teneral period because of warmer ambient temperatures, and attain flight-maturity sooner after emergence than temperate species.

Adults that have hibernated may need a second period of flight maturation before they can resume migration in spring. In England, most over-wintering females of *Limothrips cerealium* lose their ability to fly during

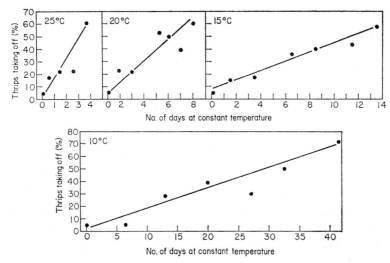

FIG. 45. The effect of temperature on the development of flight maturity in a population of *Limothrips cerealium*. Thrips hibernating in bark were exposed to standard conditions and the percentage taking-off on each occasion recorded. The period of flight maturation ranged from about 4 days at 25°C to 42 days at 10°C (after Lewis, 1962).

hibernation and regain it only after a temperature-dependent maturation period, even though their flight muscles remain intact throughout the winter (see p. 180 and Plate XI). In experiments made in early spring only 5% of individuals hibernating in pine bark emerged from it and took-off when slivers of bark were placed in a glass box at 24–25°C and at a light intensity of 1,080 lux (100 f.c.). However when batches of bark from the same trees were stored at 10, 15, 20 and 25°C and samples from each batch taken weekly for testing in the same box, a greater proportion emerged and took-off when the treatment was longer and the temperature higher (Fig. 45) (Lewis, 1962).

Take-off

Thrips usually take-off from an elevated point such as the tip of a leaf, a grass awn or the edge of a scale of bark. Launching is often preceded by abdominal and wing flexing, waving of the front legs and wing combing, sometimes lasting several minutes. Both wings on each side are combed at the same time with the adjacent hind tibia. This is placed over the wings just proximal to the mid-point and moved gradually towards the tips to straighten the setae on the leading edges. Meanwhile, the tibia of the opposite hind leg gently combs the setae on the hind margins of the same wings (Fig. 46a). Immediately before launching, thrips stand

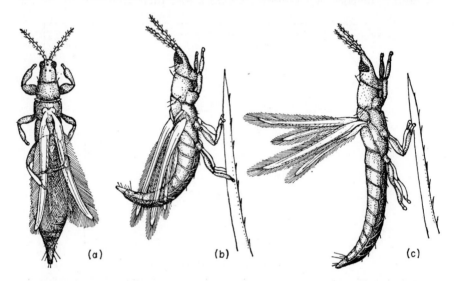

FIG. 46. Preparation of the wings before take-off by a thripid (original). (a) Combing wing setae. (b) Abdominal flexing to separate the wings. (c) Launching.

poised with wings held backward at a steep angle and abdomen curled upwards to assist spreading of the wings (Fig. 46b); then they finally launch themselves with a kick of the hind legs (Fig. 46c). In contrast to this description typical of many Terebrantia, Putman (1965b) believed that the tubuliferan *Haplothrips faurei* depended on the lift from its beating wings to free its tarsi from the resting surface. Often launching attempts fail and weaken into a mere hop. Once airborne, the abdomen hangs downwards and the insects fly in weak, irregular spirals or zig-zags, usually towards light. Bailey (1933) claimed that *Caliothrips fasciatus* could fly only about a metre without being blown, but *Limothrips* spp. can fly much further than this in still air and individuals of many species must be capable of sustained flight for several hours to permit them, with the aid of the wind, to attain the great heights and travel the long distances recorded (see pp. 152, 155).

PHYSICAL FACTORS AFFECTING TAKE-OFF

The most important microweather factors affecting take-off are temperature and light; wind is less important and relative humidity, atmospheric pressure and electricity have slight or doubtful effects.

Temperature

Thrips cannot take-off if they are too cold, but unlike larger insects no warming-up process is detectable and their threshold temperature for take-off is usually distinctly defined in terms of ambient temperatures. The threshold temperatures for take-off in populations living in temperate regions is usually between 17 and 20°C, depending on the species. In the field, take-off thresholds can be determined by comparing hourly catches in suction traps with maximum hourly temperatures plotted as scatter diagrams or cumulative frequency distributions (Taylor, 1963). It is

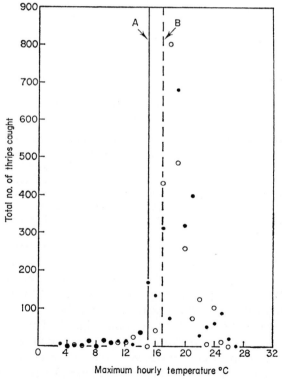

FIG. 47(a)

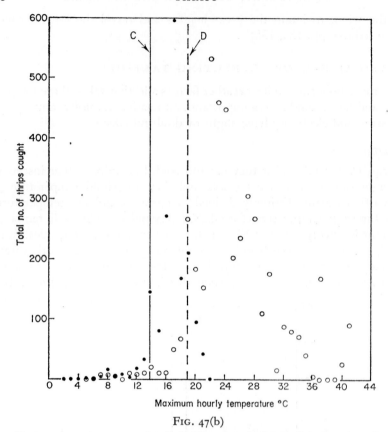

FIG. 47(b)

FIG. 47. Apparent temperature thresholds for take-off of *Limothrips cerealium*. Scatter diagrams of numbers caught per hour plotted against hourly maximum temperature show sudden rises as thresholds (vertical lines A–D) are passed. The values of the apparent thresholds depend on the position of the temperature sensor (after Lewis, 1963). A (●), 15°C; 60 cm (2 ft) above grass, mercury-in-steel thermometer. B (○), 17°C; 60 cm (2 ft) above grass and gravel, thermocouple. C (●), 14°C; 0·25 cm (0·1 in) above bark, thermocouple. D (○), 19°C; in bark, thermocouple.

essential to measure the temperature very close to the insect to obtain the true threshold temperature for take-off; otherwise misleading values are obtained. For example, the apparent threshold in populations of *Limothrips cerealium* leaving pine bark ranged from 14°C–19°C when temperatures were measured over grass near the trees, above or within the bark, or when different instruments were used (Fig. 47; Lewis, 1963). These values compare with the laboratory-measured threshold of 18·4°C for this species, so the temperature measured in the bark crevice, closest to the actual point of take-off, gave a threshold nearest to this laboratory value.

Many factors, such as sexual condition or the rate of maturation of flight muscles at temperatures experienced previously, might modify the threshold, and it would be wrong to regard individuals or populations of each species as always having one characteristic threshold. For example, a population of *Limothrips cerealium* taken from hibernation and tested in the laboratory showed two distinct thresholds for take-off, one at 18–19°C and another at 23–24°C (Fig. 48). This was probably because

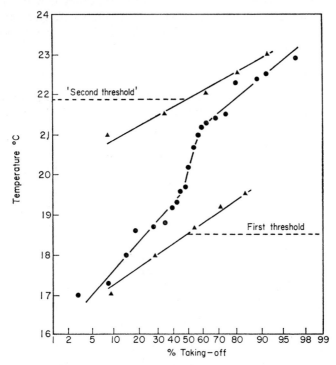

FIG. 48. Thresholds for take-off in a population of *Limothrips cerealium* emerging from hibernation in bark. The dots represent the cumulative percentages of thrips taking off plotted on arithmetical-probability paper. The results of the analysis of the two separate parts of this curve are represented by the triangular points, which show that the values for the first and second thresholds were 18·5 and 21·9°C respectively (after Lewis, 1963).

some individuals in the experimental population had not completed the necessary period of flight maturation when the temperature passed the lower threshold, and "matured" only as temperature continued to rise. Thus the higher "threshold" appeared because the population contained both flight-mature and flight-immature individuals. When only flight-mature individuals, known to be capable of take-off, were tested, there was a single threshold at 18·4° ± 1·1°C (Lewis, 1963).

Nevertheless, small differences in the true threshold temperatures for

take-off probably do occur between species and may produce small differences in the diel periodicity of exodus flights from breeding sites.
Some species are habitually more noticeable earlier in the day than others.
For example, *Stenothrips graminum* is more active on vegetation earlier in
the morning than *Limothrips cerealium* (von Oettingen, 1942). The
threshold temperatures for take-off of young adults of these two species,
measured with a mercury-in-steel thermometer as they flew from cereals,
was 14°C and 20°C respectively (Lewis, 1958) so *S. graminum* could
clearly fly in the mornings some hours before it was warm enough for
L. cerealium (see also p. 150).

Light

Thrips do not take-off in the dark and most species probably need a light
intensity of at least 1080 lux (100 f.c.) before they will do so. In temperate
regions low light intensity is unlikely to prevent take-off in the mornings

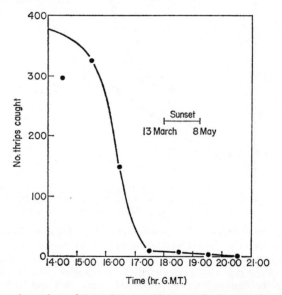

Fig. 49. The total number of *Limothrips cerealium* caught in a suction trap on spring
afternoons and evenings in 1960 and 1961 when it was warm enough to take-off.
Take-off ceased abruptly 1–2 h before sunset (after Lewis, 1963).

because it is seldom warm enough before sunrise, whereas in tropical
regions it is usually warm enough for take-off at dawn but dim light may
cause a short delay. In temperate and tropical regions it is often warm
enough for flight well into the evening, but take-off usually ceases 1–2 h
before sunset when the light fades (Fig. 49; Lewis, 1963).

Other factors

Winds up to 6·5–8·0 k.p.h. (4·5 m.p.h.) measured at about 10 cm above vegetation do not prevent take-off, and sometimes *Limothrips* spp. will take-off in winds as fast as 14 k.p.h. (9 m.p.h.). In England, days with such strong winds at vegetation level are usually cool or wet, and it is probably these factors rather than the wind that discourage take-off on such occasions. It is not surprising that thrips are undeterred from taking off by strong winds, for two reasons; they actually take-off into air that is partly sheltered by surrounding vegetation, so are probably unaware of wind speeds in exposed situations above, and also they are clearly able to tolerate the fast winds encountered during high-altitude dispersal.

Occasionally an insect may be caught flying in rain, but it is more probably one that was airborne before rain started to fall than one that had taken off since, because wet thrips cannot arrange their wing setae for flight. Relative humidity has little effect on take-off in *Limothrips cerealium*, which will take-off in the laboratory at all values tested between 46% and 85% at temperatures between 24°C and 29°C (Lewis, 1958), and it is doubtful whether it has much effect on take-off in other species (see p. 144).

BIOLOGICAL FACTORS AFFECTING TAKE-OFF

Sexual differences

The response of thrips to weather favourable for take-off differs between sexes. For example, both sexes of *Stenothrips graminum* are winged but females take-off more willingly than males. In England during 1957, the sex ratio in a population of this species on oats was 1♂:19♀♀, compared with 1♂:47♀♀ in the air 50 m away. By contrast, males and females of *Melanthrips fuscus*, *Aeolothrips tenuicornis*, *Ae. fasciatus* and *Taeniothrips vulgatissimus* take-off in approximately equal proportions (Lewis, 1961). Most species mate before take-off, and presumably this is always so in species with wingless males.

Ovarian development

Sexually immature females are the most abundant thrips in migrating populations, and also probably fly the longest distances.

Take-off and migration in relation to the sexual cycle and ovarian development was studied in *Limothrips cerealium* in southern England (Lewis, 1959a). The females only are winged and those of the new generation mate and fly from host plants in late summer to hibernate in litter and bark. About 90% of all females caught on this exodus flight, or collected from vegetation before hibernation, had undifferentiated ovaries and the ovaries of the remainder were developed only slightly. In the following spring females flew from hibernation sites to feed on grasses and young cereals, and they were ready to lay after about six weeks. By the time all the overwintering populations had emerged from hibernation, 82% of those already feeding on wheat were still maturing and 18% were laying. A second, longer, period of flight activity occurred in late June when adults flew to fresh hosts as the grass inflorescences dried. Of these, 88% had partially developed ovaries and only 12% were laying though it is uncertain whether they were a new generation. But, the point to stress is

TABLE 14

Ovarian development in *Limothrips cerealium* caught immediately over a wheat crop in England and in the air 200 m away from it during June and July (after Lewis, 1965a)

	Total examined	Immature (%)	Maturing (%)	Laying (%)
Crop level in wheat field	44	32	18	50
7·5–15 m high, 200 m away from field	53	55	30	15

TABLE 15

Numbers of female thrips at different stages of ovarian development trapped in the air (after Lewis, 1965a)

	Spring: from hibernation			Summer: between hosts			Late summer and autumn: to hibernation		
	I	M	G	I	M	G	I	M	G
*Anaphothrips obscurus**	—	—	—	—	—	—	29	0	1
Chirothrips manicatus	40	0	0	17	3	8	—	—	—
Limothrips cerealium	47	0	0	5	22	28	91	2	1
Limothrips denticornis	8	0	0	5	3	4	10	0	0
Stenothrips graminum	—	—	—	8	8	8	—	—	—
Frankliniella intonsa	28	2	0	—	—	—	—	—	—
Melanthrips fuscus	7	2	0	—	—	—	—	—	—
Taeniothrips vulgatissimus	—	—	—	21	22	18	—	—	—
Thrips major	—	—	—	16	13	10	—	—	—
Aeolothrips fasciatus ⎱ *Aeolothrips tenuicornis* ⎰	—	—	—	8	2	6	—	—	—
Percentages (from totals)	97	3	0	34	31	35	97	1·5	1·5

I = ovaries undifferentiated or undeveloped; M = ovaries maturing; G = gravid females.   * Parthenogenetic.

that relatively few laying thrips flew in late May or early June even though the weather was suitable. Similarly, sexually immature *Haplothrips faurei* take-off willingly, but laying females do not unless oviposition is prevented or delayed by a shortage of food (Putman, 1965b). This behaviour has interesting implications for the population dynamics of this predator. The inhibition of flight in well-fed, gravid females ensures that they remain to lay in areas where prey is abundant, but if prey becomes scarce, their ability to fly returns enabling them to "search" for food elsewhere.

Not only are immature females of *Limothrips cerealium* more willing to fly than gravid ones, but they also disperse more widely. A comparison of ovarian development in individuals of this species caught on a wheat crop and 200 m away from it at 7·5 and 15 m above the ground, showed that laying females predominated in the air around the wheat ears, but immature ones at a distance from the field (Table 14). A small proportion (15%) of gravid females were caught far away from host plants, perhaps through being blown from their breeding sites accidentally by gusty winds. Table 15 shows emphatically for eleven species common in southern England that immature females are the predominant migrants in either spring, autumn or both, whereas immature, maturing and gravid females all fly locally between hosts in summer (Lewis, 1965a).

Quality of food

Drying food plants may also stimulate thrips to take-off, but because many thrips become adult at a time when plants are drying, it is uncertain whether the unsuitable food causes take-off or is merely correlated with it.

SPEED AND MECHANISM OF FLIGHT

Once airborne, the airspeed of individuals probably depends on their size. The mean flight-speed of *Limothrips cerealium* measured in still air in the laboratory over 2–3 m was 14·7 cm/sec compared with 21·5 cm/sec for *L. denticornis*, a slightly larger species. The maximum speeds recorded were 24·1 and 33·1 cm/sec respectively (Lewis, 1958). These two species of *Limothrips* represent medium-sized Terebrantia, so the range of airspeeds throughout the Order probably extends from about 10 cm/sec for the smallest Terebrantia to 50 cm/sec for the large Tubulifera. Shull's (1911) record of 90 cm/sec for the medium-sized species, *Ctenothrips bridwelli*, seems unlikely. Even in "calm" weather the speed of air currents usually exceeds this range of unaided flight-speeds, so airborne thrips will almost always be blown by the wind along a track over which they have incomplete, and often minimal control.

The structure of the minute, copiously fringed wings of thrips and other small insects such as Mymaridae, small Staphylinidae and Ptiliidae suggests that they function differently from the larger, sparsely fringed or completely membranous wings of other insects. The drag on such tiny wings might exceed the lift that they could generate if used as aerofoils in the manner common to larger insects (Weis-Fogh and Jensen, 1956; Pringle, 1957), and instead, these very small insects probably "swim" in the air in much the same way as a fish would if it merely flapped its tail (Horridge, 1956). The absence of strong veins on the hind edges of the wings of most thrips probably increases the flexibility of the trailing edge, which would help the flapping action necessary for such movement.

Settling

The act of settling is seldom seen in the field because thrips are difficult to observe approaching a landing site, and are usually visible only after alighting, when they fold their wings quickly and remain still a moment while sensing the landing site with quivering antennae. Their great dependence on wind for dispersal probably means that even in light breezes they are blown to within a few centimetres of an object before they can orientate for landing, and in strong winds they appear to be impacted directly on to a surface and struggle to gain a foothold only after contact with it. There is no doubt that in fairly still air some species can see, orientate towards, and land on a surface of their choice; the specific composition of the catch of airborne thrips on coloured surfaces differs with the colour, confirming this (Lewis, 1959b; see p. 119). However, the distance over which they can make this choice is unknown, and, except in light winds, a large proportion of individuals attempting to reach an observed point probably fail. Based on evidence that walking *Limothrips cerealium* can distinguish between odours emitted by different parts of plants (see p. 98), Holtmann (1963) suggested that the final stages of orientation to a plant depended on scent. However, this assumes a degree of control over flight more precise than any observed in the laboratory and seems to be refuted by the vast number of thrips that alight on surfaces with unnatural smells, such as domestic washing, synthetic glues and water containing detergents and even formalin.

Nothing is known of the interdependence of flight and settling in thrips. Some individuals will take-off immediately after landing on an apparently suitable host plant, whereas others remain, but whether those that settle have flown more than those that take-off again is not known. The complex alternation of take-off and settling responses observed in aphids over periods of a few hours (Kennedy and Booth, 1963a, b) and many other insects (Johnson, 1969) suggests that in thrips too, the short-term willingness to settle is probably preconditioned by the duration of the preceding flight. Certainly over much longer periods their willingness to settle changes. For example, *Limothrips cerealium* is much more willing to settle in late autumn than in early spring. In autumn, presumably after flying for a considerable time, alighting females search for hibernation sites and are often reluctant to take-off again, whereas in spring, after 6–7 months inactivity, the same individuals become restless and anxious to fly.

Collective Aspects of Migration

MASS FLIGHTS

Most winged species have a flight period lasting for a few weeks each year. Within this period some thrips fly on most days, but there are usually a few days when sudden, mass flights occur. When the parental breeding population is large these may be spectacular. In Europe and central Asia

TABLE 16

Maximum densities of airborne thrips recorded during mass flights in southern England 1957–1962 (after Lewis, 1965a)

Species	Density		Situation
	(per 10^6 m³)	(per 10^6 ft³)	
(a) Above cereals—at probable source			
Limothrips cerealium	2,066,000	58,500	Above wheat
	1,978,000	56,000	Above oats
	479,600	13,580	By pine trees
Limothrips denticornis	144,800	4,100	By oats
	74,100	2,100	By wheat
Aeolothrips spp.	49,400	1,400	By beans
(b) Unknown distance from source			
Limothrips cerealium	427,300	12,100	Above young wheat
	287,800	8,150	Above beans
Chirothrips manicatus	190,700	5,400	Above young wheat
Anaphothrips obscurus	176,600	5,000	Above wheat
Stenothrips graminum	134,200	3,800	Above wheat
Aeolothrips spp.	5,300	150	Above wheat
Phlaeothrips coriaceus	41,000	1,160	In woodland

the most noticeable mass flights are produced by cereal and grass thrips when they leave ripening crops towards the end of the growing season (Bold, 1869; Körting, 1930; Tansky, 1958b). In many areas farm workers are annoyed by the countless numbers of thrips that land on bare skin and irritate nose and eyes, and corn thrips have been known to occur in such numbers as to enter watches and block the mechanism (Blunk in Körting, 1930). Sudden increases in flight activity are not confined to grass-feeders, but also occur among flower-dwellers and even predatory thrips, though in these groups they often pass unnoticed because the parent populations are relatively small. However, if a mass flight is defined as a doubling of the average daily aerial density measured over the whole flight period, then many of these less abundant species also have several mass flights each year. For example, eight of the common species occurring in southern England were abundant in the air on 19 of the 92 days between 1 June and 1 September 1956 (Lewis, 1964).

During a mass exodus flight from the breeding site individual thrips can be seen as they spiral upwards into the air, and when viewed against a bright sky they appear collectively as a cloud of drifting black specks just above the vegetation. The maximum density of *Limothrips cerealium* recorded over wheat in England is $2,066,000/10^6$ m³ ($58,500/10^6$ cu ft) near the source (Table 16). Further away, densities are less and mass flights usually pass unobserved. It is worth noting that aerial densities exceeding these measured for thrips have only been recorded for aphids (Johnson, 1952) and locusts (Rainey, 1958), so thrips clearly rank as one of the most migratory orders of insects.

Mass flights and weather

Mass flights occur sporadically within the flight period and in many species are reputedly associated with thundery weather. Indeed, in many parts of Europe thrips are called "Thunder Flies" or "Storm Flies". However, Körting (1930) showed that the occurrence of mass flights of cereal thrips in Germany depended more on temperature than on atmospheric pressure or humidity, and in Sweden, Johansson (1946) also stressed the importance of temperature, claiming that thrips flew most readily at 19–22°C and at 70–75% relative humidity. A careful study of the effect of weather on mass flights in southern England showed no correlation with thunder (Hurst, 1964; Lewis, 1964). In the summer of 1956, 27 mass flights of eight common species were recorded, and 24 of these occurred on the first or second day after a period of cool weather, when temperatures rose for a few hours above the threshold for take-off. The three other flights occurred on days when the maximum temperature rose above that of the day before (Fig. 50). During the spring and summer of 1956, 1957 and 1958, a total of 40 mass flights were recorded, and the weather on the day each occurred was analysed. Six criteria were common to 31 of these days, and the only occasions when flights occurred without all six being fulfilled were when food plants were drying rapidly. The criteria were:

No rain or drizzle during the day.

Day maximum temperature at least 20°C.

Day mean temperature above that of the previous day.

Low dry adiabatic lapse rate with no convection above about 1,600 m (5,000 ft).

At least 1 h of sunshine.

Dew point between 5 and 15°C.

Using this information, an appraisal was later made of days when mass flights would have been expected in these three years (Hurst, 1964), and

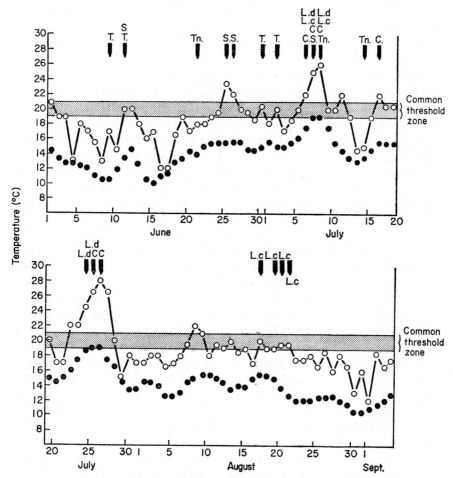

FIG. 50. Mass flights of thrips in 1956 in relation to temperature. O———O Maximum temperature; ● ● 3-day running mean temperature; arrows indicate mass flight (after Lewis, 1964). C. *Chirothrips manicatus*, L.d. *Limothrips denticornis*, S. *Stenothrips graminum*, T. *Thrips* spp. (mainly *Thrips major* and *T. fuscipennis*), L.c. *Limothrips cerealium*, Tn. *Taeniothrips* sp. (mainly *Taen. vulgatissimus* and *Taen. atratus*).

on nearly two out of every three days when these weather conditions occurred together, there were mass flights. The probability that the relationship between these meteorological factors and mass flights occurred by chance was less than 0·001. When mass flights occurred in different weather, they were usually of *Thrips* or *Taeniothrips* spp. which sometimes flew in more humid and cooler conditions than most of the other common species, such as *Anaphothrips obscurus*, *Chirothrips manicatus*, *Franklinella intonsa*, *Limothrips cerealium*, *L. denticornis* and *Melanthrips fuscus*.

Thus, allowing for small specific preferences, mass flights were encouraged by dry, settled weather, a maximum temperature of at least 20°C, a stable atmosphere above about 1,600 m with an unstable one below, and by sunny rather than dull weather. Frontal conditions and thundery weather generally discouraged mass flights.

By contrast, Kittel (1958), working mainly with *Limothrips denticornis*, *Chirothrips manicatus* and *Taeniothrips pini* in the Harz Mountains, central Germany, insisted that mass flights were associated with thunder because he believed that only rapidly changing factors such as electrical potential, temperature or light, could stimulate the thrips to behave collectively in this way. Unfortunately, no convincing experimental or meteorological evidence was produced to support this claim. There are two aspects of atmospheric electricity that might affect insects; one is the potential gradient or electric field, and the other the density of unipolar ions in the air (Maw, 1962, 1963a, b, Johnson, 1969). In fine weather the earth is usually negatively charged and the atmosphere positively charged. The potential thus produced increases with height and the rate of change is called the potential gradient. In a thunderstorm this may be more than 10^6 watts/m, resulting in a current of positive charges towards the earth's surface caused by the unipolar ions in the atmosphere. There is no confirmation that the thrips respond in any way to changes in either the electric field or the density of air ions. Experiments with *Drosophila melanogaster* Mg. suggest that this fly might respond to natural changes in the potential gradient near the earth during the passage of clouds or rain storms (Edwards, 1960a, b), but there is no comparable evidence for thrips and the most searching analysis of field records so far undertaken (Hurst, 1964) refutes the suggestion. Probably thrips are simply more obtrusive in sultry weather because humans perspire freely then, and the thrips irritate the skin when they stick to the sweat.

There are a few indications that the migration of some species over large continental land masses might be associated with frontal winds. In the U.S.A. *Frankliniella tritici* may be blown northwards every year. It has never been found hibernating in Illinois, yet large numbers of this species appear suddenly in spring (Stannard, 1968), and similarly in Maryland it was suddenly abundant in June 1957 (Henneberry et al., 1961). In view of the major displacements of leaf hoppers, bugs and aphids by large-scale weather systems in North America, it seems probable that flying thrips could likewise be carried by frontal winds. In the Gezira, Sudan, vast numbers of *Caliothrips indicus* sometimes appear suddenly, perhaps similarly blown from heavily infested cotton to the north (Johnston, 1925).

The way in which the weather in temperate regions causes young adult thrips to accumulate on vegetation so that large numbers are ready to migrate together was clearly shown by the development of a population of *Limothrips cerealium* in southern England in 1957 (Lewis, 1964). The

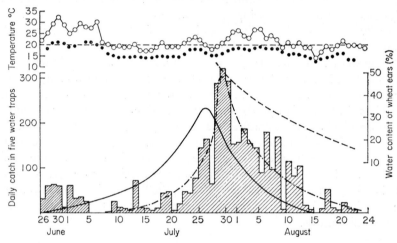

FIG. 51. The development of a population of *Limothrips cerealium* at Silwood Park, England, 1957 (after Lewis, 1964). O——O daily maximum temperature; ●● 3-day running mean; —— expected flight curve; ·-·-·· actual flight curve; ----- water content of wheat.

number of non-migrating, wingless males on a wheat field showed when the migratory females of the same brood were developing. The total number of young females caught in the air above the field during the 8-week flight period was fitted to a curve of similar shape to the production/mortality curve obtained for males, to give the *expected* flight curve for young females had they flown within a day or so of emergence (Fig. 51). However, the *actual* flight curve drawn through the daily catches represented by the histogram in Fig. 51, was steeper and occurred later than expected from the emergence pattern of the population on the crop.

This effect could have been produced had the males emerged before the females, but there was no evidence of this. It seems more likely that migrants accumulated and were suddenly released because of two separate effect of weather. In late June and the first few days of July, the thrips caught were mostly old, overwintered females and a few individuals of the new generation that had developed elsewhere. From 6 to 27 July, the adult populations on the crop increased because the daily mean temperature was about 15°C, which was warm enough for development but, except for occasional short periods, too cold for take-off. On 28, 29 and 30 July the mean and maximum temperatures increased and the migrants, all immature females, left the crop in large numbers. By this time also the wheat ears were ripening and quickly becoming unsuitable food, but it is uncertain whether this was a contributory factor though superficially it appeared to be so. Thus the weather, acting on the breeding population and possibly on the host plant, influenced the size and time of migration. Migrants accumulated during a period suitable for development but generally unfavourable for flight, and mass migration followed when the temperature increased and the host plant deteriorated.

In the wet tropics mass flights are much less common than in temperate regions. This is not because thrips are less abundant, but probably because the temperature is almost always warm enough for them to take-off and fly. Thus, instead of populations of young flight-mature adults accumulating on plants at temperatures warm enough for development but too cool for flight, individuals can usually take-off as soon as they are ready. This is illustrated by the low densities of *Taeniothrips sjostedti* measured over heavily infested cowpeas in Nigeria. The maximum density recorded was $110,567/10 m^3$ $(3,134/10^6 ft^3)$ (Taylor, 1969), only about one-twentieth the density that occurs over some temperate cereal crops supporting lighter infestations.

Diel periodicity of exodus flights

The timing of the exodus flight each day depends on the weather. Thrips will not fly if it is too cold or rainy, so on cool, cloudy days flight-mature females may remain on plants until temperatures rise in mid-afternoon. In warm, showery weather take-off may occur sporadically during intervals between showers. Figure 52 shows the diel periodicity of flying *Limothrips cerealium* caught in a segregating suction trap over a wheat crop whilst a new generation developed and migrated. Maximum aerial density usually occurred near noon or in early afternoon. No thrips flew at night. Take-off started each day when the field (mercury-in-steel) threshold temperature (20°C) was passed, and on 14 days maximum aerial density was reached *before* maximum temperature occurred, while on 7 other days maximum density and temperature coincided. Density never increased after the hottest time of day, even though it remained warm enough for take-off, suggesting that the initial decrease of aerial density each day occurred because there were fewer flight-mature females available, rather than because it was too cold.

Measured over a period of a few weeks the fluctuations caused by

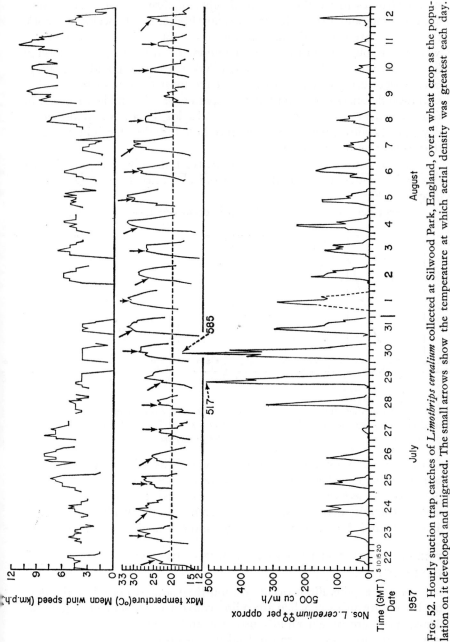

FIG. 52. Hourly suction trap catches of *Limothrips cerealium* collected at Silwood Park, England, over a wheat crop as the population on it developed and migrated. The small arrows show the temperature at which aerial density was greatest each day. This usually occurred soon after the field threshold temperature for take-off (20°C) was passed, but *before* the maximum air temperature was reached (after Lewis, 1958).

minor changes in weather and availability of thrips from hour to hour are obscured, and a smooth curve results (Fig. 53). All species of Aeolothripidae, Thripidae and Phlaeothripidae caught over a five-year period were most abundant in the air near midday or in the early afternoon at light intensities of 55,000–104,000 lux (5,000–9,500 f.c.), and their flight curves were usually symmetrical about the most favoured time. No important differences between predatory species, flower- and leaf-feeders, or fungus-feeders were observed (Lewis and Taylor, 1964), but on individual days species with a low-temperature threshold for take-off might fly earlier than others with a higher one.

A similar pattern of diel flight periodicity, initiated by a slightly higher threshold temperature, occurs in the tropical species *Taeniothrips sjostedti* (Taylor, 1969).

Some species are more active than others on vegetation at different times

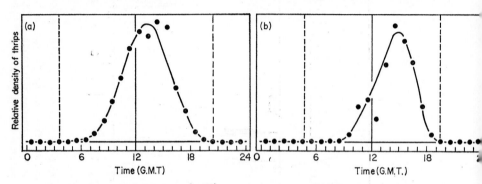

FIG. 53. Mean diel flight periodicity curves obtained from catches in segregating suction traps operated over 2–3 weeks (after Lewis and Taylor, 1964). (a) *Limothrips cerealium*; (b) *Phlaeothrips coriaceus*.

of the day but no comparisons of the periodicity of terrestrial and airborne activity have been made. Von Oettingen (1942) swept a greater proportion of *Stenothrips graminum* from meadows earlier in the day, and this terrestial activity may be reflected in differences in the diel flight curves of *Stenothrips* and, for example, *Limothrips* and *Chirothrips* (see p. 138).

VERTICAL DISTRIBUTION AND LONG-DISTANCE DISPLACEMENT

Because the exodus flight occurs during the warmest period of the day, the insects are airborne when ascending convective movement is greatest. Thus, many are blown rapidly upwards and away from their breeding site. For example in southern England at a site only 200 m from the nearest large breeding area, three common species were trapped at five heights between vegetation level and 15 m, and many individuals were caught in the upper trap (Lewis, 1959b). The mean aerial density of all species decreased with height, but the vertical gradient differed between them (Fig. 54). This difference between gradients fairly near the ground may reflect the size and flight-speed of the three species considered. *Limothrips denticornis*, the largest and the strongest flier of the three, had the shallowest

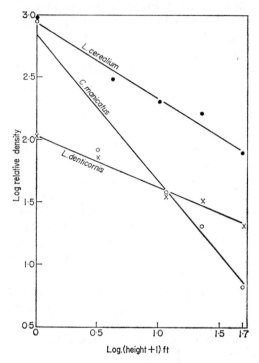

Fig. 54. Linear profiles of relative density on height (log scales) for three species caught over exposed grassland at Silwood Park, England. The density of all populations decreased with increasing height but the rate of decrease differed between species (after Lewis, 1958).

gradient, which means that it was more evenly distributed up to 15 m than the other species. By contrast, *Chirothrips manicatus*, the smallest of the three insects, and the weakest flier, had the steepest gradient, showing that it was more concentrated near the ground than *L. denticornis*. *Limothrips cerealium*, with an intermediate flight-speed, had an intermediate gradient. An explanation of these differences might be that the more powerful the flight, the higher can the insects climb by their own efforts, and thus encounter strong ascending currents to carry them even higher.

This evidence refutes the claim that thrips prefer to fly above a height of about 1 m (Körting, 1931; Johannson, 1946). An apparent concentration of flying thrips above a certain height may be recorded unless catches on sticky traps at different heights are corrected for different wind speeds at each height. The catch on an impaction trap is a product of aerial density and wind speed past it, so graphs plotted from uncorrected figures, as Körting's and Johansson's were, often show a peak in the vertical gradient, wrongly interpreted as a chosen height of flight.

The upper limit to which thrips are carried is unknown but, using nets on aircraft near Paris, Berland (1935) caught Thripidae up to a height of 2,000 m (6,600 ft) and Aeolothripidae and Tubulifera between 500 and 1,000 m (1,650–3,000 ft). In Louisiana, U.S.A., using sticky traps fitted to aircraft, Glick (1939) caught one *Frankliniella tritici* and one *Haplothrips graminis* above 3,100 m (10,000 ft) and 89 specimens representing at least 16 species at lower altitudes above 6 m (20 ft) (Table 17). At Darango, Mexico, *Stomatothrips flavus* and *Caliothrips phaseola* were also caught at 1,200 m (4,000 ft). An interesting and surprising feature of the data from Louisiana is the number of thrips caught at night, representing 35% of the total when catches from 150 m (500 ft) are included and standardized to a trapping time of 10,000 minutes. In England, suction traps at ground-level catch a much smaller proportion of the total daily catch during darkness (<0·5%) and these few thrips are considered to be stragglers that enter the trap earlier, but remain lodged accidentally on the sides until they die and fall in at night. Five out of the six night-flying individuals caught by Glick were at 150 m, but a *Sericothrips* sp. was caught at over 1,500 m (5,000 ft). Night flight can prolong the migration of other insects, for example aphids (Berry and Taylor, 1968), and these authors suggest that small insects may often fly at night in continental regions where warm, low-level jet streams develop above nocturnal inversions enabling individuals carried upwards earlier in the day to remain airborne during darkness.

At lower altitudes, six species were caught in daytime catches with two nets on a kite 450–600 m (1,500–2,000 ft) in Yorkshire, England (Hardy and Milne, 1938); in Lincolnshire, in similar nets at heights between 3 and 85 m (10 and 277 ft), five species occurred of which *Limothrips cerealium*, *L. denticornis*, and *Kakothripis robustus* were the most common (Freeman, 1945). At least six species were caught at 300 m (1,000 ft) in suction traps suspended on balloon cables in Bedfordshire, England (see Table 18) (Lewis, 1965), and *Idolothrips spectrum* was caught at 450 m (1,500 ft) near Adelaide, Australia (Mound, *in litt.*).

Vertical profiles of aerial density constructed from these data, corrected as appropriate for wind speed at different altitudes, all approximate to the linear log density × log height relationship established for a great variety of insects up to about 1,500 m by Johnson (1957, 1969) (Fig. 55). The gradient of the density × height profile for each species probably changes during the day, as for other small insects (Johnson *et al.*, 1962),

TABLE 17

Thysanoptera collected by aircraft at Tallulah La., U.S.A., August 1926–October 1931 (after Glick, 1939)

	60 m (200 ft)	150 m (500 ft)	300 m (1,000 ft)		600 m (2,000 ft)		900 m (3,000 ft)		1,500 m (5,000 ft)		Over 1,500 m (5,000 ft)	Total	
	Day	Night	Day	Night	Day	Night	Day	Night	Day	Night	Day	Day	Night
Thripidae													
Sericothrips sp.	1	—	1	—	1	—	1	—	—	1	—	4	1
Sericothrips cingulatus	1	—	1	—	—	—	—	—	—	—	—	2	0
Frankliniella tritici	8	5	4	—	—	—	4	—	1	—	1	18	5
Frankliniella fusca	2	—	2	—	2	—	—	—	—	—	—	6	0
Microcephalothrips abdominalis	—	—	—	—	2	—	—	—	—	—	—	2	0
Mycterothrips longirostrum	—	—	1	—	—	—	—	—	—	—	—	1	0
Phlaeothripidae													
Hoplandothrips pergandei	1	—	—	—	—	—	—	—	—	—	—	1	0
Neurothrips magnafemoralis	4	—	—	—	—	—	—	—	—	—	—	4	0
Leptothrips mali	4	—	1	—	2	—	—	—	—	—	—	7	0
Elaphrothrips tuberculatus	4	—	1	—	—	—	—	—	—	—	—	5	0
Elaphrothrips sp.	1	—	—	—	—	—	—	—	—	—	—	1	0
Liothrips caryae	7	—	4	—	—	—	—	—	—	—	—	11	0
Liothrips castaneae	1	—	—	—	—	—	—	—	—	—	—	1	0
Liothrips citricornis	1	—	3	—	—	—	—	—	1	—	—	5	0
Liothrips sp.	2	—	2	—	—	—	—	—	—	—	—	4	0
Haplothrips graminis	1	—	—	—	2	—	—	—	1	—	1	5	0
Undetermined	7	—	1	—	1	—	2	—	1	—	—	12	0
TOTAL												89	6

N.B. The duration of trapping differed for different heights.

TABLE 18

Percentage of different species of Thysanoptera in the air at 3 heights
on one day (4 August 1955) at Cardington, Bedfordshire, England
(after Lewis, 1965a)

	3 m (9 ft)	75 m (250 ft)	300 m (1,000 ft)
Limothrips cerealium	42·5	66·4	62·0
Limothrips denticornis	14·2	11·1	14·1
Frankliniella tenuicornis	8·7	2·2	4·2
Anaphothrips obscurus	1·9	—	1·4
Aeolothrips albicinctus	0·2	—	—
Thrips spp.	16·4	15·9	15·5
Taeniothrips spp.	4·3	3·2	2·8
Aeolothrips tenuicornis and *fasciatus*	0·5	—	—
Frankliniella intonsa	0·4	—	—
Undetermined	10·9	1·2	—
Total catch	1,193·0	315·0	71·0
Percentage grass dwellers	67·5	79·7	81·7

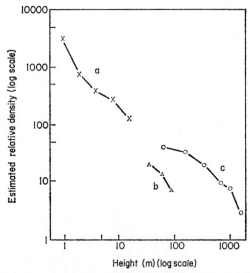

FIG. 55. Profiles of estimated relative density on height (log scales) for catches of total
thrips from (a) Berkshire, England, in water traps (Lewis, 1959b); (b) Lincolnshire,
England, in nets (Freeman, 1945); (c) Louisiana, U.S.A., by aircraft (Glick, 1939). The
relative density is plotted because catches on impaction traps and in nets were converted
to relative, not absolute, densities at different heights, because absolute wind speeds
were unknown.

being shallower around noon, when ascending currents are strong, than later in the afternoon when convection is less and airborne populations descend.

While reaching these high altitudes, thrips are also blown long distances from their original breeding site. This is clear from the many that have been caught on ships or occasionally by aircraft after being blown accidentally over the sea (Yoshimoto and Gressitt, 1961; Yoshimoto et al., 1962). These individuals probably represent merely the persistent stragglers of vast numbers that migrate from breeding sites. In a series of cruises through the Pacific, Indian and Atlantic Oceans sponsored by the Bishop Museum, Hawaii, between 1957 and 1964, 94 thrips, mostly Thripidae, were caught. As an example of the long distances they can travel there is a record of a specimen caught in the Yellow Sea 460 km from the nearest land, the Wei-hai-wei Peninsula, in wind speeds of 28–32 km/h (18 m.p.h.). The insect must have remained airborne about 16 h to travel so far in this wind (Holzapfel and Harrell, 1968). On a southern voyage from Melbourne, Australia, to Macquarie Is. three thrips were caught at 40°S and two more at 44°S (Gressitt et al., 1961), and one was caught as far south as 58° between New Zealand and McMurdo Sound, Ross Is. (Gressitt et al., 1960), all some hundreds of kilometres from land. In fine weather long-distance travel over sea may be assisted by air turbulence above the waves. Insects that have reached a considerable altitude over land tend to fall as they are blown seawards, but as they approach the surface of the water they may be prevented from touching it by a cushioning layer of turbulence (Holzapfel and Harrell, 1968).

There is no comparable direct evidence for such long-distance displacement of thrips over land, because it is difficult to be certain of the source of individuals caught in high-altitude traps. They probably cover similar distances in appropriate weather systems, as the sudden appearance of Frankliniella tritici in the northern U.S.A. each year suggests (see p. 146). Two years after Liothrips urichi was released on Fiji to control the weed Clidemia hirta, the insect was abundant at a distant site which could only have been reached by travelling 48 km over land or across 32 km of ocean (Simmonds, 1933).

Another example of the ease with which thrips are spread by wind is provided by the rapidity with which they can colonize an isolated habitat. After killing, by fumigation, all insects on seven very small islands of red mangrove (Rhizophora mangle) in the Florida Keys (Wilson and Simberloff, 1969) the interval before arthropods invaded them was measured. Five species of thrips appeared, four of which (Neurothrips magnafemoralis, Liothrips n. sp., Haplothrips flavipes and Pseudothrips inaequalis) were widespread among the islands. Almost all invasions occurred within 4–5 months of fumigation. Large populations were occasionally produced but they rarely persisted (Simberloff and Wilson, 1969). At the end of two years, four species were still present in the islands, suggesting that not only colonization, but also equilibrium between the thrips and other fauna, was achieved rapidly (Simberloff and Wilson, 1970).

Specific composition of airborne populations and habitat

Often several species migrate at the same time, so the population in the air is mixed. In Europe, grass- and flower-feeding species usually predominate. *Limothrips cerealium* and *Kakothrips robustus* were the commonest species in catches from eastern England (Freeman, 1945), and Körting (1930) estimated that mass flights in Germany in late summer contained over 95% *L. cerealium*, except when the cereal-feeding *Haplothrips aculeatus* and *H. tritici* also flew.

In Britain, migrating Tubulifera are uncommon. In Louisiana, by contrast, Glick (1939) found Tubulifera more common than Terebrantia though the most common single species caught was a terebrantian, the flower-feeding *Frankliniella tritici*. Ten of the twelve identified species he caught above 900 m (3,000 ft), which suggests them to be the more vigorous migrants, were grass-feeders. There are probably three reasons why cereal- and grass-feeding thrips predominate in mass flights in southeast England and other parts of Europe (*cf* Table 18); cereal crops provide uniform breeding sites over large areas and all ripen more or less simultaneously; wild Gramineae are common and widely distributed; and insects breeding in temporary habitats must migrate to survive (Southwood, 1962). The behaviour of cereal thrips contrasts with that of the winged females of *Taeniothrips ericae* which may likewise be abundant in the more permanent habitat provided by *Calluna* and *Erica* spp. over large areas of moorland in Britain, but which rarely migrate.

This tendency for species living in temporary habitats to be more migratory than species from permanent ones was emphasized by classifying many thousands of thrips caught in water traps in southern England according to their habitat (Table 19). Species feeding on trees and woody shrubs, including those that feed on fungi or live under bark, and predatory species confined to trees, were classified as dwellers in permanent habitats (P), while species that feed on annual flowers or grass inflorescences were considered as living in temporary habitats (T). A second classification separated the thrips into abundant (A), common (C), occasional (O) or rare (R), according to the frequency with which they usually occur on their host plants. Species in which short-winged or wingless forms predominate in both sexes and roaming predatory thrips were excluded from the analysis; catches of species strongly atttracted to white were corrected (see p. 118) to permit a valid comparison of the abundance of different species in the air.

"Abundant" and "Common" species were separated from "Occasional" ones, then the corrected catches divided by the number of species in each habitat type (Table 20). The mean catch per species, which can be interpreted as a crude index of "migratoriness", was much greater for thrips

TABLE 19

Total thrips caught during 1956, 1957 and 1958 in water traps at Silwood Park, England. The catches of species attracted to white were corrected (end column) (from Lewis, 1961)

	Total catch	Habitat type	Abundance group on hosts	Approximate corrected catch
Melanthrips fuscus	393	T	O	393
Chirothrips manicatus	3,560	T	A	3,560
Chirothrips hamatus	23	T	O	23
Limothrips denticornis	399	T	C	399
Dendrothrips ornatus	1	P	C	1
Anaphothrips obscurus	196	T	C	196
Odonothrips cytisi	.	P	C	.
Odonothrips ulicis	.	P	C	.
Odonothrips meridionalis	.	P	C	.
Taeniothrips atratus	11,998	T	A	120
Taeniothrips vulgatissimus	14,687	T	A	147
Physothrips consociatus	1	P	C	1
Thrips physapus	.	T	C	.
Thrips hukkineni	2,700	T	C	90
Thrips praetermissus	4,500	T	O	150
Thrips fuscipennis	3,500	T	C	117
Thrips major	22,000	T	C	733
Thrips tabaci	2,700	T	C	90
Thrips flavus	480	T	C	16
Stenothrips graminum	853	T	C	853
Frankliniella intonsa	3,423	T	C	3,423
Frankliniella tenuicornis	145	T	O	145
Kakothrips robustus	.	T	O	.
Hoplothrips ulmi	60	P	O	60
Hoplothrips fungi	2	P	O	2
Hoplothrips corticis	.	P	O	.
Haplothrips subtilissimus	10	P	O	10
Haplothrips distinguendus	2	T	R	2
Haplothrips setiger	15	T	O	15
Phlaeothrips coriaceus	1	P	R	1
Hoplandrothrips annulipes	1	P	C	1
Hoplandrothrips bidens	.	P	O	.
Megathrips lativentris	.	P	R	.

from temporary habitats than from the permanent ones. Of the temporary habitats considered, grass inflorescences usually remain favourable for longer than flowers. This persistence is perhaps reflected in the behaviour of grass-dwellers, such as *Limothrips* spp. and *Chirothrips manicatus*, whose

TABLE 20

The relationship between mean catch per species and habitat for thrips from different abundance groups (from Lewis, 1961)

Abundance group	Habitat group	Number of species	Corrected total catch	Mean catch per species
Abundant and common	P	6	3	0·5
	T	13	9,744	750
Occasional	P	5	72	14
	T	6	726	121

migration is confined to a few short periods in the annual cycle with longer periods of inactivity between, whereas many flower-dwellers fly throughout the late spring and summer (Lewis, 1961).

SETTLING AND PATTERNS OF DISTRIBUTION ON CROPS

While the settling of individuals is difficult to observe in the field, the collective result of settling or "deposition" over a period of time produces distinct patterns of infestation on some crops. These patterns are largely determined by local air currents formed when the free wind encounters obstacles such as the edges of crops or sheltering hedges and trees. The nearness of hibernation sites also affects the patterns in spring (Koppa, 1967).

Even in unsheltered fields a crop only a few centimetres tall can create enough turbulence in the wind along its windward edge to cause flying thrips to accumulate there. For example, when winds in spring blew steadily from one direction, the initial infestation of a wheat crop with *Limothrips cerealium* was heaviest on the windward edge of the field, and the thrips gradually spread downwind into it as the season progressed (Fig. 56). In years when winds were variable during the migration period no marked edge effects occurred (Lewis, 1959a). Similarly, the initial distribution of thrips in hibernation sites such as hedge-bottoms or along the edges of woodland largely depends on wind directions during the autumn migration. The pronounced accumulations of *Taeniothrips laricivorus* which often occur along the edges of larch stands in spring arise partly because wind blows the thrips there, but the numbers present

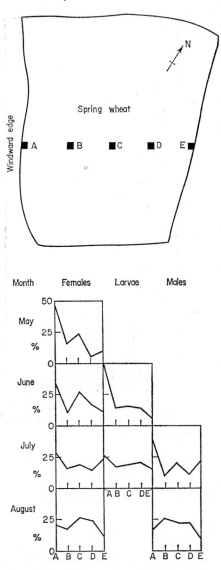

FIG. 56. The distribution of *Limothrips cerealium* in a field of wheat. The line of sampling positions, A–E, was perpendicular to the windward edge of the field during the spring migration. The arriving females accumulated along the windward edge. This initial pattern of infestation was reflected in the subsequent development of the population and persisted throughout the breeding season (May–July) (after Lewis, 1958).

also depend on the proximity of the larch to spruce, where the thrips hibernate (Fig. 57) (Maksymov, 1965; Zenther Møller, 1965).

Barriers to the wind such as hills and ridges, buildings, trees and even

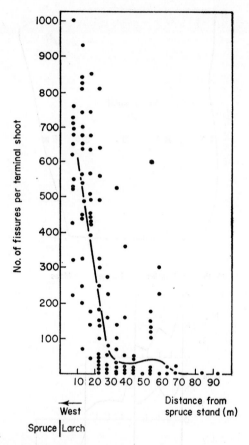

FIG. 57. Distribution of damage by *Taeniothrips laricivorus* in a Danish larch stand. Such a distinct edge effect only occurs when the larch is very near to spruce on which the thrips overwinter. When the distance between larch and the nearest spruce is greater than about 100 m, no edge effects are detectable (after Zenther-Møller, 1965).

irregular canopies of crops create sheltered zones in which many insects accumulate, and zones where faster winds blow them away. Experimental work in the field using first artificial windbreaks and eventually natural ones has shown accurately how they affect the distribution of most flying insects, and thrips respond in a typical manner.

The height and permeability of windbreaks affects the shelter they produce. The more open the barrier, the less intense is the shelter but the further it extends to leeward. For example, a solid wooden fence gives most shelter (80%) at a distance to leeward equal to the height of the fence (1 H), whereas a fence of wooden laths with 45% open area gives maximum shelter (60%) at a distance three times the height of the fence

(3 H) to leeward. The patterns of distribution of airborne thrips, measured with suction traps over bare ground, reflect the patterns of shelter, and most thrips accumulate in the calm zones (Fig. 58a, b; Lewis and Stephenson, 1966). Similar patterns occur in the air behind low hedges and windbreaks of tall trees. Details of the pattern of distribution depend on the speed of the wind, the angle at which it strikes the windbreak, and whether the thrips originate from a distant source or have bred in

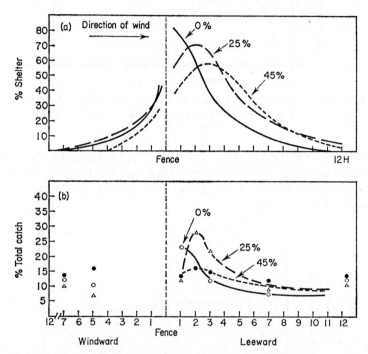

Fig. 58. The pattern of shelter behind lath fences and its effect on the accumulation of airborne thrips (after Lewis and Stephenson, 1966). (a) Mean shelter profiles behind solid (0%), 25%, 45% and 70% permeable fences measured in direct incident winds. The fences were 1 m high, and distance from them is expressed in multiples of their height (1H = 1 m). (b) Relative density profiles of thrips to leeward of the fences, closely reflecting the pattern of shelter.

the windbreak itself. In winds between 3 and 13 km/h migrating *Limothrips* spp. accumulated in the air behind a hawthorn hedge about 2 m tall, with most thrips at a distance of 2 to 4 times the height of the hedge to leeward. By contrast, in mean wind speeds slower than 2·5 km/h the distribution behind the hedge was uniform because there was rarely enough wind to blow thrips towards it (Fig. 59). In the zone between the hedge and 10 times its height to leeward, thrips were 14 times more abundant in windy weather than in calm (Lewis, 1969a). Accumulations

G

are greater when the incident wind blows against the windbreak directly than when it strikes it obliquely. The patterns of distribution behind a windbreak of pine trees about 20 m tall in direct (90° ± 40°) and oblique

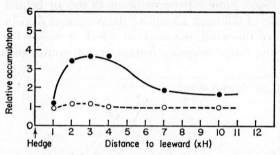

FIG. 59. Comparison of patterns of distribution of airborne thrips to leeward of a 2 m-high hedge, in fast and slow winds (after Lewis, 1969). ●——● 3–13 km/h; ○ - - - - ○ <2·5 km/h (1H = 2 m).

(>±40°) winds show this (Fig. 60; Plate IXb). In situations where thrips are actually breeding in the windbreak itself before migrating, the pattern to leeward shows an even more distinct peak almost immediately behind the barrier (Lewis 1970a).

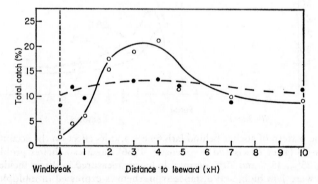

FIG. 60. Changes in patterns of accumulation of airborne thrips behind a windbreak of tall trees caused by changes in the direction of incident winds (after Lewis 1970). ○——○ direct winds; ● - - - - ● oblique winds (1H = 20 m).

These aerial patterns are important because they are reflected on the crop beneath. This can only be established for insects that alight to breed, feed or rest long enough to be sampled. The distribution of a population of *Limothrips cerealium* on a wheat crop sheltered by the tall pine trees, illustrated in Plate IXb, provided an excellent example of how the pattern of primary migrants and the subsequent distribution of the new generation depended on the shelter produced by the trees. When the females blown from hibernation sites in spring find suitable host plants on which to feed and lay, they become less willing to fly. Populations of these primary migrants on wheat crops are usually sparse, so it is difficult to collect large enough samples to plot their distribution. However, later in the season it is easier to detect where they alighted because the males of the new generation that develops from eggs laid by the migrants are wingless, so largely confined to the individual plants on which they have fed and grown.

In 1969, 72% of spring migrants flew while winds blew towards the trees, including 32% in direct incident winds. Figure 61 shows the distribution pattern of migrants *above* the crop during this period compared

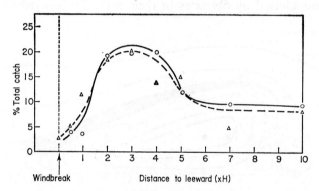

FIG. 61. The pattern of airborne *Limothrips cerealium* to leeward of a windbreak of pine trees, 20 m high, compared with the population of males that developed on the crop four weeks later (after Lewis, 1970). O————O airborne females; △----△ males on crop.

with the distribution of males *on* the crop measured 28 days after the migration ended, when the eggs the migrants laid had developed. The pattern of shelter, the density of migrants over the crop, and the new generation produced on it, corresponded closely. The heavy infestations of *Diarthrothrips coffeae* that occur on leeward slopes and behind windbreaks in coffee fields on Mt Kilimanjaro (Notley, 1948) probably arise in a similar way.

An explanation of how many small insects with body area <3 mm², including thrips, accumulate in sheltered zones has been suggested by Lewis and Dibley (1970) following detailed measurements of air movement near windbreaks. The accumulations are believed to arise largely from physical causes for two reasons. First, the patterns depend greatly on the structure of barriers, as do the patterns of horizontal wind speed. If alternatively, insects accumulated largely because they saw obstacles and flew towards them, patterns would be similar on both sides of barriers, but they are not. Also, if these patterns were produced in response to visual stimuli, the insects would need to distinguish between fine structural differences in fences and hedges from a distance of 3–60 m or more, probably far beyond their range of visual discrimination. Secondly, size and airspeed of different taxa have little effect on the patterns of distribution. For example, patterns for weak-flying Thripidae and stronger-flying Staphylinidae (Coleoptera) are very similar (Lewis and Stephenson, 1966). If the insects' airspeed had much effect on the accumulations, patterns would differ greatly in incident winds of different speeds, but they do not.

The explanation is based on the way inert particles drifting in an air-stream are affected by obstacles in their path. To leeward of any long

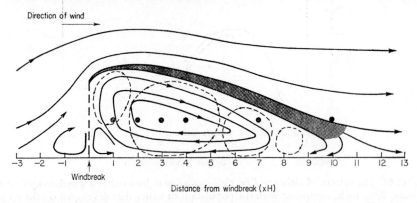

FIG. 62. Diagram of the approximate wind flow likely to cause accumulations of thrips near slightly permeable barriers. Solid lines show the general pattern, and dotted lines stress the erratic movement within this. The shaded area represents the turbulent shear layer where wind passing over and beyond the barrier separates from air that recirculated behind it (after Lewis and Dibley, 1970).

narrow barrier there is a relatively calm zone resembling a bubble extending 7–20 times the height of the barrier downwind, the distance depending on the angle of the incident wind and the permeability of the barrier. In this "bubble" the wind fluctuates in all directions but on the average follows a circulatory motion in planes normal to the barrier around a calm region near its centre. The bubble is limited below by the ground and above by a turbulent shear layer, which forms where the wind passing over and beyond the barrier separates from air that recirculates in the bubble (Fig. 62). Thus, some of the insects blown towards the barrier will be carried over and beyond it; others will be diffused into the bubble by the turbulence in the shear layer, or convected into it with the re-entrant flow generated near the downstream end of the bubble.

Although the mean air movement in the bubble is towards the barrier near the ground, and away from it near the shear layer, minor erratic eddies occur in all other planes. Depending on the direction of air currents encountered within the bubble, an insect may be detained there a long time, or escape quickly; but the net effect will be to increase the aerial density of insects in the sheltered zone. The accumulation will tend to increase until a steady state is reached, when the number entering the bubble from above or from populations on the vegetation beneath it, is balanced by the number escaping upwards through the shear layer under the action of turbulent diffusion or by directed flight, or the number alighting on the vegetation.

The concentration of inert particles to leeward of a solid elongated rectangular sheet can be estimated in order of magnitude by equating the ratio of convection into the bubble and diffusion out of it.

Expressed mathematically:

$$\frac{\Delta c_b}{\Delta c_f} \approx \frac{u}{\sqrt{(\bar{q}^2)}} \approx 10$$

where Δc_b = concentration in the bubble

Δc_f = concentration in free air

u = wi speednd

$\sqrt{(\bar{q}^2)}$ = fluctuating velocity in root mean square.

The concentration of particles would be less behind a permeable barrier where the convective re-entrant flow is weaker.

For most taxa of small insects, and for thrips in particular, the increases in aerial density behind solid barriers, lath fences and hedges range from about 2- to 14-fold (Lewis and Stephenson, 1966; Lewis, 1970). The similarity between the magnitude of accumulations measured for thrips behind many types of windbreaks, patterns of paper particles behind fences (Lewis, 1965b), and the accumulations theoretically probable for purely physical reasons, strongly suggest that the process whereby thrips accumulate is basically similar to that described. The hypothesis no doubt oversimplifies a very complex process, difficult to quantify, but it offers a better understanding of how thrips and insects of similar size accumulate in the air and on vegetation near barriers than a simple statement that they are "deposited". Though basically a physical process, insects within the bubble can presumably retain a degree of control over their track and eventual alightment rather than accumulate by inert "deposition". This hypothesis could also explain the patchy distribution of thrips that sometimes occurs in fields where there are shallow depressions or isolated trees to create wind shadows.

There are economic advantages in knowing beforehand where pests will accumulate on crops. Species harmful even when sparse, such as virus-transmitting thrips, need to be detected on crops as soon as possible.

Searches should therefore be concentrated in sheltered areas. Also, insecticides could perhaps be limited to the edges of fields, thus saving money and preserving beneficial insects. Because thrips accumulate in wind shadows created by obstacles, crops with gaps between plants or with uneven canopies will probably receive more migrants and heavier infestations than other crops in the same locality that present a smoother profile to winds blowing over them. Therefore, apart from cultural reasons, even growth should be encouraged to lessen losses from harmful species.

8 Survival in unfavourable seasons

COLD AND HIBERNATION

Stages and forms

Few species remain active outdoors throughout the year where the mean temperature in the coldest month falls below 5–6°C. Winter may be survived by adults of both sexes, adult females, females and larvae or pupae, or larvae only, depending on the species. Most commonly, as in *Limothrips* spp., *Chirothrips* spp. and *Thrips physapus*, adult females only survive (Lewis and Navas, 1962; Wetzel, 1963). Late-developing larvae and males of such species may persist into late autumn on fading host plants but they are killed by starvation and early frosts; just a few males of *Chirothrips manicatus* or *Limothrips denticornis* may survive for longer (Wetzel, 1963, Koppa, 1969b). Less often, adults of both sexes survive, as in *Thrips angusticeps*, *Oxythrips ulmifoliorum* and *Haplothrips aculeatus*; or females, larvae and occasionally eggs, depending on the mildness of the winter, as in *Aptinothrips rufus* (Sharga, 1933a) and *Thrips tabaci* (Sakimura, 1937a). In South Australia *Thrips imaginis* breeds throughout the winter, but the stages best adapted to survive are adults on vegetation and pupae in the soil; in cold weather the eggs and larvae have not time to complete development on individual flowers before these fade (Andrewartha and Birch, 1954). Only larvae of *Thrips minutissimus*, *Kakothrips robustus* and of many European phlaeothripids overwinter (Morison, 1943, 1948). *Aeolothrips tenuicornis* usually overwinters as larvae, but in warm seasons, when development is rapid, the adults hibernate (von Oettingen, 1942). In areas with warm equable climates like California or South Carolina, many thrips, such as *Frankliniella* spp., breed throughout the mild winter and all stages are present (Watts, 1936; Bailey, 1938), but in the more severe winter of sub-tropical continental areas, like southern New Mexico, the immature stages of the same species die (Faulkner, 1954). The citrus thrips, *Scirtothrips citri*, is one of the few species which overwinters only in the egg stage (Bailey, 1938).

Structure and colour may differ between individuals of the overwintering generation and their parents of summer generations. Structural differences between generations are common in species overwintering in turf or soil; in *Thrips angusticeps* in Holland (Franssen and Huisman, 1958) and *Sericothrips bicornis* in England (Ward, 1966) the overwintering generation is brachypterous and flightless, whereas both sexes of summer generations are macropterous and can fly. Females of *Thrips nigropilosus* show all degrees of wing development and long- and short-winged individuals overwinter in England (Morison, 1957). In northern Europe, II stage larvae of *Anaphothrips obscurus* may hibernate as well as long- and short-winged adults (Weitmeier, 1956; Koppa, 1969b). Where winters are cool, brachypterous *Thrips tabaci* are rarely found, but in warmer areas, for example in Louisiana, U.S.A., brachypterae and macropterae are common, though brachypterae predominate (Newsom *et al.*, 1953). This is surprising as brachypterous forms of other species usually appear in the cooler areas of their range and in the cooler seasons.

In polyvoltine species, generations developing at higher temperatures usually consist of smaller, paler-coloured individuals than those developing at lower temperatures, so the appearance of some species may differ slightly in different seasons and geographical latitudes (Priesner, 1964a). In England (latitude 52°N), overwintering *Thrips tabaci* may be darker than the yellowish-brown summer generations (Mound, 1967b), but in Japan (latitude 36°N), where this species may complete up to 10 generations a year, there is a definite gradation of coloured forms from light yellow in July to dark brown in January (Sakimura, 1937a). By contrast, individuals of species overwintering as young adults in the soil like *Thrips angusticeps* and *Stenothrips graminum*, often stay pale throughout the winter and darker pigmentation does not develop until spring.

SITES AND POPULATION DENSITY

In late summer or autumn, thrips usually move away from light and seek cracks and crevices, eventually coming to rest with as large a part of their body surface as possible touching the surrounding surfaces. Some species enter the soil beneath their host plant as fully grown II stage larvae and crawl downwards through cracks, old root channels and worm burrows searching for suitable sites. They penetrate deeper in friable sands and loams than in compact clays. Larvae of *Stenothrips graminum* usually concentrate 50–70 cm below the surface (Fig. 63) depending on the compactness of the soil, though a few may burrow as deep as 100 cm. Each larva moulds a cell in the soil and within this changes to the adult. Most adults emerge from the soil the following spring but in Holland 20–30% remain diapausing within the cell for a further year. *Thrips angusticeps* is another deep-burrowing species in which usually about 5% of the overwintering adult generation remain in the soil for 20 months (Franssen

and Huisman, 1958). Such prolonged inactivity would only be possible
at depths where the seasonal temperature fluctuations were small. Second
stage larvae of *Kakothrips robustus* burrow slightly less deeply to 20–50 cm
(Kutter and Winterhalter, 1933; Buhl, 1937; Franssen, 1960) but do not
pupate until spring when they complete development and emerge from
the soil as adults (Williams, 1915). As the II stage larvae do not crawl
far before entering the soil, large populations accumulate near to host

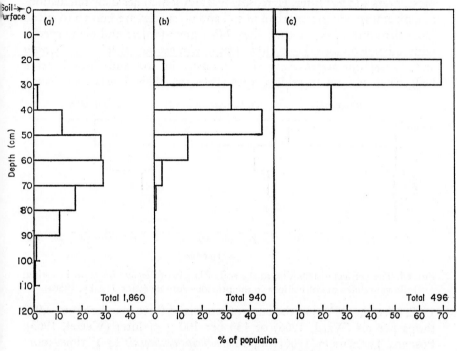

FIG. 63. The distribution with depth of 3 populations of thrips overwintering in soil.
(a) *Stenothrips graminum* (Franssen and Mantel, 1965); (b) *Thrips angusticeps* (Franssen
and Huisman, 1958); (c) *Kakothrips robustus* (Buhl, 1937).

plants. In Dutch flax fields 1 m² of soil 70 cm deep may contain up to
6,000 hibernating *Thrips angusticeps* (Franssen and Huisman, 1958), and
on harvested pea fields there may be 800 *Kakothrips robustus* larvae per
m² (Franssen, 1960). In soil beneath uncultivated, mixed vegetation,
populations are usually much smaller.

   Larvae rarely penetrate beyond a ploughsole or hard pan (Bailey, 1938)
but the natural vertical distribution of larval populations in soil may be
changed by cultivations. For example, hibernating larval populations of
Haplothrips tritici, a common European species, concentrate in the top
10 cm of uncultivated soil, but cultivations redistribute the larvae to

greater depths (Fig. 64) (Tanksy, 1958a; Shurovenkov, 1961). In arid parts of south-eastern European Russia, in early autumn, larvae remain near the surface on damp, cool irrigated plots but penetrate deeper in hot dry soil. There is also evidence of movement within the soil in early winter, the larvae ascending to shallower layers in October and November, presumably as the soil becomes wetter (Grivanov, 1939).

Many species overwinter just beneath the soil surface or among turf, moss, grass and leaf litter. Here, seasonal and sometimes daily fluctuations in temperature are greater than in soil and no species are known to spend more than one winter in such sites. Most grass-feeders and many species with dicotyledonous hosts, like *Thrips nigropilosus, Taeniothrips atratus* and *Taeniothrips vulgatissimus*, are common in soil and litter beneath hedgerows and on wayside verges (Priesner, 1925; Lewis and Navas,

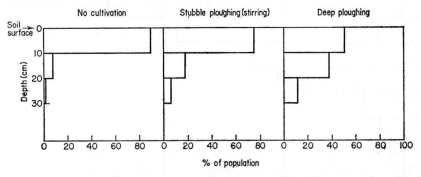

FIG. 64. The vertical distribution in the soil of larvae of *Haplothrips tritici* in spring, following different cultivations in the previous autumn (after Tansky, 1958a).

1962; Wetzel, 1963). On the edge of woodland there may be up to 3,000 thrips per m² (Ward, 1966) or 130 per 100 g of litter (Wetzel, 1963); Post and Thomasson (1966) found *Limothrips denticornis* 13–17 times more abundant in shelterbelt turf than in roadside turf. In more open situations, vegetation which produces compact ground cover provides the best conditions for overwintering. For example, in New Mexico, winter temperatures in litter beneath dense grass swards where *Frankliniella* spp. are abundant, may be up to 5°C higher than among less dense cover where fewer thrips survive (Faulkner, 1954). Similarly in Germany, stubbles of grass-seed crops contain large overwintering populations only where the inter-row spaces have become overgrown to produce a dense ground cover (Wetzel, 1963). Crop remains left on and around fields also provide suitable overwintering sites for many pest species and encourage large infestations in the following season (Horsfall and Fenton, 1922). Some cereal-feeding species can overwinter in shoots of winter cereals; populations of *Frankliniella tenuicornis* in winter rye may be as dense as 220/m² (Koppa, 1969b).

A hibernation site entered in early autumn may deteriorate with the onset of winter and some species move to more protected sites as winter progresses. In England *Aptinothrips rufus* accumulates in upright hollow grass stalks and between leaf sheaths in November, but later moves deeper into the flattened stalks and surface litter, presumably on occasional warm days (Fig. 65). On wayside verges up to 300 *A. rufus* may collect in a single square metre (Lewis and Navas, 1962) and up to twice as many in the same area of woodland litter (Healey, 1964). Other species survive best in drier, exposed sites like the upstanding stalks of dead grasses. *Chirothrips manicatus* is a tiny species that feeds on *Dactylis glomerata*, and many females remain through the winter in dead inflorescences of their host plant, deeply embedded between the shrivelled pales

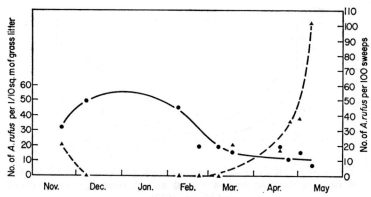

FIG. 65. The number of *Aptinothrips rufus* collected from standing grasses (by sweeping) and from surface litter (in quadrats) showing change of site during the winter. ▲---▲ Number per 100 sweeps; ●——● Number per ft² (0·1 m²) of litter (after Lewis and Navas, 1962).

(Lewis and Navas, 1962). *Frankliniella intonsa* and *Thrips physapus* may also overwinter on the shrivelled inflorescences of their hosts and on neighbouring plants (Morison, 1957; von Oettingen, 1942).

Where plant growth continues throughout the winter, many species migrate from harvested or withered summer hosts to green crops and weeds, where they continue to feed and breed, though more slowly than in summer (Bailey, 1934a; Davidson, 1936; Newsom *et al.*, 1953; Boyce and Miller, 1954).

Still higher above the ground, cracks and crevices in tree bark provide an abundance of hibernation sites, but because bark temperatures and humidities fluctuate so much [in maritime European winters from about −5°C in heavy frost to 35°C in direct sunlight and from 30% to 80% R.H. (Lewis 1962)], usually only adults are hardy enough to survive there. Trees with loose, flaky bark, like pine (*Pinus* spp.) have more suitable crevices than those with compact bark like poplar (*Populus* spp.) or beech (*Fagus sylvatica*), and even on one species of tree the age and scaliness of the bark affects the number of suitable sites available (Plate Xa). Young trees have usually fewer cracks per surface area of the trunk than older ones.

Limothrips denticornis and *L. cerealium* are especially common in much of western Europe (Maltbaek, 1932; von Oettingen, 1942) where they overwinter in bark, but the proportion of each species in the hibernating population depends on the predominant local vegetation and crops. For example, in barley-growing districts of southern England *L. denticornis* is usually the more abundant, but in wheat-growing districts, *L. cerealium* predominates. This species generally occupies higher and drier sites than

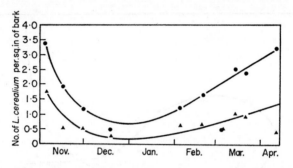

FIG. 66. The numbers of living *Limothrips cerealium* per in^2 (6·25 cm^2) in the outer crevices of bark on Austrian pine ●———● and chestnut ▲———▲ (after Lewis and Navas, 1962).

L. denticornis and prefers bark to litter. Many other species hibernate in bark including some flower-dwellers, but usually they are less abundant than *Limothrips* spp. (Table 21). There are usually more thrips near the base of the trunk than higher up it. This is probably because when thrips fly to overwintering sites the greatest aerial density occurs nearest the ground (Lewis, 1959b; Taylor, 1960; see p. 151), and partly because the bark at the bottom of trunks has more scales and crevices than higher up.

The distribution of thrips over the trunk is often patchy. A suitable crevice 1 cm^2 may contain up to 60 individuals, whereas another much larger but less suitable area nearby may contain few. Average numbers of all species found per 10 cm^2 of bark range from less than 1 on unsuitable trees like poplar to about 20 on large pines. Thus mature trees with flaky bark may contain 100,000 to 500,000 hibernating thrips in cereal-growing areas in England, but relatively unsuitable trees like poplar only 5,000 to 25,000.

As with grass stalks, the suitability of bark crevices changes during the winter and *L. cerealium*, and probably other species, penetrate into the

TABLE 21

The proportions (%) of different species of thrips in overwintering populations in soil, litter and bark in south-east England (after Lewis and Navas, 1962, and Ward, 1966)

Species	Soil	Litter				Bark				
	Chalk down	Chalk down	Hedge-bottoms	Wayside verge	Yew woodland	Old Austrian pine	Young Scots pine	Horse chestnut	Poplar	Yew
Chirothrips manicatus	—	1·4	—	7·3	<0·1	—	—	0·3	—	0·9
Chirothrips ruptipennis	—	0·3	—	—	—	—	—	—	—	58·3
Limothrips denticornis	26·1	35·0	22·7	11·7	95·1	<0·1	4·8	—	—	29·6
Limothrips cerealium	—	1·4	2·7	2·1	1·3	99·3	93·2	99·3	98·7	0·9
Aptinothrips rufus	10·9	19·4	71·4	78·2	1·5	0·2	0·3	0·2	1·3	—
Sericothrips bicornis	2·2	1·1	—	—	<0·1	—	—	—	—	—
Anaphothrips silvarum	4·3	0·3	—	—	<0·1	—	—	—	—	1·9
Taeniothrips vulgatissimus	—	—	2·9	—	<0·1	—	0·5	—	—	2·8
Taeniothrips atratus	—	—	—	—	1·5	—	—	—	—	1·9
Thrips physapus	—	—	—	—	<0·1	<0·1	—	—	—	0·9
Thrips fuscipennis	—	—	—	—	<0·1	—	0·5	—	—	—
Thrips major	56·5	40·3	—	—	<0·1	—	—	—	—	—
Thrips nigropilosus	—	—	—	—	<0·1	—	—	—	—	2·8
Thrips praetermissus	—	—	0·3	0·7	—	0·2	—	0·2	—	—
Thrips (other spp.)	—	—	—	—	<0·1	—	—	—	—	—
Platythrips tunicatus	—	0·8	—	—	—	—	0·7	—	—	—
Phlaeothrips coriaceus	—	—	—	—	—	0·3	—	—	—	—

deeper crevices to spend the coldest months (Fig. 66). There must be many occasions when it is warm enough for thrips to change positions within the bark, for *L. cerealium* walks fairly readily at 5°C, and when the sun shines directly on to tree trunks temperatures in crevices often reach 15–17°C, even in mid-winter (Lewis, 1962). *Thrips fuscipennis* can also crawl at near freezing temperatures to avoid unsuitable conditions, particularly moisture (Speyer, 1938). Individual *L. cerealium* prefer crevices

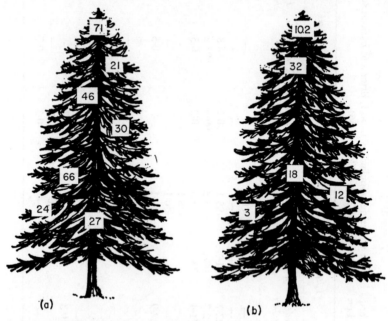

(a)

(b)

FIG. 67. The distribution of overwintering *Taeniothrips laricivorus* on two young spruce trees in a predominantly larch stand. Tree (a) was isolated from other spruce, so the thrips were distributed over it more evenly than on tree (b), which stood among a small group of spruce and was therefore unevenly exposed to migrants, the top receiving by far the greater proportion. The figures give the number of thrips per 2-litre sample of twigs (redrawn after Vité, 1956).

between 0·2 and 0·4 mm wide, so that the upper and lower surfaces of their bodies can touch the crevice walls; wider crevices are unattractive and narrower ones inaccessible (Lewis and Navas, 1962). The texture of crevice walls may also influence the selection of suitable sites, the rougher surfaces being most attractive (de Mallman, 1964).

The aspect of hibernation sites in relation to the direction of prevailing winds during the autumn migration and their exposure to cold winds in winter, affects the numbers of thrips present. Isolated trees, and trees in hedgerows or on the edges of woodland, usually contain more thrips than trees within woodland.

The larch thrips, *Taeniothrips laricivorus*, hibernates mainly on spruce beneath the previous year's bud scales, but the density of overwintering populations on young trees depends on the composition of the surrounding stand, and, on each tree, on the amount of foliage in the crown. When spruce trees occur in a predominantly larch stand, populations of

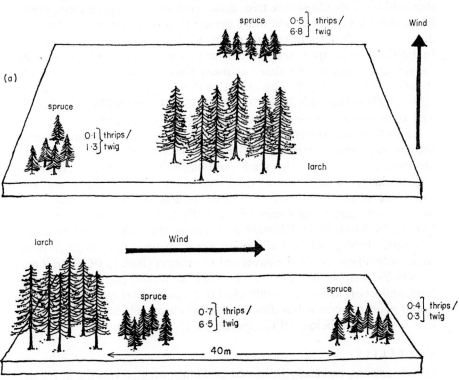

FIG. 68. The effect of wind direction on the density of overwintering populations of *Taeniothrips laricivorus* on young spruce trees. Each pair of bracketed figures represents the mean number of thrips on a terminal and second uppermost whorl of twigs. (a) The population on spruce trees downwind of larch trees, the summer host, was much greater than on spruce trees cross-wind to larches. (b) Spruces adjacent to larches had 0·7 and 6·5 thrips per uppermost and second uppermost whorl of twigs respectively; 40 m downwind the densities were 0·4 and 0·3 respectively. Thus thrips seeking hibernation sites may be blown a considerable distance, and where suitable sites are relatively scarce, the density of hibernating populations remains similar for many metres from the source of migrants (after Maksymov, 1965).

thrips per tree are greater and more evenly distributed on spruce trees standing alone than on trees standing in small groups (Fig. 67) (Vité, 1956). This is presumably because the isolated trees are exposed to flying thrips from all directions whereas the grouped trees have some sides and their lower branches, sheltered from sources of thrips. The effect of wind on distribution is well illustrated by the disposition of hibernating *Taen. laricivorus* in Swiss spruce plantations. Spruce trees around the edges of plantations and downwind from the host larches contain most thrips, partly because this species like well-illuminated sites, but largely because thrips drifting in the air are more likely to alight at the edges of plantations than within them. On spruce trees downwind of a larch plantation, population density was about five times greater than on other spruce the same distance away but transverse to the wind (Fig. 68) (Maksymov, 1965). In England, vegetation and litter beneath the southern side of wayside hedges may contain four times as many thrips as on the northern side (Lewis and Navas, 1962). This is true of the wingless species *Aptinothrips rufus* which presumably crawls short distances to overwintering sites, as well as of winged species that are blown there (see p. 158).

As well as the common hibernating sites mentioned, many diverse natural and artificial crevices shelter thrips in winter. Out of doors they hibernate in haystacks, sedge stacks, stubble, hollow grass stalks, thatches, logs, in clumps of moss, among dead insects, inside lepidopterous cocoons and even beneath the scales of scale insects (Coccidae). *Phlaeothrips coriaceus* may enter old mines of the holly leaf miner (*Phytomyza ilicis* Curtis), in Yugoslavia *Liothrips oleae* enters the galleries of Scolytids (Tominic, 1950), and in Californian citrus groves, *Caliothrips fasciatus* commonly hibernates in the navel end of oranges (Bailey, 1938). Indoors, *Limothrips cerealium* squeezes into cracked beams and picture frames, behind wallpaper, and even inside clocks; *Taeniothrips simplex* survives on stored gladiolus corms but dies on unlifted corms when the temperature remains for long below 10°C (Speyer, 1951).

MORTALITY

In indoor sites, especially in heated buildings, many thrips die in the dry atmosphere, but out of doors, winter mortality depends largely on the weather and the site occupied. Many species can withstand a wide range of temperatures including freezing; very dry or wet conditions probably cause more deaths than low temperatures, especially among species overwintering in turf and litter.

Overwintering forms in soil can survive wide fluctuations in air temperature, extreme cold, and often flooding. *Thrips angusticeps* can withstand experimental temperature regimes of two months at 20°C, or one month at 2°C and one month at 20°C; even one month at −20°C and one month at +20°C caused little mortality. During the severe 1955/56 winter in Holland no deaths were detectable in an undisturbed field population of this species between November and March (Franssen and Huisman, 1958). It can also survive flooding for weeks, perhaps because air is retained in the larval cells. However, soil-dwellers may be killed by autumn and winter cultivations which disturb the soil and redistribute the thrips. For example, in North Khazakhstan, autumn ploughing lessens by 30–60% the number of *Haplothrips tritici* emerging in spring, because it buries the larvae so deep that young adults cannot easily reach the soil surface. Breaking the stubble in autumn causes 65–75% mortality because without the stubble to retain it, snow is blown from the open fields in winter leaving the larvae exposed to prolonged severe frost (Tansky, 1958a). Snow covering the branches and foliage of conifers lessens the daily fluctuations in temperature there so thrips overwintering on snow-covered branches may experience about 8°C less change per day than insects on exposed ones (Wellington, 1950).

Prolonged dry weather in spring may sometimes kill thrips in soil before they emerge. Larvae of the pear thrips, *Taeniothrips inconsequens*, construct a cell within the earth in which to hibernate and complete development (Jary, 1934). In clay or loam soils the cell walls harden as the soil dries in spring and the larva inside is protected, but on coarse sandy soil the walls crumble as the soil dries and the larva may also dry and die (Bailey, 1944).

Providing the air is moist, hibernating *Limothrips cerealium* survived in the laboratory for about 18 days at −5°C, 50 days at −1 to +1°C and 60 days at 20°C, extreme temperatures they are unlikely to experience for such long periods during maritime winters. They survived even longer when temperatures fluctuated between −1° and 20°C (Lewis, 1962). However, many were found dead in Holland after the severe 1962–63 winter (Franssen and Mantel, 1965), and in Germany after a cold dry February when the average temperature was −7·5°C (Holtmann, 1962). This evidence, and the tolerance of this thrips to cold temperatures mentioned above, support Maltbaek's (1932) belief that the species occurs only where the average temperature of most winter months exceeds −1°C.

This species can also acclimatize to rapid changes in temperature. Cooling from 31°C to −7°C in about 45 minutes and heating from −10° to +7°C in 15 minutes were harmless provided the thrips were not chilled for too long (Lewis, 1962). Such efficient acclimatization is essential for species overwintering in bark or other exposed habitats where direct sunlight may produce temperature increases as great as 25°C in an hour. In mid-latitudes this occurs especially on the eastern side of tree trunks in clear weather when the rising sun, low in the late winter sky, shines directly on to the bark.

These fluctuating temperatures within crevices are accompanied by rapid changes in relative humidity. During the night the air within them is nearly saturated (R.H. >80%) but it quickly dries to about 20% R.H. in direct sunlight, and once dry it stays drier than the surrounding air throughout the day: on cloudy days it remains saturated. Hibernating

Limothrips cerealium would die if the most severe saturation deficiencies recorded in bark in the late winter (23 mm Hg at 30°C and 18 mm at 25°C) persisted for 14–24 h (Lewis, 1962) but in north-western Europe 2–4 h per day is usually the maximum time that such severe conditions continue. The high temperatures causing very dry conditions occur mostly in early spring after emergence from the crevices has started and when warmth stimulates the thrips to come to the bark surface and fly away. So, although hibernating *L. cerealium* in cool temperate zones are much more susceptible to injury by desiccation in early spring than by frost in mid-winter, they are unlikely to stay in hot crevices long enough to be killed, unless they are trapped within the bark.

This may sometimes happen when bark crevices contract at high temperatures. *L. cerealium* usually chooses crevices 0·2 to 0·4 mm wide so a slight contraction is sufficient to squash individuals that are loosely wedged between the crevice walls. In spring, crevices may contract by 0·1 mm in direct sunlight, and expand again at night, and the local pressures produced by this movement probably kill some thrips, and temporarily seal others within crevices. Loose flaky bark often contains groups of dead individuals (Plate Xb), suggesting that many die in bark each winter. This is a false impression created by the slow rate at which bark is shed from tree trunks. Most older trees shed their bark gradually and some, like pine, shed little before the trees are about 20 years old. Therefore the dead thrips found in old pine bark have accumulated for 10 to 20 years and the mean annual mortality is probably less than 10% (Lewis and Navas, 1962; Ward, 1966). In less persistent hibernation sites winter mortality can be assessed more easily. In Switzerland, 20–26% of *Taeniothrips laricivorus* hibernating beneath the dried bud scales of spruce, Douglas fir and Weymouth pine died in 1963/64 (Maksymov, 1965). In turf, mortality may be much greater; in North Dakota up to 60% of *Limothrips denticornis*, 80% of *Frankliniella tritici*, 40% of *Anaphothrips cameroni* and 40–66% of two tubuliferan species in roadside turf died during the winter (Post and Thomasson, 1966).

Where winters are milder and thrips survive exposed on plants, rainfall probably causes more deaths than cold. In California, many overwintering *Thrips tabaci* and *Caliothrips fasciatus* are washed from plants by rain and the survivors are found in protected places beneath leaves and in flowers and buds (Bailey, 1934b, 1937). Twenty per cent of adult c. *fasciatus* hibernating in the navel end of oranges, where rain could reach them, died by mid-winter (Woglum and Lewis, 1936), whereas in dry places dormant adults of this species survive temperatures as low as −9°C (Bailey, 1937).

Immature stages of species that feed throughout mild winters may sometimes die because food is short. For example, in South Australia nymphs of *Thrips imaginis* develop so slowly during winter that they may fail to complete development before the flowers in which they are living wither (Andrewartha and Birch, 1954).

Individuals of the larger species are eaten by insectivorous birds, but the proportion taken from an overwintering population is probably too small to affect the size of the population in the following season.

OVARIAN DIAPAUSE

Adult thrips arrive at hibernation sites with their fat-body distended and conspicuous. Their gut soon empties and the fat reserves nourish them throughout the winter. By spring, the fat-body is shrunken and inconspicuous.

In most species there is little ovarian development before hibernation, and in most individuals the egg rudiments are not differentiated. In species that mate before hibernation like *Limothrips* spp. and *Chirothrips manicatus*, some differentiation occurs in about 10% of females, especially in a mild autumn, but their longest egg rudiments rarely exceed 50 μ. In a minority of *Thrips fuscipennis* in England, developing eggs may be 120–150 μ long by February and they continue to grow slowly even before emergence from hibernation, but do not reach full size until thrips have fed (Speyer, 1939). The ovaries of young adults produced from larvae that hibernate in the soil probably remain completely undifferentiated until after they have emerged and mated in spring. Where winter temperatures fluctuate, a few parthenogenetic species like *Aptinothrips rufus* and *Thrips tabaci* may lay in occasional warm spells and their egg production seems only to be delayed by cold weather; in warm latitudes breeding continues throughout the year (Sakimura, 1937a).

In contrast to this temperature-induced reproductive quiescence in *Aptinothrips* and *T. tabaci*, the delayed ovarian development in *Limothrips cerealium* seems to be a form of reproductive diapause. In experiments, females collected from hibernation at approximately monthly intervals from September to February and placed on young cereals for 20 days at 25°C did not lay and not until March were they able to resume ovarian development when placed in favourable conditions. Constant temperatures of $-5°$, $+1°$, $5°$, $10°$ or $15°C$ and daily fluctuations of 8 h at 20°C and 16 h at 10°C failed to break ovarian diapause before then (Lewis, unpublished observations).

The inception and duration of diapause may be influenced by the photoperiod, temperature, and perhaps the inferior quality of the food in the period before the thrips enter hibernation. Reproductive diapause in adults of the parthenogenetic *Anaphothrips obscurus* was induced experimentally by exposing larvae to short days (10 h light:14 h dark). The ovaries of adults produced by these larvae remained undeveloped after

30 days, whereas ovaries of adults produced from larvae that had experienced long days (16 h: 8 h) were well developed after the same period. Diapause was maintained for about 60 days, by keeping adults in a short day at 21°C, but thereafter diapause terminated and reproduction started. Short days at 15°C delayed but did not prevent the termination of diapause. In the field, the influence of photoperiod on the maintenance of diapause probably declines during the winter, temperature regulating the resumption of ovarian development in spring (Kamm, 1972).

The reason for the two-year diapause in a minority of *Stenothrips graminum* and *Thrips angusticeps* is unknown; the basis of the difference between diapausing and non-diapausing individuals may be genetical and there may be some selective advantage in spreading emergence over more than one season.

INHIBITION OF FLIGHT

Once they have arrived at hibernation sites many macropterous species lose the urge to fly until the following spring. In experiments made in southern England in early autumn, about 80% of *Limothrips cerealium* females collected from grass flew towards a bright light at 25°C, compared with only 23% of those that had already entered bark (Lewis, unpublished observations). This was not due to degeneration of the wing muscles which appear to remain intact throughout the winter (Plate XI), in contrasting to hibernating Colorado beetles [*Leptinotarsa decemlineata* (Say)], whose wing muscles degenerate to thin muscle fibres within 6 weeks of entering hibernation sites and do not regenerate until diapause has almost ended. However, though there is no visible degeneration in flight muscles in thrips, flight may be nevertheless inhibited by slower muscle respiration and "loosely coupled" oxidative phosphorylation, as happens in Colorado beetles when their muscles have regenerated after diapause but before they have completely regained their function (Stegwee, 1964).

Whatever environmental conditions and physiological processes induce thrips to enter or leave hibernation sites, in autumn they produce strong negative phototactic and positive thigmotactic responses so the insects generally seek dark places and maximum bodily contact with their surroundings. When emerging in spring, however, they temporarily seek light and avoid crevices, usually crawling to the extremities of bark flakes before take-off. This reversal of responses to contact and light in the two seasons may be influenced by temperature differences. A similar change occurs in the bark beetle, *Blastophagus piniperda* L., which is more strongly photopositive in spring than in autumn, the intensity of the response depending on temperature. In spring the beetles are photopositive at temperatures between 10° and 35°C but photonegative at other temperatures (Perttunen, 1959, 1960). Adults of the second generation of *Limothrips cerealium*, *L. denticornis* and *Haplothrips aculeatus*, are photopositive at 20°C in summer (Holtmann, 1963), but as temperatures fall in early autumn this positive response may gradually reverse and dark crevices become attractive to them. If this explanation is correct, the changes in behaviour in autumn and again in spring must develop gradually, perhaps over a period of weeks, because the high temperatures occurring suddenly on occasional sunny days do not usually stimulate thrips to emerge from bark crevices. Such a gradual change in behaviour would serve to protect overwintering individuals from emerging on occasional warm days very early in the year, before their food plants had started to grow.

EMERGENCE FROM HIBERNATION

Thrips are stimulated to leave overwintering sites in spring by rising temperatures. Towards the end of winter, species which have survived deep in the soil, such as *Thrips angusticeps*, gradually move upwards

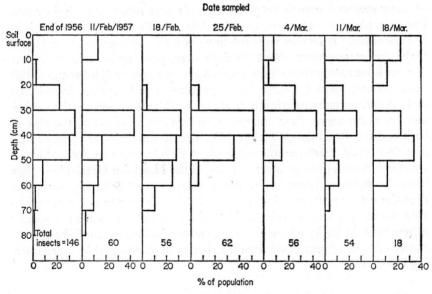

FIG. 69. The upward movement of brachypterous *Thrips angusticeps* in the soil from early February to mid-March 1957, in Holland (after Franssen and Huisman, 1958).

(Fig. 69), and in Holland this species usually emerges from the soil when the average air temperature over a 10-day period exceeds 8°C (Franssen and Huisman, 1958). Similarly, *Taeniothrips inconsequens* in British Columbia starts to emerge when mean air and soil temperatures reach 10°C and 7°C respectively (Cameron and Treherne, 1918), though further south in California significant emergence (i.e. 5 adults/sq m/day) begins only after the mean soil temperature has exceeded 12°C for 2–3 consecutive days (Bailey, 1934a). There is a large discrepancy between these two estimates of the soil temperatures necessary to stimulate emergence in the pear thrips. Perhaps it arises because the warmer temperature given by Bailey is the one necessary before *most* individuals will emerge; whereas movement *starts* at the lower temperature quoted by Cameron and Treherne.

Among species which overwinter in litter, *Aptinothrips rufus* becomes active when air temperatures reach 15°C for a short time, so in northern Europe this species usually appears on young grass from early to mid-March (Lewis and Navas, 1962; Wetzel, 1963) but adults return to litter during dry spells later in spring (von Oettingen, 1942). *Chirothrips manicatus* and *Aeolothrips albicinctus* need a longer period of fine weather with temperatures above 20°C before they emerge, so they do not usually appear until late April or early May.

Because the time of emergence depends on temperature the appearance of many species is associated phenologically with other biological events. For example, the emergence of *Limothrips cerealium* and *L. denticornis* in Europe usually coincides with the flowering of *Anemone nemorosa*; *Haplothrips aculeatus* appears about 10 days later, possibly because it favours damper hibernation sites which take longer to warm up (Holtmann, 1962; Franssen and Mantel, 1965). As temperatures rise in spring, individuals of *L. cerealium* overwintering in bark gradually move towards the surface and on warm sunny days emerge from crevices to crawl to the tips of bark scales before take-off (see p. 135). Before the thrips are ready to do this they need a temperature-dependent development period which is shorter at warmer temperatures. Thus, in the Northern Hemisphere they leave the warmer eastern and southern sides of tree trunks slightly earlier than the cooler northern sides (Lewis, 1963). Air temperatures are probably therefore a less reliable guide to the time of emergence for species hibernating in bark, where the temperature in intense sunlight greatly exceeds the air temperature, than for species in more equable overwintering habitats.

HEAT AND AESTIVATION

Much less is known about survival in hot seasons than in cold ones, but in climates where habitats dry out in the hot season many thrips aestivate to avoid desiccation. Where the dry season is prolonged, as in deserts and steppes, many species have only one generation per year. Most Thripidae that aestivate do so as adult females, the males surviving only for a short time in spring, but phlaeothripid males are generally more resistant to hot dry weather than thripid males and a few often survive the unfavour-able season. The most vulnerable stage in dry conditions is usually the pupa which, unlike larvae and adults, cannot feed to replenish lost water, neither is it embedded in moist plant tissue like the eggs. However, Putman (1965b) found the egg stage of the predatory *Haplothrips faurei* more vulnerable to dryness than the pupal stages; in the laboratory at 24°C, all eggs kept at 65% R.H. survived, about half survived at 40% and none at 30% R.H., whereas prepupae and pupae were unexpectedly resistant to the drier conditions. In some orchards in Ontario, Canada, humidity falls to these lethal levels, but the air probably does not remain dry long enough to kill the eggs.

Sites

The sites entered for aestivation in hot climates are similar to those used for hibernation in colder ones, and flower heads, hollow branches, dead stalks, tree bark and soil are all suitable. A few desert species produce galls on their hosts and remain protected within them (Priesner, 1964a), but even though they enter protective sites, millions of thrips die in hot dry seasons (Andrewartha and Birch, 1954). They are especially vulnerable to sudden hot, dry winds like the Khamsin that blows in Egypt and along the Levantine coast, where temperatures may rise rapidly to 42°C and relative humidity fall to 10% (Rivnay, 1938).

In hot dry climates, Thysanoptera avoid direct sunlight until just before take-off, and even in hot humid areas species living on the surface of plants usually avoid excessive exposure to sunlight by remaining on the shaded underside of leaves. On cashew, *Anacardium occidentale*, the cacao thrips, *Selenothrips rubrocinctus*, shows a slight preference for the stomatal-bearing axial surface, but if under natural conditions this surface is inverted and exposed to sunlight, the thrips move to the other surface to avoid heat and undue dryness (Fennah, 1963). Phlaeothripidae usually rest on the shaded side of tree trunks. *Retithrips syriacus*, a Middle-eastern thripid, selects vertical leaves for oviposition sites to protect its eggs from excessive heating (Priesner, 1964a), though adult females of other species may require brief exposure to direct sunlight to complete maturation of the eggs during the preoviposition period (Ghabn, 1931).

Mortality

Total radiation from sun and sky is unlikely to raise directly the body temperature of such small insects more than a fraction of a degree (Digby, 1955), except perhaps in the largest Phlaeothripidae, or when large numbers of smaller thrips are clustered together. It does however heat their surroundings, especially dark, dull surfaces like bark, and produces a shallow zone of hot, dry air around such surfaces (Lewis, 1962) in which many species could probably not survive for more than a few hours. In the laboratory, mortality of *Thrips tabaci* larvae was no greater after 2 days at 38°C than after exposure to lower temperatures, provided the air remained moist (Macgill, 1937). Thus for species living exposed on vegetation, dryness rather than heat alone is probably the greatest danger.

9 The size and natural regulation of field populations

In all temperate regions the reproduction and growth of individual thrips are retarded or cease completely during cold weather, and in hotter regions dry weather has a similar effect. Usually only one stage survives the unfavourable season (see Chapter 8). These survivors form the basis of populations that develop later when weather and other environmental factors improve. This chapter shows how the physical and living components of the environment, whose separate effects on individuals are described in Chapters 2, 3 and 4, combine to control the growth and size of field populations. The effect of these components within a season is considered first, then the evidence examined to see what might determine the long-term trends of abundance around which populations fluctuate from one year to the next.

The rate of increase during the breeding season and the eventual size of a population in an area depends partly on the number of thrips breeding in the area and partly on the number entering and leaving it. The mechanism and scale of aerial dispersal was considered in Chapter 7 but with such immense numbers of small insects living on and flying above their habitats, it is difficult to tell whether immigration or local breeding contributes most to each population. This problem, and the interdependence of dispersal and population density, is discussed at the end of the chapter.

The Size of Populations and Seasonal Changes in Population Density

When populations reach their peak many individuals may live on each plant, resulting in large numbers per acre (1 ac = 2·47 ha), especially in species infesting crops. As many as 3,600 nymphs of *Thrips tabaci* may occur on a single onion inflorescence (Carlson, 1964a), which is equivalent to 300–650 million/acre, while on younger onion plants, Harris *et al.* (1936) recorded 160 individuals of all stages per plant representing 15–30 million/acre. In England individual wheat plants support about 8 adult

Limothrips cerealium, equivalent to 6–7 million/acre (Lewis, 1958), and in Trinidadian cacao fields there may be 500,000–4 million *Selenothrips rubrocinctus*/acre (Fennah, 1955). Muma's (1955) record of 2·2 predatory *Aleurodothrips fasciapennis*/citrus leaf represents 2–6 million individuals/ acre depending on the age and type of tree; Laughlin (1970) estimated that the peak population of *Isoneurothrips australis* on an isolated gum tree was about 55,000 and that during a season 300,000–400,000 larvae developed on it.

On mixed vegetation the density of each species is probably much less. For example, on chalk downland, Ward (1966) found only 32,000 *Thrips angusticeps*, 6,000 *T. physapus* and 1,500 *T. validus*/acre.

All populations fluctuate between seasons, but the changes are usually least in species that do not disperse far and in regions where the weather throughout the year is equable. In Trinidad, where mean monthly temperatures vary by only a few degrees, populations of *Selenothrips rubrocinctus* on cashew increase about 9-fold, and on cacao 20- to 24-fold, during the breeding season (Fennah, 1955, 1963). Many overlapping generations develop, so the small increases per generation are difficult to measure. In cooler regions populations fluctuate much more, sometimes increasing a 100- or even a 1,000-fold in a breeding season. Also, there are only a few discrete generations each year, so it is easier to measure the increase in numbers per generation. Most of the cool temperate species studied move from hibernation to host plants in spring, so the difference between the maximum number of winter survivors reaching the breeding sites and the peak number of adults living there after the first generation has developed gives the actual increase per generation. This net rate of multiplication (R_0) can be conveniently expressed as a ratio of the number at the *end* of a generation to the numbers at the *beginning* of it. The most reliable estimates of R_0 obtained in the field are for species whose males are produced in equal numbers with the females and are wingless, and can be counted easily. One such species, *Limothrips cerealium*, was observed in England in two consecutive years and R_0 was found to be 26 and 74 respectively (Lewis, 1958), while in Germany, Holtmann (1962) recorded values of 16–22 for this species and 8–32 for *L. denticornis*, living on wheat. Both sexes of *Taeniothrips laricivorus* are winged but only females survive the winter and fly to larch trees in the spring to lay their eggs. For this species values for R_0 ranging from 4 to 40 have been recorded in Germany (Vité, 1956) and 30 in Denmark (Zenther Møller, 1966).

Where the climate in summer is suitable for very rapid breeding, and the limits between generations become obscured, the cumulative increase during the whole breeding season over the small overwintering population, or over the first migrants to an area, may be much greater than the increases for single generations cited above. For example, near Tokyo,

Japan, where *Thrips tabaci* may complete at least ten generations a year, the peak populations on onion fields may be 300 to 3,000 times greater than the overwintered population at the beginning of the year (Sakimura, 1937a). In England, some flower-dwelling thrips also show very large increases in numbers during the breeding season, though it is difficult to make reliable estimates since these species spread quickly over the flowers available to them, and the number of these is always changing. However, in populations feeding on wild Compositae, Ward (1966) measured an approximate 1,600-fold increase in *T. angusticeps* over two generations and a 500-fold increase in *T. physapus* over three generations.

Factors affecting Abundance

Short-term changes

The effects of separate components of weather on individuals are described in Chapter 2, but in the field the effects of temperature and rainfall, the two most important weather factors affecting the numbers of thrips, are interdependent. In temperate regions, warm, sunny, dry summers encourage reproduction and survival of most species until drought causes plants to wither and food becomes scarce. Cloudy, wet weather, when the foliage is often moist, discourages breeding of species that feed on flowering plants, but fungal-feeders living under dead bark or wood are harmed less. Survival during unfavourable seasons when it is too cold or too hot for activity is considered in Chapter 8.

In warm springs, hibernating species emerge earlier and start to breed sooner. Populations of parthenogenetic species are usually greater in warm seasons because less time is needed to complete development so more generations are produced in the time available (see p. 32). For example, infestations of *T. tabaci* on onions remain small and do not become economically important until the daily mean temperatures rise above 14·5°C (Harding, 1961).

In temperate climates, warm weather is usually dry so mortality caused directly by heavy rain washing individuals from plants is less in warm summers than in cool ones. When rainfall is heavy, the effect on populations is easy to see. Figure 70 shows how two days of driving rain with hail washed about 70% of the population of *T. tabaci* from an onion crop (Harris *et al.*, 1936). In South Africa, *Scirtothrips aurantii* usually shelters on the underside of citrus leaves when clouds obscure the sun, but when a sudden heavy rain shower falls while the sun is still shining, many thrips, especially larvae, are washed from the upper surface and die (Hall, 1930). Larvae of pear thrips, *Taeniothrips inconsequens*, are also washed from leaves by heavy rain (Bailey, 1934a), and larvae and adults of *Haplothrips verbasci* tend to be washed from the exposed side of mullein inflorescences in beating rain (Shull, 1911). In India, populations of thrips on tea decline suddenly with the onset of the monsoon (Dev, 1964). In Europe, heavy rains in spring and summer destroy a large proportion of macropterous *Thrips augusticeps* and *T. linarius* (Bonnemaison and Bournier, 1964), and Franssen and Huisman (1958) found 90% fewer brachypterous *T. augusticeps* hibernating in the soil after cool rainy summers than after hot dry ones. The number of adults in the soil depended on the amount of rain falling when larvae were boring into the ground. In

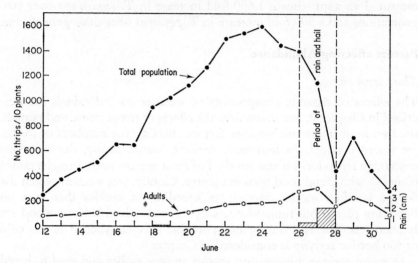

FIG. 70. Changes in abundance of a population of *Thrips tabaci* on onions showing particularly the adverse effect of driving rain and hail (after Harris *et al.* 1936).

heavy rains many were drowned as the crevices in the soil filled with water.

Soil-dwelling stages are vulnerable to changes in soil moisture. Larvae of pear thrips cannot live longer than three days at temperatures between 5·5° and 43·5°C if soil contains less than 9% water. In experiments made at 25°C using soil saturated or wetted to field capacity, a high proportion of larvae also died within 24 days. The greatest number survived when soil moisture was between 10 and 13%, near to the permanent wilting point (Bailey, 1938). Similarly, pupae of some species are killed if the soil is too dry or too wet. When the temperature is 23°C and the moisture content 25 to 85%, most pupae of *T. imaginis* become adults but few survive in soil drier or wetter than this (see Fig. 37) (Andrewartha, 1934).

By contrast, a few species, such as *Anaphothrips obscurus*, thrive better in damp than in dry weather, and others can even survive immersion in water for long periods. *Aptinothrips rufus* often lives on maritime grasses submerged twice daily by normal high tides (Morison, 1957). Moreover, larvae of this species do not usually appear in spring until a few days after warm rain (von Oetingen, 1942). Adults of *Euchaeothrips kroli* can survive for at least four days in the leaf sheaths of *Glyceria* plants flooded by fresh water (Lewis, 1955; see p. 17). *Iridothrips mariae* enters the leaves of *Typha* even when they are under water and *Organothrips bianchi*, which normally lives in slime at the bases of the stems of the taro plant can adapt to a completely aquatic life in fresh water (Titschack, 1969; see p. 61).

When blown accidentally on to water, many species float temporarily on the surface film and may survive until they drift to land. Adult *Haplothrips verbasci* are kept afloat by air bubbles trapped by their setae, and larvae of this species can also survive total immersion for about 30 h (Shull, 1911). Adults of many Phlaeothripidae are resistant to thorough soaking and can often be collected unharmed from the sodden litter that accumulates in drains after tropical storms. Indeed, Priesner (1964a) suggests that gall formation, which is most common in tropical areas, was originally a protective adaptation against frequent torrential rains, even though the few gall-formers that occur in deserts use the same adaptation as protection from drought.

The mortality that occurs during a whole season or generation, in contrast to mortality during a single stage of the life-cycle, is more difficult to assess. One approach is to prepare a budget based on absolute or relative estimates of as many stages of a population as possible, and in temperate regions where the generations of many species are distinct, an age-specific life-table based on the members of a single generation is the easiest type to construct. Holtmann (1962) attempted this for cereal-dwelling thrips in Germany. Using populations on rye, wheat and winter barley, he calculated the number of eggs laid per female by dividing the number found on five plants by the number of gravid females caught there. Samples of plants were collected at intervals during the development of each generation, and the mortality in the population entering each instar was calculated. For example, in a population of *Limothrips denticornis* on rye in 1958 the total mortality between the egg and adult stage was 89·7% apportioned between the stages as shown in Table 22. The mortality on wheat, a less suitable host, was 83·4%. Similar proportions of *L. cerealium* on the same hosts died according to the suitability of the food plant. In both species it was found that the eggs were much more vulnerable than the other stages. Thus, in these species, mortality within a generation depended largely on the death of the eggs. Holtmann claimed that this was due mostly to mechanical damage at blastokinesis, caused by changes in the osmotic pressure of the plant cells surrounding them which, in turn, depended on weather. Although this work clearly shows that egg mortality has a great effect on the seasonal abundance of these cereal thrips, the accuracy of these estimates is unknown since the analysis was based partly on the assumption that females within a generation lay all their eggs on a single plant. This is unlikely and certainly unproven.

There are few convincing examples of *rapid* change in population density caused by natural enemies, even though high levels of parasitism are recorded (see p. 75). In many field populations the proportion of

TABLE 22

Mortality occurring during each stage in the development of a generation of *Limothrips denticornis* (data from Holtmann, 1962)

	On rye			On winter barley		
	Mean No. eggs/♀	Mean No. individuals killed	% Mortality /stage	Mean No. eggs/♀	Mean No. individuals killed	% Mortality /stage
	122			66		
Eggs		102·6	84·1		44·1	66·8
Larvae		6·2	32·1		11·0	50·0
Pupae		0·7	5·0		0·0	0·0
TOTAL		109·5	89·7		55·1	83·4

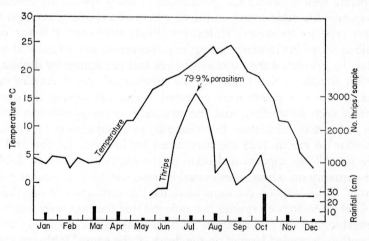

FIG. 71. Changes in abundance of a population of *Thrips tabaci* caused by parasitism by *Thripoctenus brui* in Japan. Temperature and mean monthly rainfall data show that weather was favourable for a continuing increase in abundance after mid-July, but parasitism prevented it (after Sakimura, 1937a).

deaths caused by other animals seems small, but nevertheless they may be critical in determining abundance over a long period.

The relationship between *T. tabaci* and its parasite, *Thripoctenus brui*, in Japan is one of the few that shows how rapid decreases in populations can occasionally be caused by other animals. Figure 71 shows the mean monthly curve for temperature during 1933 and the curve for relative density of thrips on onion crops. Between mid-July and mid-August the population of thrips decreased to about a quarter of its maximum, although temperature continued to rise and there was little rain to wash thrips from plants. This decrease was caused by a sudden rise in parasitism by *Thripoctenus brui*, from less than 1% in May and June, to an average of 79·9% after mid-July. By early September cooler temperatures contributed to the continuing decline in abundance (Sakimura, 1937a).

Long-term changes

Detailed quantitative studies on thrips populations are rare and have seldom lasted more than a few years, so the causes of long-term changes in abundance are little understood. Enough is known, however, to suggest which factors might cause fluctuations between years, and for a few species, even between decades.

In Great Britain, the climate for the period 1900–1940 was milder and drier than at any time in the previous century (Lamb, 1965) and thrips were probably more abundant during this equable period than since. The cooler temperatures and the slight but general increase in cloudiness and precipitation, in the form of rain and mist, during the last three decades, has produced wetter and cooler foliage for a longer period each year, and the abundance of many species that feed on flowering plants has declined. For example, aeolothripids which bask in sunlight in adult and larval stages, have been much less common since about 1960 than before. *Aeolothrips ericae*, the most common aeolothripid in north-east Scotland, has its peak adult emergence about two weeks later now than 20 years ago. The range of some British species does not extend as far north as it used to. *Limothrips cerealium* used to be so abundant that in north England and parts of Scotland, it often annoyed people harvesting cereals from the end of July to September, but it is gradually becoming scarcer in these areas and is now rarely a nuisance (see pp. 79, 143). It has never established in Caithness, west Sutherland, Skye, Outer Orkney and Shetland, presumably because the climate there is too cool and wet (Morison, *in litt.*).

In north Germany, von Oettingen (1942) tried to relate the annual abundance of grassland species to changes in temperature, rainfall and humidity. Populations were sampled by sweeping vegetation and dissecting stems, and the numbers of thrips counted were plotted on graphs alongside climatic data appropriate to the breeding season of each species (Fig. 72). This simple presentation gave a crude general impression of the relationship between abundance and weather, and generally supported the theory that cool, wet years "discouraged" most species, but the validity of some conclusions drawn from such unreliable sampling methods and elementary analysis are doubtful. For example, von Oettingen surprisingly found *Limothrips denticornis* more common in wet years, which is contrary to observations in other parts of Europe (Morison, *in litt.*; Priesner,

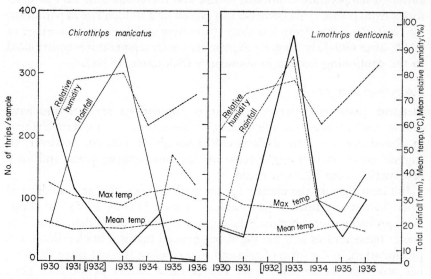

Fig. 72. Annual abundance of two species (heavy lines) compared with broad climatic data (after von Oettingen, 1942).

1964a) and which contrasts with his observations on *Chirothrips manicatus*, a species occupying a similar niche in grass inflorescences.

A much longer and incomparably more detailed quantitative study of the effect of weather on abundance was made on *Thrips imaginis* in South Australia. But, despite the immense amount of background knowledge of the biology and ecology of the insect collected in the early stages of this work, including the effect of diet on longevity and fecundity (Andrewartha, 1935), the duration of the life-cycle at different temperatures (Andrewartha, 1936), and the effect of soil moisture on the viability of pupae (Andrewartha, 1934), as well as the detailed analysis that followed (Davidson and Andrewartha, 1948a, b; Andrewartha and Birch, 1954), the effect of weather on the abundance of this insect remains controversial.

Thrips imaginis breeds and develops continuously in South Australia, although the birthrate and rate of survival are much lowered by heat and drought during the hottest part of the summer and by low temperatures during the winter. In the warm part of the year a generation is completed in a few weeks, and maximum numbers occur in November and December (Fig. 73) (Davidson, 1936). The adults and larvae live on a wide variety of weeds and garden plants, including apple and pear blossom, and the larvae pupate in litter and soil. The adults are especially attracted to rose flowers even though these are unsuitable for extensive breeding (Evans, 1933b) and in this study populations were sampled by collecting 20 roses almost daily for 14 years (see p. 125).

Using multiple regression analysis, Davidson and Andrewartha (1948a)

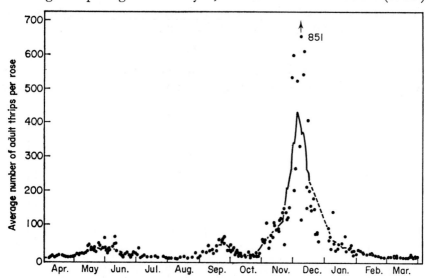

FIG. 73. The individual daily records and the trend throughout one year (1932–33) of the number of *Thrips imaginis* per rose. The points represent daily records. The curve is a 15-point moving average (after Davidson and Andrewartha, 1948a).

related maximum populations each year to weather factors likely to affect the development, survival, longevity and food supply of the thrips. Four independent variates were chosen to represent:

(i) opportunities for the growth of food in the previous autumn, winter and spring (x_1),

(ii) the length of time the host plants remained succulent and suitable for breeding (x_2),

(iii) the period in spring warm enough for breeding (x_3),

(iv) the effect of the abundance of thrips, or seed production of their annual host plants, in one season on populations of thrips in the next (x_4),

each variate being defined as follows:

H

(x_1) Total effective day-degrees (in units of 100 d°) from the break of the season to the end of August, usually a period of 5–6 months before the population began to increase. The effective temperature was calculated from the formula:

$$\frac{\text{Max. daily temperature} - 48}{2}$$

(x_2) Total spring rainfall (in inches).
(x_3) Total effective day-degrees in spring (in units of 100 d°).
(x_4) As for (x_1) but for the year before the samples were collected.

The equation derived from the analysis, in which Y represents the populations of thrips and x_{1-4} the variates listed above, was:

$$\log Y = -2\cdot390 + 0\cdot1254_{x1} + 0\cdot2019_{x2} + 0\cdot1866_{x3} + 0\cdot0850_{x4}$$

from which theoretical values of Y were calculated for each year. These closely resembled observed values of abundance (Fig. 74). These four

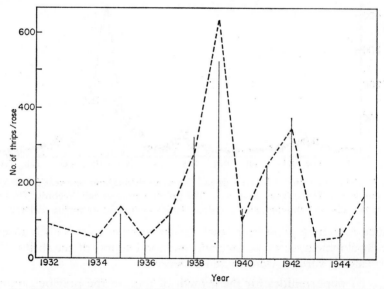

FIG. 74. Observed (solid vertical lines) and theoretical (dotted line) values of the numbers of *Thrips imaginis* per rose. The observed values are geometric means of the daily counts of thrips per rose, and the theoretical values calculated from the expression on p. 194 (after Davidson and Andrewartha, 1948b).

quantities, obtained solely from meteorological records, accounted for 78·4% of the variance, and Andrewartha and Birch (1954) concluded that little room was left for any other systematic cause of variation. In other words, they claimed that the abundance of *Thrips imaginis* depended entirely on temperature and rainfall affecting the thrips directly and through their food and host plants, without the involvement of density-dependent factors.

This interpretation has been disputed by many population dynamicists (Kuenen, 1958; Nicholson, 1958; Smith, 1961; Huffaker and Messenger, 1964; Lack, 1966; Clark *et al.*, 1967). The consensus of their criticisms is that the original method of analysis was inappropriate, and in fact demonstrated the degree to which the population conformed to a predicted level, hence the degree to which it *was* regulated by a density-dependent system. The method of analysis is likely to remain controversial, but two other straightforward biological points overlooked in the study also detract from the validity of the original conclusions. One is that the populations of adults on roses served only as an index of a population that largely bred elsewhere, and it is impossible to distinguish between abundance and activity by the sampling method used; the other is that populations of immature stages were not studied, so a source of density-dependence regulation such as egg mortality, shortage of larval food, or parasitism of the immature stages, would not necessarily be revealed by samples of adults taken from roses.

Considering the wide variety of parasites and predators that attack thrips (see p. 68), the high incidence of larval parasitism sometimes recorded in other species, e.g. 68% in pea thrips (Kutter, 1936) and 80% in cacao thrips (Cotterell, 1927), and the clear density-dependent relationship between *Thrips tabaci* and its larval parasite *Thripoctenus brui* (see p. 76), it would be surprising if no similar density-dependent factors affected the populations of *Thrips imaginis* and other thrips from year to year. In fairness to Andrewartha and Birch, their critics have not suggested what the density-dependent factors might be. Perhaps the scarcity of natural enemies of *T. imaginis* occurs because it is not native to Australia, and was introduced there without them. The surprising absence of closely related species, there being no other representatives of the genus known from Australia (Mound, *in litt.*), supports this suggestion. Bailey (1933) found that 60–62% of all mortality in the bean thrips, *Caliothrips fasciatus*, occurred before the adult stage, including 0·9% caused by larval parasitism, so a study on populations that disregards the immature stages is unlikely to reveal all the causes of changing population density. Nevertheless, it is worth stressing that thrips are delicate insects whose development and dispersal are affected greatly by temperature, rain and wind, and superficially these appear to be the dominant factors affecting long-term trends. Most of the scanty evidence suggests that the weather usually affects populations of thrips directly, or through their host plants, rather than through other animals. One indisputable example of density-dependent regulation occasionally occurs in the predatory *Haplothrips subtilissimus*, a species that is usually communal but when food is scarce the larvae become cannabalistic (Putman, 1942).

Gradual changes in agricultural practice almost certainly affect the abundance of some species by changing the amount of food available, but reliable information is difficult to collect. The utilization of grass for grazing, ensilage and hay probably severely checks populations of grass-dwellers each year as they approach a peak, because the breeding and feeding sites among the upper leaf sheaths and inflorescences are removed at a critical time. Over many years these practices probably lessen abundance and may be the reason for the decline in numbers of *Anaphothrips*

TABLE 23

Species probably affected by forestry practice in Great Britain
1940–1964

Species	*Period of increased abundance*
Acanthothrips nodicornis	1952–60
Cryptothrips nigripes	1944–61
Hoplandrothrips annulipes	1943–64
Hoplandrothrips bidens	1945–57
Megathrips nobilis	1949
Hoplothrips corticis	1945–53, 1956–64
Maderothrips longisetis (an inquiline)	1942–63

obscurus in Great Britain. By contrast with grass crops, cereals are allowed to develop to maturity and thus carry a heavier infestation, though their harvesting with combine-harvesters followed by artificial drying may have reduced numbers slightly. In Britain, *Limothrips denticornis* has become generally more common since about 1950 when barley, its preferred host, started to be grown widely in preference to oats and wheat. Some British species that feed on fungi under dead wood or bark seemed to benefit temporarily from changes in forestry practice during and after the Second World War, and from the effects of the severe storm of 31 January 1953. Due to the wartime increase in felling and the later storm damage, many branches and trunks were left to rot in the forests. Some species of fungus-feeding thrips, present in small numbers in these wooded sites before felling, seem to have flourished on the saprophytic fungi that developed on the rotting wood. No other explanation is known for the sudden increase, over the periods specified, in numbers of the species listed in Table 23. They are now scarce again and their breeding restricted by the clean forestry methods now practised (Morison, *in litt.*). Under natural conditions it takes one to three years for the fungi on which these thrips feed to develop, then they continue to produce spores yearly as long as the wood or bark remains nutritious and the climate suitable. When trash is cleared promptly from the forest, the fungi have no time to establish.

Fire is also a hazard for species living in litter and among the crowns of prairie grasses. Burned prairies are especially poor in numbers of species, particularly in flightless ones (Stannard, 1968).

Dispersal and the Size of Populations

In many winged species the scale of airborne dispersal is so great that changes in abundance within a habitat depend greatly on immigration and emigration, as well as on breeding and mortality. It is extremely difficult to estimate the quantitative effects of such dispersal on populations. No such comprehensive studies have been made on thrips, but work on different aspects of dispersal, density and abundance, can be pieced together to show how they are interrelated.

Population density and dispersal within habitats

Flower thrips need to disperse because the individual blooms they feed on are usually ephemeral. Frequent flights ensure that thrips spread quickly through their habitat. It is unlikely that all individuals of the same population remain in a prescribed area from day to day, since their restlessness and spiralling flights mean that some individuals are blown from and others into it (see p. 141). However, at certain times of the year in temperate regions, especially before breeding, the density of populations within an area appears to depend on the number of suitable flowers there, either because long-distance flight is discouraged, perhaps by calm weather, or because immigration and emigration more or less balance each other. Ward (1966) counted the numbers of *Thrips validus* and the number of host plants on 7 acres of chalk downland in England. The main hosts were species of *Taraxacum*, *Leontodon*, *Sonchus*, *Crepis* and *Hypochaeris*. In early spring, when the first overwintered females appeared, only a few *Taraxacum* were in flower and the thrips concentrated on them at an average density of 3·0 per flower head. Within two weeks the number of flowers had greatly increased, but the number of thrips remained about the same, so the average number per flower decreased to 0·3, as they spread over the greater number of blooms available. Later in the season the number of flowers, mainly of *Leontodon hispidus*, increased quickly, so the average number of thrips per flower was even lower, though it represented a larger total population in the area (11,000). At the end of the season the number of flowers and thrips decreased, but the flowers disappeared more rapidly than the insects, so density per flower

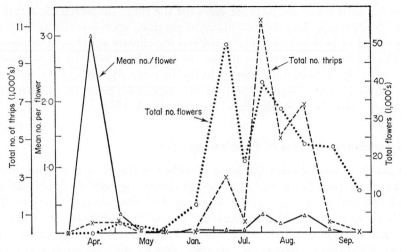

Fig. 75. The relationship between the mean number of *Thrips validus* per flower and total numbers in a 7-acre area of chalk down in 1964 (after Ward, 1966).

was slightly greater than when the thrips were most abundant (Fig. 75). Similarly, on a single Australian gum tree, larvae of *Isoneurothrips australis*, another flower thrips, crowded into the young remaining flowers as old ones fell (Laughlin, 1970).

Mortality during dispersal between habitats

In uncultivated areas, many phytophagous species live on small, discrete patches of their food plants scattered throughout a larger area. Similarly, in cultivated areas many of the crops that support heavy infestations of thrips are not continuous over large areas, but grow in small fields surrounded by fields of non-host plants. Thus, between suitable habitats, there are often large areas of territory unsuitable for feeding or breeding, and a large proportion of emigrants from breeding sites must die because they fail to find a host plant. It is not known what proportion of thrips migrating between habitats is lost, and there are few estimates of losses for other small flying insects. For aphids, there is general agreement that of the millions that are produced and disperse, only a minute fraction, perhaps of the order of 0·001% or less, survives (Hughes, 1963; Johnson, 1969). The scanty evidence available for thrips suggests that losses incurred in movements between habitats are great but not as great as this.

In north-west Germany, overwintered females of *Limothrips cerealium* and *L. denticornis* produce an early first generation on rye, and these young females disperse to produce a second generation mainly on spring oats when rye becomes less succulent. Holtmann (1962) estimated the maximum density of females on rye towards the end of the first generation, and found that only 10% of the first generation were lost during dispersal; this is probably an underestimate, since teneral *Limothrips* usually take-off as soon as they become flight-mature, so some thrips must have left the rye on non-sampling days. His estimate for losses between peak populations in summer and the number reappearing next year after hibernation was between 86 and 94%. There is usually little mortality of this species during hibernation (Lewis and Navas, 1962), so most of this loss must also have occurred during flight to and from overwintering sites. Similarly, studies on *L. cerealium* on wheat, showed that there was a 98·2% loss incurred in flight to and from hibernation, even though the field was surrounded by easily accessible overwintering sites (Lewis, 1958).

These estimates of mortality during dispersal are considerably lower than the probable losses of aphids. Thrips are generally less prolific breeders than aphids, so populations could only maintain themselves if a smaller proportion died, but the reason for their greater hardiness during dispersal is unknown. Perhaps, compared with aphids, a smaller proportion of individuals in a population of thrips fly persistently after take-off, and so are less likely to be blown far from their original habitat. Nevertheless, losses during dispersal are probably as great as, or may even exceed, the other mortality factors discussed above. Indeed, as such a large proportion of migratory populations die during dispersal, young adults may be more vulnerable than any other stage in the life-cycle. Evaluation of losses in the immature stages would be needed to confirm this. Studies on population dynamics of migratory species that fail to include dispersal as one of the major mortality factors, are therefore unrealistic and give false impressions of abundance.

Immigration, emigration and breeding on a single tree

The influence of dispersal on population dynamics of thrips is clearly shown by Laughlin's (1970) study of a population of *Isoneurothrips australis* on a single gum tree (*Eucalyptus calycogona*) in South Australia. The life history of this species is simple and similar to many other flower thrips. When adult females arrive at flowers, they feed, mate, and lay eggs beneath the surface of the flower tissue. The larvae hatch and feed on the flowers, pupate in the soil under the tree, and the adults return to flowers to start the cycle again (Steele, 1935).

When a tree begins to flower it is quickly colonized by adults. They initiate the population of larvae and are the only source of eggs until adults begin to emerge from the soil beneath the tree. From then onwards until flowering ends, eggs are laid by local and immigrant adults. Numbers of local adults are low at the beginning of the breeding season, but eventually they form most of the breeding stock.

The number of flowers and thrips per tree, and larvae in the soil beneath it, were sampled weekly (see p. 114). The number of adults per favourable flower increased to between 10 and 20 at the beginning of the flowering period, then remained at this density. Thus, the total number of adult thrips in the tree depended on the number of suitable flowers open. This increased for the first 6 weeks of flowering, then decreased for the last 14 weeks. The movement of thrips to and from the tree reflected this cycle of flower production, so there was a net gain in numbers for 6 weeks followed by a net loss for 14 weeks. The intriguing, but unsolved, problem is how the relatively constant number of thrips per flower was maintained when the supply of thrips to the tree from immigration and local emergence changed so enormously during the season.

The first curve in Fig. 76 derived by multiplying the number of flowers in bloom each week by the average number of adults per flower, shows

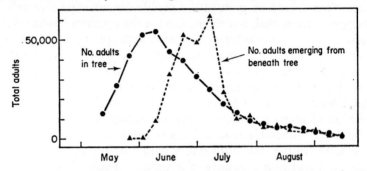

FIG. 76. Weekly totals of the number of adult *Isoneurothrips australis* in, and emerging from, a gum tree during the flowering season (after Laughlin, 1970).

the estimated number of adults in the tree. This curve is a function of the number of adult immigrants and emigrants, and the mean duration of their stay on the tree.

Adults on the tree at any one time, may have come from the soil beneath the tree, or have flown from elsewhere. The number of adults emerging locally was estimated from funnel catches of larvae (see p. 114) and plotted on Fig. 76 one week to the right (the time taken for mature larvae to pupate) to produce the second curve. This assumes that few pupae died in the ground and that all emerging adults flew up into the tree. The number of immigrants from elsewhere was not counted directly. During the first three weeks there were no adults emerging from the ground but the number on the tree increased by about 40,000. If no adults left the tree during this period, the mean rate of arrival must have been at least 2,000/day, but as individuals probably arrived and departed, the actual rate of immigration was probably greater. Such a rate of arrival would pass unnoticed by an observer watching the air around the tree; aerial densities of thrips measured in England ranged from about 0·005 to $2.0/m^3$ of air, and at the lesser density flying thrips were rarely seen (Lewis, 1965a). The mean wind run over this particular gum tree for the first three weeks was 65.4 km/day and the cross section of the tree about 7 m^2. Thus, 2,000 individuals would have been blown on to the tree each day from an aerial population containing as few as 0.0043 thrips/m^3.

For the next month an estimated 25,000 to 60,000 adults also emerged from the ground weekly, equivalent to between 3,000 and 8,000 per day. Even if many larvae, say 90%, died in the ground, 300–800 adults would still have emerged daily, and some adults probably continued to arrive from elsewhere. Despite this potentially great influx of thrips, numbers in the tree continued to decline.

It is not known how the number of adults per flower, hence the population on the whole tree, was regulated. Perhaps most of the emerging adults flew away from the tree before settling to feed and reproduce. This would help to maintain constant numbers per flower by lessening the number of local immigrants, but the number of immigrants from distant trees would presumably increase as the season progressed. If, by contrast, most of the locally produced adults flew first into the tree but, unable to tolerate overcrowding some then flew away, the density per flower would similarly remain constant.

The evidence presented in this section shows the great mobility of thrips within and between habitats. It stresses the influence on their population dynamics of dispersal which is probably as important as natality and mortality. There seem to be behavioural mechanisms which adjust the size of local groups, and which might lead to density-dependent processes capable of regulating numbers. For example, if crowding enhanced dispersal it would provide such a mechanism. Unfortunately there is no direct information on this for phytophagous thrips, but populations of some predatory thrips stabilize at fairly constant prey/host ratios (see p. 75), suggesting that crowding might be critical.

10 Habitat, specific diversity and spatial disposition

There are many lists of species of Thysanoptera and their host plants from different parts of the world (see Appendix 1) but few relate the number of individuals and species living in different habitats to the vegetation there, or describe the spatial disposition or grouping of individuals within their habitats. For example, von Oettingen (1954) listed the species of Swedish thrips characteristic of sand dunes, heath-land, damp meadows, dry steppes and woodland (see Appendix 4), but such general surveys do not show the diversity of communities or their spatial patterns; these aspects are examined in this chapter.

Almost all known Thysanoptera are terrestrial and most species dwell and feed on living higher plants (Phanerogams). Most well-defined habitats occupied by thrips are therefore determined by vegetation, though where the flora is fairly uniform, local physical differences may also determine which species are present. As there are so many migratory species and ubiquitous feeders it is often difficult to separate the true denizens of a habitat from species straying into it from elsewhere. For example, Morison (*in litt.*) has found *Taeniothrips atratus* on 566 species of plant though it breeds on only about 20. Nevertheless, the *number of species* in a habitat largely depends on the diversity of the plant community, whereas the *abundance of individuals* is affected mostly by weather and other organisms in the environment (see Chapter 9).

Specific Diversity within Habitats

Tables 24 and 25 give examples of the number of species and individuals found in communities living in natural and semi-natural habitats and in crops. Some of these communities contain more species than others, but the richer ones (i.e. those containing the greater number of species) have not necessarily more individuals per unit area. The number of individuals caught in each of these habitats differed greatly, because of different population densities, and different methods and duration of sampling, so it is difficult to appreciate or to compare the specific diversity of the communities directly from the raw data. Neither can their diversity be compared using expressions based on proportions (e.g. number of species/total catch) because such values change with sample size (Williams, 1949, 1950). This is because at first species are collected almost as rapidly as individuals, but subsequently new species are collected progressively less often, although many individuals may be caught. A measure of diversity must take account of this and Williams's (1944) index of diversity (α) obtained from the formula

$$S = \alpha \log_e \left(1 + \frac{N}{\alpha}\right)$$

where S = number of species, and N = number of individuals, fulfils this requirement, providing that sampling is random and that the frequency distribution of species with different numbers of individuals conforms to a logarithmic series. Most of the samples listed in Tables 24 and 25 were collected by sweeping vegetation, which probably removed thrips at random, especially from uniform stands, such as grassland and low-growing herbaceous crops. Unfortunately many samples were either too small, contained too few species, or were incompletely identified to show clearly whether the distribution of species and individuals fitted a logarithmic series. However, this was established for three samples collected by different methods from three contrasting communities. They included suction-trap samples of the aerial population at Lilongwe, Malawi (Mound, *in litt.*); samples collected by beating and sweeping vegetation in different habitats in Morocco (zur Strassen, 1968); and samples collected by beating, shaking and searching larch trees in Czechoslovakia (Kratochvil and Farsky, 1942). The distribution of individuals within species in all these communities approximated to a logarithmic series (Fig. 77). The poorest fit was for the community on larch perhaps due partly to non-random searching or to the fact that one species, the larch thrips, completely dominated it (see p. 217). Nevertheless, the fit for these communities from such different habitats and regions is good enough to justify the use of α to describe diversity in communities of thrips generally. Samples from 22 other communities living on mixed vegetation or crops, fitted to a logarithmic series by computer, confirmed this. Values for α are therefore given where possible in Tables 24 and 25; the greater the value, the more diverse, or richer, the community.

There is no absolute standard with which to compare indices of diversity from different habitats. For most communities α is affected by the duration of the sampling period and the time of day when samples were taken. The longer the sampling period the greater the diversity because species that appear in different seasons are more likely to be included in the samples; sampling at different times of day also changes α because different species are active at different times. Most published

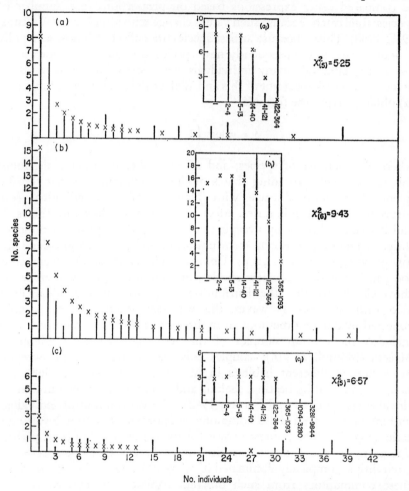

FIG. 77. The frequency distribution of species of thrips with different numbers of individuals, in 3 populations. (a) An aerial population from Malawi sampled over 1 month by a suction trap (data from Mound, *in litt.*, after Farrell, unpublished). (b) Population on mixed vegetation in Morrocco sampled over 6 weeks by beating and sweeping (data from zur Strassen, 1968). (c) Populations on larch trees in Czechoslovakia sampled over 3 years by beating, shaking and searching (data from Kratochvil and Farsky, 1942). The crosses show the appropriate fitted logarithmic series. a_1 b_1 c_1 show the same data grouped more conveniently into $\times 3$ classes of abundance with the fitted logarithmic series again represented by crosses (Williams, 1964). The values for χ^2 show that there were no significant differences (P $= 0.05$) between the observed frequency distribution and the appropriate logarithmic series.

records of thrips swept from vegetation refer to samples taken on a few days each season when thrips were abundant, and they rarely extend over more than one or two years. Thus, values for α obtained from these short-term samples are probably underestimates.

One of the most intensively sampled thrips communities is that living in the artificially enriched parkland habitat at Silwood Park, Berkshire, in south-east England, where there is a mixture of coniferous and deciduous woodland, sandy heath and scattered cereal and fruit crops. This community was sampled for three consecutive years from early spring to late summer (Lewis, 1961 and unpublished observations). White-painted water traps sampled the aerial population throughout each day and frequent sweeps were made of local vegetation. The water traps were highly atttractive to a few common flower-dwelling species, but after correcting catches for this (see p. 157) the two methods together yielded about 18,900 individuals and 33 species (though a few species probably remained unrecognized), giving an overall index of diversity of 4·2. Neither sampling method gave a completely representative sample of the community, though both were random. Tree- and bark-dwelling species were missed by sweeping which gave a value for α of 2·8 with 8,300 individuals in 22 species. Water traps missed wingless species, but with 10,600 individuals in 29 species indicated a richer community with $\alpha = 3·7$. This was because the traps sampled continuously, thus catching individuals of the rare species, and because the airborne population probably included a few species blown from other habitats; by contrast, sweeping sampled the populations from discrete areas for relatively short periods.

The unreliability of α calculated from samples collected by non-random sampling, especially over short periods, is illustrated by comparing Sęczkowska's (1963) catches obtained by sweeping spread over two years, with those obtained by searching for thrips in wild flowers in different seasons. The fauna swept from a *Carex–Inula* dominated sward in the upper part of the Stawska mountain in Poland, an area near farmland but remote from all but scattered trees, had an index of diversity of 4·8, whereas the indices for collections from flowers in the same area in spring, summer and late summer were only 1·2, 2·6 and 2·4 respectively.

COMMUNITIES IN NATURAL AND SEMI-NATURAL HABITATS

Some of the most intensively sampled natural habitats are different parts of the Khazakhstan Steppes, east and west of the Ural Mountains (Tansky, 1965). Unfortunately, although all species caught in these areas were identified, only "typical" species (see p. 213) in each habitat were counted. Thus only the approximate numbers caught are known, but

TABLE 24

Specific diversity in natural and semi-natural habitats

Area	Vegetation	Soil	Sampling method	Duration of sampling	No. of species	No. of individuals	Approx. index of diversity (α)	Author
S.E. England Silwood Park	Farmland, heath, meadow, woodland (see p. 205)	Podsol	Trapping Sweeping Searching	Daily April–Oct. 3 yrs	33	19,900 (corrected)	4·2	Lewis (1961)
Poland Stawska Gora Reserve	Mainly *Carex bumilis* or *Inula ensifolia*, near farmland, deciduous trees and shrubs	—	Sweeping	Fortnightly April–Oct. 2 yrs	24	685	4·8	Sęczkowska (1963)
U.S.S.R. N. Khazakhstan Kustanay Karabalyskii district	Moderately dry steppe, rich multi-grass flora dominated by *Stipa pennata*	Cherno-zoem	Sweeping	Intermittently over 6 yrs	40	?1,017	8·4	Tansky (1965)
Semiozernyi district	Dry steppe, dominated by *Festuca ovina* and *S. pennata*	Chestnut	Sweeping	1 yr	26	?262	6·5	
Orenburg district	"High" steppe	—	Sweeping	1 yr	28	?3,569	4·4	
Volgograd district	"High" steppe	—	Sweeping	1 yr	18	?374	4·1	

Northern plain		Loamy	Sweeping	1 day	16	436	3·4	von Oettingen (1942)
	Open meadow	Loamy	Sweeping	1 day	16	436	3·4	von Oettingen (1942)
	Woodland glade	Loamy	Sweeping	1 day	11	83	3·4	
	Coastal meadow	Sand	Sweeping	1 day	10	266	2·1	
	Woodland glade	Sand	Sweeping	1 day	12	316	2·5	
	—	Dry bog	Sweeping	—	18	192	5·2	
	—	Damp bog	Sweeping	—	13	532	2·5	
Egypt Desert valleys and dunes	*Zilla*, *Alhagi*, *Zygophyllum*, *Haloxylon*, *Panicum*, *Penniseta* and Annual Cruciferae	Sand	Collecting	—	23 "typical"	—	—	Priesner (1964a)
Coastal plain	*Thymelaea hirsuta*, *Asphodelus*, *Inula*, *Phlomis*, *Salsola*, *Suaeda*	—	Collecting	—	13 "typical"	—	—	
Higher valleys of Mt Sinai	*Alkanna*, *Achillea*, *Phlomis*	—	Collecting	—	3 "typical"	—	—	
Salt swamps	*Atriplex*, *Tamarix*	—	Collecting	—	3 "typical"	—	—	
Nubian mountains	Annual grasses: *Eragrostis*, *Panicum*; *Acacia* forests in valleys	—	Collecting	—	7 "typical"	—	—	
Canada Ellesmere Island	Tundra high arctic flora	—	?Collecting	2½ months (in 1 yr 1 growing season)	3	—	—	Oliver (1963)
Alaska Point Barrow	Tundra	—	Soil samples (Berlese extracted)	2 yrs	2	25	1·0	Hurd and Lindquist (1958)

TABLE 25

Specific diversity in crops

Area	Crop	Sampling method	Duration of sampling	No. of species	No. of individuals	Approx. index of diversity (α)	Author
U.S.S.R. N. Khazakhstan Kustanay Karabalykskii district	Wheat	Sweeping	6 yrs	23	6,885	3·0	Tansky (1965)
Semiozernyi district	Wheat	Sweeping	1 yr	13	1,354	2·1	
Orenburg district	Wheat	Sweeping	1 yr	10	1,728	1·5	
Volograd district	Wheat	Sweeping	1 yr	11	8,646	1·2	
Finland	"Cereals", winter and spring wheat, rye barley, oats	Sweeping	6 yrs	29	163,725	2·8	Koppa (1967)
N. Germany	"Cereals", winter and spring barley, winter wheat and oats	Sweeping	3 yrs	11	35,777	1·0	Holtmann (1962)
Czechoslovakia Slavice	Bird's foot trefoil (Lotus corniculatus)	Sweeping	1 yr (over 5 mths)	12	1,519	1·9	Obrtel (1965)

Location	Crop/habitat	Method	Duration	No.	No. of specimens	Ratio	Reference
S. Moravia	Lucerne (*Medicago sativa*)	Sweeping	2 yrs	19	4,976	2·6	Obrtel (unpublished)
Central Bohemia	Red clover (*Trifolium pratense*)	Sweeping	2 yrs	14	5,623	1·7	Skuhravý *et al.* (1959)
Brno	Larch forests	Beating, shaking	3 yrs	23	9,870	2·8	Kratochvil and Farsky (1942)
New Zealand S. Island	Cocksfoot seed crop (*Dactylis glomerata*)	Random samples of vegetation	3 yrs	7	—	—	Doull (1956)
N. Island 113 localities	Pasture: mainly ryegrass (*Lolium perenne*) cocksfoot (*Dactylis glomerata*), *Paspalum*, *Danthonia*, white clover (*Trifolium repens*)	Sweeping	1 yr	12	12,000 approx.	1·3	Cumber (1958, 1959)
Kenya	Pyrethrum (*Chrysanthemum cinerariaefolium*)	Random samples of vegetation	—	7	4-5,000	<1	Bullock (1963, 1965 and *in litt.*)
Holland	Flax (*Linum usitatissimum*)	Sweeping and collecting	4 yrs	26	26,366	2·8	Franssen and Mantel (1961)
Great Britain	Flax (*Linum usitatissimum*)	Collecting	—	18	—	—	Morison (1943)
U.S.A. N. Utah	Tomatoes (*Lycopersicon esculentum*)	—	—	5	—	—	Maddock (1949)

assuming that "atypical" species were represented by singletons or a few individuals, and that about 95% of the total number of individuals caught belonged to "typical" species, the index of diversity in the different areas can be estimated (Table 24). On moderately dry steppes where the grasses *Stipa pennata* and *S. speldens* were dominant among a rich variety of other grasses and flowering plants, especially Leguminosae, Labiatae and Umbelliferae, α was about 8·4. It decreased to 6·5 in drier steppes characterized by *Festuca ovina*, *S. pennata* and fewer flowering plants. West of the Urals on the high steppes where much of the vegetation withers in June (Schimper, 1903) and in the Volga basin, diversity was less. In north Khazakhstan, Tansky (1961) recorded 29 species on virgin steppes and 32 species on reverted arable land, but α cannot be derived from this data because species were grouped only into categories of relative abundance. Knechtel's (1960) record of 67 species from the Dobrogea Steppes in Rumania is not comparable with the Khazakhstan data because the Rumanian collections were from a wide variety of habitats including wood and coastal vegetation.

Von Oettingen (1942) studied diversity in smaller semi-natural grassland habitats in north Germany. He compared communities on different types of meadow growing on similar soils, with others from grasslands having similar floras but growing on different soils. Samples were mostly small and taken on one day. The coefficient of similarity (Jaccard, 1912) used by von Oettingen to compare the different populations is less appropriate than α, and details of some of his conclusions are doubtful. Nevertheless, his suggestion that the richness of the grassland fauna is affected by the microclimate produced by the soil as well as by the diversity of the flora, is supported by the values for α derived from his data (Table 24); there were greater differences between α for communities in botanically similar grasslands on different soils, than for communities in different types of grassland on similar soils.

Although there are many national and regional faunal lists (see Appendix 1) these are neither confined to well-defined habitats nor compiled from randomly sampled data so α cannot be derived from them. However, Priesner's (1964a) lists of "typical" species in five desert habitats in Egypt (Table 24) show clear relationships between the richness of flora and the number of species of thrips present. In this work "typical" species were not defined, but presumably they were those that could complete their life-cycle in one recognizable habitat. The size of a habitat probably affects the number of "typical" species in it because the greater its area the less likely are species from other habitats to spread through it. Communities in low-lying desert valleys with many perennial plants and spring annuals were richer in species than those in other arid habitats, though the 23 species "typical" of this habitat did not occur in all valleys;

the species present differed at different latitudes and distances from the sea. Communities in the higher valleys of Mt Sinai and in salt swamps were poor, each with only three "typical" species, and on the higher slopes in the Arabian and Libyan desert, which remain bare except for a few bushes, thrips cannot survive. The cold desert of the tundra also supports few species (Table 24).

The specific diversity of thrips communities in natural habitats resembles the diversity of other small groups of phytophagous insects in similar habitats. In Finland the index of diversity for leafhoppers (Homoptera–Auchenorrhynca) in 30 different habitats, sampled over three summers, ranged from 1·4 to 5·8 (Williams, 1964, after Kontkanen 1950), and caddis fly (Trichoptera) communities sampled with a light trap for three years in a habitat similar to Silwood Park (see p. 205) had an index of diversity of 7·6 (Crichton, 1960). The heteropteran community sampled by light trap over eight years at Harpenden, southern England, was more diverse with $\alpha = 18$ (Southwood, 1960).

The richest thysanopteran communities studied in single natural or semi-natural habitats, as distinct from national faunas, thus contain 30–40 species with indices of diversity between 4·5 and 8·5. In temperate zones, grassland seems to support richer communities than other habitats, which is not surprising since the Gramineae contains more species of host plant than any other single botanical family. The greater diversity of some steppe faunas is probably due to the rich profusion of plant species that grow where the ground is free of *Stipa* sward. In such places the vegetation on comparatively small areas is extremely varied and much more diverse than in north European meadows (Rehmann, in Schimper, 1903). Natural sub-tropical grasslands like those in parts of India also support fairly rich faunas, but in areas where many of the grasses have been introduced, as in Illinois, U.S.A., or where grass is not the dominant natural flora, as in the wet tropics, the thrips fauna on grassland is poor. The indices of diversity for the vegetation of temperate heaths, grasslands, and forests range only from 2–7, compared with 11–15 for sub-tropical forest and 21–54 for tropical rain forest (Williams, 1964). Thus the diversity of thrips communities in these tropical forest habitats is probably far greater than in the richest temperate grassland faunas. Data available for other groups of insects support this; for example, the specific diversity of North American Drosophilidae is generally greater in sub-tropical Mexico than in the U.S.A. (Williams, 1964, after Patterson, 1943).

The poor fauna in hot and cold deserts is presumably a result of the fewer plants present to provide food and habitats. But in parts of the tundra many non-woody species flourish (e.g. about 110 at Lake Hazen, Ellesmere Is. (82°N) (Oliver *et al.*, 1964), and a more fundamental reason that insect species in general, and Thysanoptera in particular, are poorly represented there, is that the environment approaches the physiological limits of insect life, with cold as the limiting factor (Downes, 1964). There are few species of thrips adapted to very cold environments (Priesner, 1964a) but the short wings of the three species found at Lake Hazen are probably a morphological adaptation to the high arctic, where wings would be an encumbrance because temperatures, with a July mean of 6·5°C (Oliver *et al.*, 1964), rarely exceed the threshold temperature for flight long enough for them to be used. In contrast, sparse vegetation, high temperature and water-loss are physiological barriers to most species in hot deserts. As most thrips have a volume of only about 0·5–1·0 mm³

TABLE 26

The basic species in north German meadows and additional species associated with them especially in wet and dry places (after von Oettingen, 1942)

Basic meadow	Hydrophiles	Xerophiles
Aeolothrips fasciatus	*Aeolothrips fasciatus*	*Aptinothrips elegans*
Chirothrips hamatus	*Aeolothrips albicinctus*	*Chirothrips manicatus*
Chirothrips manicatus	*Chirothrips hamatus*	*Oxythrips brevistylus*
Limothrips denticornis	*Limothrips denticornis*	*Frankliniella tenuicornis*
Aptinothrips rufus	*Aptinothrips stylifer*	*Taeniothrips firmus*
Aptinothrips stylifer	*Anaphothrips obscurus*	*Thrips physapus*
Anaphothrips obscurus	*Dictyothrips betae*	*Haplothrips dianthinus*
Frankliniella tenuicornis	*Odontothrips phaleratus*	*Haplothrips acanthoscelis*
Frankliniella intonsa	*Frankliniella intonsa*	*Haplothrips arenarius*
Stenothrips graminum	*Taeniothrips atratus*	
Haplothrips aculeatus	*Euchaetothrips kroli*	
Haplothrips acanthoscelis	*Rhopalandrothrips annulicornis*	
	Bolacothrips jordani	
	Haplothrips caespitus	
	Cephalothrips monilicornis	
	Nesothrips dentipes	

and larvae especially have a thin cuticle, they are extremely vulnerable to desiccation in dry air (MacGill, 1937; Ghabn, 1948; Lewis, 1962; Cederholm, 1963) though some desert dwellers may drink dew to supplement water sucked from plants (Priesner, 1964a). Another adaptation for survival evolved by many species living in dry habitats where drought persists for part of the year, is the completion of but a single generation in the cool season, and aestivation for the rest of the year when it is hot and dry (see Chapter 8).

Different communities may have similar indices of diversity but contain different species; if they are qualitatively different the index of diversity of the combined samples is greater than for the communities separately.

Thus, for the separate communities in open and woodland meadows sampled by von Oettingen (1942), the index of diversity was 3·4 (Table 24), but combined, the samples contained 519 individuals and 20 species giving $\alpha = 4\cdot1$. Qualitative differences between thrips faunas in different habitats are largely caused by differences in the species of plants present and possibly by soil-induced differences in microclimate. Von Oettingen (1942) found that meadows contained a "basic" fauna of species having either a wide host range, or breeding on a few ubiquitous species of plants. In addition, species with fewer host plants, or with more exacting microclimatic requirements, occurred wherever plants and humidity permitted (see Table 26). Table 27 shows that the composition of communities in pure grassland habitats depends partly on the soil and associ-

TABLE 27

The effect of soil type and associated humidities on the proportion (%) of different genera in thysanopteran communities in north German grasslands
(after von Oettingen, 1942)

Genera	Lowland moor	Peaty loamy sand	Dry sand	Heavy dry clay	Heavy wet clay
Aeolothrips	0·5	1·1	9·4	11·2	8·0
Anaphothrips	2·0	0·7	0·8	32·0	8·0
Aptinothrips	70·0	—	—	3·0	—
Chirothrips	1·0	18·0	78·0	4·3	9·0
Frankliniella	2·2	1·5	0·3	13·0	34·0
Limothrips	1·8	2·2	—	26·5	6·0
Haplothrips	20·0	66·0	6·0	7·1	23·0
Others	2·5	10·5	5·5	2·9	12·0

ated humidity, as well as the constituent grasses, though Koppa (1967) found no significant difference between the species present on oats grown on three different soils. Generally, the more flowering dicotyledons included in a grass sward the greater the diversity of the thrips population, and leguminous plants especially encourage species of *Odontothrips*.

Of the 90 species von Oettingen recorded from grassland habitats in north Germany, 24 (26%) fed solely on Gramineae, 19 (22%) lived on species of plants from other families and 47 (52%) were "non-typical", feeding on plants not normally found in meadows, having probably originated elsewhere. The average estimate of "non-typical" species from the steppes is less (39%) (Tansky, 1965), probably because the habitats studied covered a much larger area than individual meadows, so fewer species strayed into them from other habitats. The true proportion of "non-typical" species on the steppes may be even less than 39%; Tansky regarded a "typical" species as one that occurred in more than about 20% of samples, rather than one that could complete its life-cycle in one recognizable habitat. His criterion is unsatisfactory because rare species which could develop only on steppe flora, would still be classified as "non-typical" simply because they were uncommon. *Chirothrips manicatus* was the most widespread species on the steppes, occurring in 63–100% of all samples taken.

COMMUNITIES ON CROPS

Within similar lattitudes and climatic zones, communities of thrips in cultured or managed habitats such as crops or commercial forests contain fewer species but many more individuals per unit area than communities in more natural habitats. Indices of diversity range from less than 1 to about 3·0 (Table 25) with an average of about 1·7, compared with an average of about 4·3 for mainly temperate natural and semi-natural habitats. Tansky (1965) found 51 species on virgin steppe but only 23 on wheat in the same districts. In most natural habitats there are usually a few co-dominant species each constituting about 15% or occasionally more of the community, but in most cultivated habitats 1 species generally overwhelmingly predominates and is often a pest. On wheat fields in north Khazakhstan, Tansky (1961) found an average of 4,994 individuals per m², of which 89–99% were *Haplothrips tritici*, compared with 351 per m² on reverted grassland where *Chirothrips manicatus* constituted 15·6–58·8% and *Nesothrips icarus* 8·5–16·8% of the community, and only 254 per m² on virgin steppe where *C. manicatus* dominated although *Haplothrips acanthoscelis*, *H. reichardti*, *H. distinguendus* and *Aeolothrips intermedius* were also common. The unrivalled success of *H. tritici* on wheat is probably because the phenology of insect and plant coincide in this temperate, mid-continental climate, and because the species has adapted to a single food plant on which its predators reproduce more slowly than on uncultivated grasses.

The number of species recorded on flax, 26 in Holland and 18 in Great Britain (see Table 25), illustrates the difficulty of determining the true diversity of the thrips community on a crop when many of the species found there are probably chance arrivals settling from the general aerial fauna. No more than three or four species breed on flax (Morison 1943) and Franssen and Mantel (1961) found that the two most abundant breeding species constituted about 87% of the population; the oats thrips, which does not breed on flax, accounted for 12%, and the remaining 23 species together made up 1%. Some of these species may breed on weeds associated with the crop and therefore truly belong to the habitat, but the estimate of diversity for many communities living in small areas is probably unavoidably exaggerated by accidental arrivals from elsewhere. One of the poorest faunas occurs on Kenyan pyrethrum crops where only seven species are recorded (Bullock, 1963b). This may be because the genus *Chrysanthemum* does not occur naturally in Kenya (Gillett, *in litt.*) where pyrethrum was not introduced until 1928, and many indigenous species may not have adapted to it yet. However, the speed with which some species can exploit a suitable niche (see p. 216) suggests that this explanation is incomplete, and the fauna may be poor because pyrethrum is toxic and possibly resistant to some phytophagous insects.

Cultivation of natural vegetation quickly alters the structure of the thysanopteran community living in it. Table 28 shows that within a year

TABLE 28

Number of grass thrips per 100 wheat plants on established arable fields, compared with numbers in the first year after ploughing reverted grassland and virgin grassland (after Tansky, 1961)

Species	Vegetation in previous year		
	Wheat (established arable)	Reverted grassland	Virgin land
Haplothrips tritici	541·0	415·0	403·5
Haplothrips aculeatus	32·5	24·5	18·0
Limothrips denticornis	7·0	18·5	13·5
Frankliniella tenuicornis	5·0	3·5	—
Chirothrips manicatus	4·5	4·0	57·8
Aptinothrips elegans	0·2	0·6	—
Rhipidothrips elegans	0·1	0·5	—

of cultivating virgin land or reverted grassland most species characteristic of old arable soils become established. Ploughing and cultivation deplete the rich plant community of the steppe to a few species which are most often wheat and associated weeds. Species of thrips needing turf for food, hibernation and protection during part of the life-cycle are destroyed and cannot re-establish as long as the land is cultivated annually. A few of the flower-dwelling species persist after cultivation but these are mostly species that were polyphagous in their natural habitat and have persisted because one or other of their host plants has survived as an arable weed. If land is ploughed carelessly and the turf incompletely turned, most of the original species persist for longer. When arable land is left unculti-vated, but mown for hay or moderately grazed, grassland of a different character develops. It has a rich flower layer spread evenly throughout the sward, not patchily distributed as in virgin land, and the thrips population becomes as diverse as the original. In New Zealand, well-managed pasture has far fewer species (Cumber, 1958, 1959) but this is perhaps partly because New Zealand has few endemic species of Thysanoptera and partly because natural grassland is not extensive. If new species were introduced they could probably easily find a niche in this environment.

Where suitable niches are available thrips can exploit them quickly. *Limothrips denticornis* was first recorded in North Dakota in 1946 (Post and Colberg 1958) and 14 years later population density in barley fields averaged approximately 5,800 m^2 (Post and Olson, 1960). *Isoneurothrips australis*, which feeds on flowers of *Eucalyptus* and weeds in Australia, was recently introduced to Israel, Cyprus and Egypt where it has quickly become a pest on cultivated plants. *Thrips imaginis*, a species that lives on many species of flowers in Australia, where it is also a pest of top-fruit trees, seems capable of establishing itself in a similar niche in many parts of the world, and *Anaphothrips sudanensis*, probably of West Indian origin, has now spread to Cyprus, western Asia, Egypt and South Africa (Priesner, 1964a). Around the Mediterranean, *Taeniothrips croceicollis* is a common, but not abundant, species living on *Asphodelus* (Liliaceae) growing among grass. Cattle avoid this lily, so on intensively grazed pastures it spreads; the thrips exploit the favourable niche created, and large populations develop (zur Strassen, 1968).

Changing the botanical composition of the sward with fertilizer also eventually changes the species of thrips present. Applications of phosphate to meadows encourage *Vicia*, *Lathyrus* and *Trifolium* in the sward and von Oettingen (1942) found that the proportion of *Odontothrips phaleratus* in a meadow so treated increased from almost nil to 60% of the thrips population in four years.

On crops, the same species of thrips usually dominates year after year. On fields of meadow foxtail (*Alopecurus pratensis*) in Finland, usually over half, and sometimes more than 90% of the thrips community is *Chirothrips hamatus* (Hukkinen, 1936). On flax in Holland and Great Britain, wheat in Khazakhstan, onions in U.S.A., and pyrethrum in Kenya, *Thrips augusticeps*, *Haplothrips tritici*, *T. tabaci* and *T. nigropilosus* respectively completely dominate the communities, but on leguminous forage crops where there may be two or three co-dominants, the relative abundance of species fluctuates. On red clover in central Bohemia, *Odontothrips loti*, *Frankliniella intonsa* and the predatory *Aeolothrips intermedius* together constitute 88–92% of the population, but in consecutive years the proportion of *F. intonsa* present changed from 24 to 64% (Skuhravy *et al.*, 1959). Similarly, on a lucerne crop in Moravia, 30% of the population in

1963 was *T. tabaci* but in 1962 only 0·4% (Obrtel, unpublished observations). In different places in the world the thrips community on the same crop may have different dominants. On red clover in Idaho, U.S.A., and Ontario, Canada, *Haplothrips leucanthemi* replaces the species listed above (Burrill, 1918; Ross, 1918). On wheat, in eastern and southern Europe and western Asia, *Haplothrips tritici* is the dominant species, but further west it occurs progressively less frequently and its niche is filled by *Limothrips cerealium*; further north, in Scandinavia *L. denticornis*, *Anaphothrips obscurus*, *Frankliniella tenuicornis* and *Haplothrips aculeatus* predominate, the proportion of these species depending on whether the crop is winter or spring sown (Koppa, 1967) (Table 29). *Limothrips cerealium* is also common on wheat in many parts of the U.S.A. (Bailey, 1948), but in Kansas and parts of Oklahoma, western Missouri and southern Nebraska, *Prosopothrips cognatus* is the more abundant species (Kelly, 1915), and in Bombay, India, *Anaphothrips sudanensis* predominates (Patel and Patel, 1953). On mixed oats and barley in Sweden, *Stenothrips graminum* constitutes up to 48% of the community (Cederholm, 1963), but in northern Germany on pure oats and barley stands only 0·1 to 0·6% (Holtmann, 1962); it is absent from oats in the U.S.A. (Bailey, 1948) and in Asiatic U.S.S.R., where *Chirothrips manicatus* takes its place (Tansky, 1961).

Studies on forest faunas are limited mostly to temperate European conifer stands where specific diversity is small. For example, only 13 out of the 49 known species of Swedish thrips are characteristic woodland dwellers (von Oettingen, 1954), and only 28 out of the 82 species living in the French Alps (Weitmeier, 1956). In Czechoslovakia, Kratochvil and Farsky (1942) sampled mature larch in pure stands, nursery stock, and isolated larch growing among other trees, from April to October for three years. Twigs and bark were searched, and branches shaken and beaten to obtain large samples, which together yielded nearly 10,000 thrips of 23 species ($\alpha = 2·9$). The larch thrips, *Taeniothrips laricivorus*, was overwhelmingly dominant, constituting 88·1% of the population; the next two most common species, the predatory *Aeolothrips fasciatus* and *Ae. vittatus* each constituted only 2·5% and almost all others less than 1%. In winter, when most of these species are hibernating in turf and litter, the population on the trees consists almost entirely of larch thrips. Autumn and winter samples from mixed woodland in Switzerland, including broad-leaf species, fir, pine, larch, spruce and cedar, yielded over 4,600 *Taen. laricivorus* and a mere six *Haplothrips aculeatus* (Maksymov, 1965). By contrast the faunas in high, wet tropical forests in Polynesian islands are dominated by fungus-feeding and predatory Tubulifera, and the flower and foliage-feeding Terebrantia are commonest in the drier, lowland regions (Zimmerman, 1948).

TABLE 29

Proportion (%) of different species on cereals in north Khazakhstan, north Germany and Finland. Dominant species in bold type (after Tansky, 1961; Holtmann, 1962 and Koppa, 1967)

Species	Winter wheat			Winter rye			Spring barley			Oats		
	Khazakh-stan	Germany	Finland	Khazakh-stan	Germany	Finland	Khazakh-stan	Germany	Finland	Khazakh-stan	Germany	Finland
Melanthrips fuscus	1·1	—	—	—	1·3	—	—	—	—	—	0·1	—
Rhipidothrips elegans	3·9	—	—	0·5	0·1	—	—	—	—	**69·8**	—	1·8
Chirothrips manicatus	—	—	6·7	4·5	—	2·0	**22·8**	—	3·1	—	—	—
Chirothrips hamatus	—	—	0·3	—	—	0·7	—	—	—	—	—	—
Limothrips cerealium	—	**77·2**	—	—	**48·1**	—	—	**45·9**	—	—	**94·5**	—
Limothrips consimilis	5·7	—	—	—	—	—	<0·1	—	—	—	—	—
Limothrips denticornis	—	18·7	14·0	**41·4**	**42·0**	**27·2**	**43·3**	10·9	**25·4**	2·1	1·1	2·8
Aptinothrips elegans	—	—	—	0·3	—	—	—	—	—	—	—	—
Aptinothrips stylifer	—	0·1	<0·1	—	—	1·0	—	—	0·4	—	—	<0·1
Anaphothrips obscurus	—	0·1	17·0	—	0·3	6·7	—	0·9	9·3	—	0·1	**24·5**
Frankliniella intonsa	—	—	—	—	—	—	—	—	—	—	0·1	—
Frankliniella tenuicornis	2·4	—	**43·2**	3·9	—	**46·1**	6·0	—	**53·3**	9·1	0·1	**61·6**
Taeniothrips atratus	—	0·1	—	—	2·1	—	—	—	—	—	0·1	—
Thrips angusticeps	—	—	—	—	—	—	—	0·9	—	—	0·5	—
Stenothrips graminum	—	—	—	—	—	—	—	0·1	—	—	0·6	—
Haplothrips aculeatus	7·7	3·8	7·3	19·5	6·1	6·4	7·6	**41·3**	1·8	11·3	2·8	1·3
Haplothrips tritici	**79·2**	—	—	**29·9**	—	—	**20·3**	—	—	7·7	—	—
Others	—	—	11·5	—	—	9·9	—	—	6·7	—	—	8·0

Microhabitats and Spatial Disposition

Primary habitats with characteristic thysanopteran communities may be uniform over hundreds or even thousands of square kilometres, as in the larch forests of Europe or on the grasslands of Asia and North America. But within these expansive general habitats there are usually less diverse communities in subordinate habitats which are determined by local physical differences or by clearly defined, small-scale differences in vegetation. This is especially true where Man has modified the local vegetation, or the geology and topography change over short distances. For example, *Aptinothrips elegans* usually occurs only in warm, dry places especially on hummocks (von Oettingen, 1942). The narrow belt of vegetation bordering water-courses often supports a highly characteristic fauna; in Europe *Euchaetothrips kroli* is found only on great water grass (*Glyceria maxima*) which grows in narrow bands a few metres wide along the banks of rivers and streams. Logs and fungi provide smaller highly specialized habitats, particularly for species of Phlaeothripidae which feed upon the saprophytic fungi flourishing on dead wood and bark. The temporary suitability of sites covering even a very small area is shown by Smith's (1955) record of 33 individuals of five species, *Taeniothrips atratus, Taen. vulgatissimus, Thrips flavus, T. fuscipennis* and *T. major* on a single stinkhorn fungus (*Phallus impudicus*). These species were absent from vegetation within at least 15 m of the fungus and all feed on green plants. Perhaps the moisture or bright sunlight shining on the decomposing phallus attracted thrips to it, but whatever the reason, a discrete group of species temporarily colonized a limited area within a more general habitat. On an even smaller scale, individual inflorescences of flowering plants provide microhabitats often having a characteristic, if ephemeral, community structure. A single capitulum of a dandelion (*Taraxacum officinale*) may contain 50 individuals of six species, a large flower of *Clematis* 12 individuals of two or three species, and hundreds of individuals of five to seven species may occur in flowers in a clump of certain species of *Senecio* or *Solidago* (Morison, 1958). Among sand-dune flora red flowers are claimed to attract more individual thrips (46·6%) than yellow (32·3%) or white (21·1%) ones (Gromadska, 1954), and, on the same species of plant at the same time and place, the larger the flowers the more individuals they support (Ward, 1966). The greatest daily average of adult *Thrips imaginis* per rose recorded by Davidson and Andrewartha (1948a) was 1,582. Thus, small, discrete but highly favourable microhabitats within a larger general habitat encourage thrips to aggregate. The distribution of individuals on plants is usually associated with food and shelter (see pp. 44, 46), but within a habitat other factors, including patchiness in the distribution of vegetation, changes in microweather and shelter, feeding, mating and oviposition behaviour, and possibly an innate urge to aggregate, affect spatial disposition.

Many Phlaeothripidae feed on fungal spores which develop in the humid environment under the bark of dead and dying twigs, decaying grass or mouldering forest litter, and within these microhabitats aggregations form. Even predatory species occasionally aggregate within a small area. Eggs, immature stages and adults often occur close together, though there does not seem to be any social organization (Stannard, 1957). Hood (1950) suggested that in some aggregations individuals communicate with each other by sounds emmitted from stridulating organs on the coxae and femora. However, all the 14 species so far described with stridulating organs feed on fungal spores, and in most of them different generations overlap, so that aggregations are likely to develop whether or not they communicate by sound. Aggregations also develop in galls (see p. 52). Hundreds of Thripidae may congregate on a few cm² of leaf- or flower-surface. On vines in Italy more than 60 larvae of *Drepanothrips reuteri* may occur per cm² of leaf (Bournier, 1957). Adult "flower thrips", *Thrips fuscipennis*, *T. major*, *T. flavus*, *Taeniothrips atratus* and *Taen. vulgatissimus* often form dense aggregations on flowers that are not hosts for their larvae, behaviour that has perhaps evolved to ensure that the sexes are brought together for mating (Morison, 1957).

The largest aggregations of individual species occur when the population density is greatest but it is difficult to assess, by observation, whether the degree of aggregation is similar at different population densities. Taylor (1961) described a method which gives an index of aggregation for a population independent of its density. Variance (s^2) is proportional to a fractional power of the mean (m) so that $s^2 = am^b$, where a and b are characteristic of a specific population. For a set of samples, a log × log plot of s^2 on m gives a straight line whose slope (b) is an index of aggregation. In a near regular distribution b approaches 0, $b = 1$ for a random distribution and values >1 indicate a progressively increasing degree of aggregation.

Most thysanopteran populations are probably aggregated even where the habitat appears uniform. In Australia, Davidson and Andrewartha (1948) sampled *Thrips imaginis* in rose flowers at a similar stage of development, just before the petals began to fall. Each sample contained 20 flowers. For this population $s^2 = 1 \cdot 0m^{1 \cdot 8}$ showing that thrips were clumped and not randomly distributed between apparently similar flowers in the hedge (Fig. 78a). In a uniform stand of pyrethrum, Bullock (1965) showed, by comparing the frequency distribution of numbers in samples with a theoretical Poisson distribution having the same mean, that a population of *Thrips nigropilosus* was aggregated; he did not measure the degree of aggregation.

Spatial distribution is a characteristic of the species rather than of a particular habitat. This is shown by the degree of aggregation in *Limothrips cerealium*, a species often found hibernating in the bark of many trees, ranging from the loose, scaly bark of pine with many cracks and crevices suitable for hibernation, to the dense, compact bark of poplar where suitable crevices are rare (see p. 172 and Plate X). Although these trees provide different microhabitats for the overwintering population, a single straight line fitted the log s^2 × log m plot for all samples taken from the trunks of different trees (Fig. 78b), showing that the degree of aggrega-

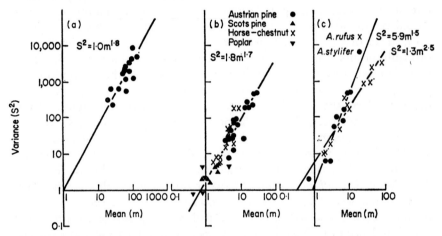

FIG. 78. s^2 × m plots for three populationa of thrips. (a) *Thrips imaginis* in roses (data from Andrewartha and Davidson, 1948a). (b) *Limothrips cerealium* hibernating in the bark of four species of trees (after Lewis and Navas, 1962). (c) *Aptinothrips rufus* and *A. stylifer* in grasses on a woodland floor (after Healey, 1964).

tion for this species was the same in each of these four habitats (Lewis and Navas, 1962). It follows that in a single habitat two species that respond differently to the same environment will aggregate differently. Healey (1964) studied the numbers and distribution of *Aptinothrips stylifer* and *A. rufus* on a woodland floor in England. These two closely related species live in the grasses *Holcus lanatus* and *Deschampsia flexuosa* which, with bluebell (*Endymion non-scriptus*) dominated the ground flora. Nine to twelve samples of litter including dead leaves, twigs, grasses and superficial roots were collected on nine occasions throughout 1960. For *A. stylifer* $s^2 = 1·3m^{2·5}$ and for *A. rufus* $s^2 = 5·9m^{1·5}$ (Fig. 78c); *A. stylifer* was the more strongly aggregated species. The difference in the degree of aggregation within this habitat was probably due to the preferences of each species for the different grasses, *A. stylifer* for *Deschampsia* and *A. rufus* for *Holcus*. Where field populations have been sampled quantitatively this method could be used more often to show differences in the grouping of species within habitats, and may reveal unsuspected responses to small changes in the environment.

Where the boundaries of different types of natural vegetation meet there is often a transitional zone that contains more species of animals and denser populations than either of the neighbouring habitats. This is because weather, shelter and the variety of food plants in the transitional zone are a mixture of those existing on either side, enabling rare species to become more abundant along the boundary. There has been no systematic study of thrips in such areas between extensive primary habitats, but on a smaller scale, populations of thrips on vegetation bordering woodland are often more dense, and the community is richer than within open areas of the wood, especially in winter. The increase in density probably occurs because flying thrips accumulate and settle in the wind shadow created by the sudden difference in height of the vegetation (see p. 158), and the richer community develops because of the greater floral diversity along the woodland edge. Similarly, where cultivated and uncultivated land meet, the population density may also increase. *Sinapis arvensis*, a common arable weed, is a host plant of *Thrips angusticeps*, and Ward (1966) found this thrips more abundant on plants of *Leontodon* near to a field infested with *Sinapis* than on *Leontodon* distant from the field. Many *Haplothrips leucanthemi* occur on rye when this thrips' host, scentless mayweed (*Matricaria* sp.), grows in or near the cereal crop, and the number of *Taeniothrips atratus* in cereals also depends on the proximity and abundance of its host plants, species of Carophyllaceae and Labiatae, to the fields (Koppa, 1967).

SECTION IV

ECONOMIC IMPORTANCE

11 Thrips as crop pests

Several hundred species of thrips are pests. The literature abounds with brief papers describing symptoms of attack on many crops and short-term insecticidal control measures. A review of all these papers is beyond the scope of this book, partly because severe infestations on some crops are rare or very local and therefore not of general interest, and partly because the insecticides available change frequently, so specific recommendations soon become out of date. This chapter therefore deals with the general principles of thrips as pests, and the cultural, chemical and biological methods that help to lessen damage to crops.

Range and Spread of Pests

Of the 5,000 or so known species only a few hundred, mostly in the sub-order Terebrantia, attack cultivated plants, but this small minority has been studied far more than the rest. Harmful species occur around the world in all latitudes between Finland and Alaska in the north (60°N) and New Zealand in the south (45°S). The damage they cause is often slight but sometimes it can be severe and result in serious losses.

Most field and plantation crops support numerous species ranging from as few as five on tomatoes to twenty-three on wheat (see Table 25), but usually only one or two are abundant enough to damage the plants severely. A few crops are attacked by the same species in widely separated parts of the world. *Thrips tabaci* is a cosmopolitan pest of onions grown between sea-level and 2,000 m, and *Chaetanaphothrips signipennis* is found on bananas from Fiji, Australia, Central and South America and Africa. More often, when the same crop is grown in different regions it is infested by different species in each place. In southern Africa, *Scirtothrips aurantii* is the most harmful thrips on citrus, but in California, Texas and Arizona, *S. citri* fills this niche. In the Sudan, *Caliothrips impurus* and *C. sudanesis* attack cotton, whereas *Frankliniella fusca* and *F. tritici* cause comparable damage in North America. Many thrips that attack crops over vast areas are polyphagous species common on many wild and cultivated plants.

I

...ies with a more limited host range, such as *Kakothrips robustus* which ...eds on peas in southern England and parts of Europe, or *Chirothrips hamatus* infesting meadow foxtail grass especially in Scandinavia, are more localized.

Some species native to an area have become pests because they have changed their feeding habits to exploit introduced cultivated plants. An example is *Taeniothrips orionis* in Alaska, which probably normally feeds on mountain lupin and columbine but which severely damages lettuces, potatoes and cabbages when these are planted in freshly cleared areas (Washburn, 1958). Others have become injurious because they have been introduced accidentally by Man to new areas and have adapted to the crops growing there. Once established they have been spread by wind, assisted by further local introductions. By contrast, airborne drift over seas has probably only rarely resulted in the permanent establishment of species in parts of the world far from their original home, despite the vast numbers of individuals that are blown long distances (see p. 155).

Such small insects easily pass unnoticed on fruit, flowers and vegetables unless these are scrutinized, and there are many instances of leaf-, flower- and bulb-feeders, and species that pupate in the soil, spreading great distances in the course of normal trade. The native land of *Taeniothrips simplex* is unknown, but it probably originated in a Mediterranean type of climate. Now, it occurs in South Australia, New South Wales, Victoria, many parts of Canada and U.S.A., Argentina, Hawaii, Bermuda, Jamaica, South Africa, Southern Rhodesia, Turkey and Europe, mostly carried to these places in soil or on the bulbs or roots of plants, especially gladioli and carnations. The original home of the grape thrips, *Drepanothrips reuteri*, was probably southern Europe, but it has been accidentally carried as far as California; and the western European corn thrips, *Limothrips cerealium*, now infests cereals in areas climatically as different from its native lands as the Seychelles and Israel (Morison, 1957).

In the last century a few species may have spread along shipping routes in firewood, soil, or in hay used for animal fodder (Mound, 1968a, 1970a). Modern transport and commerce can spread thrips great distances very quickly; thus specimens of *Caliothrips fasciatus* were present on tangerines and oranges imported into Hawaii from California (Whitney in Morison 1957); in Washington, D.C., the yellow orchid thrips, *Anaphothrips orchidaceus*, was found on orchids sent from England (Swezey, 1945) and species of *Chirothrips* on grass seed imported from Russia (Andre, 1941), all journeys of thousands of kilometres.

Many exotic species introduced to areas where they could not normally survive, persist on plants grown in glasshouses. For example *Heliothrips haemorrhoidalis* breeds outdoors between latitudes 45°N and 40°S, but although it cannot survive the winter in Great Britain, it thrives in glasshouses and is the most widespread and economically important introduced thrips living there (Morison, 1957).

Plant Injury and Crop Loss

The typical damage to plant tissue caused directly by feeding thrips is described in Chapter 3 (p. 41). The silvering, scarring and distortion of leaves and fruits caused by feeding, and their later discoloration due to excrement and the growth of moulds (Plate IIIc), are usually easy to detect, but the economic importance of these symptoms is often more difficult to assess.

Leaf-feeders

When infestations are so heavy that plants are killed the extent of the damage is clear and its cost calculable. Complete loss of crops occurs most often when seedlings are attacked, or in dry seasons when plants lose water rapidly through the damaged epidermis. Young onion plants are often destroyed by heavy infestations of *Thrips tabaci* (Plate XIIa) (Sakimura, 1937a; Schmutterer, 1969), flax seedlings can be killed by the brachypterous generation of *Thrips angusticeps* (Franssen and Huisman, 1958) and cotton seedlings by *Frankliniella* spp. (Fletcher and Gaines, 1939).

Usually plants survive attack, but damage to their leaves, fruit or seeds retards growth and lessens the final yield. When light infestations of leaf-dwellers feed on young leaf tissue the leaves continue to develop but grow distorted (Plate XIIb). The "leaf-curl disease" of chillies in India is really caused by *Scirtothrips dorsalis* feeding in this way (Plate XIIIe) (Ananthakrishnan, 1956; Fernando and Peiris, 1957). On fully grown leaves of most plants light infestations produce silvering and scarring, either in patches on the laminae or in streaks alongside veins, but the leaves usually retain their shape and persist on the plant (Plate XIIIc). Heavily infested leaves of all ages dry, shrivel, turn brown and brittle, and eventually droop or fall. Leaves on the sunny side of trees or in exposed sites usually deteriorate sooner than leaves supporting similar infestations in shade, showing that loss of water is the prime cause of damage. On grasses and cereals, injured leaves turn yellow or whitish rather than brown, before they wither.

The severity of the damage caused by leaf-feeders often depends on the type of soil and local weather. On light, dry soils and in dry seasons the symptoms develop more rapidly than on heavy moist soils and in wet weather. For example, the degree of damage to cotton in the Sudan by leaf-feeding *Caliothrips* spp. is related to the texture of the soil and hence its water content. Injury develops earlier in the season and is most severe on light soils or on high land where the plants suffer from shortage of water. On clay soils damage is less because there is no water strain on the plants, and very wet soils provide additional protection by preventing the emergence of adults from the buried pupae (Pearson, 1958).

There is much evidence showing that plants under water stress not only develop symptoms more rapidly than adequately watered plants,

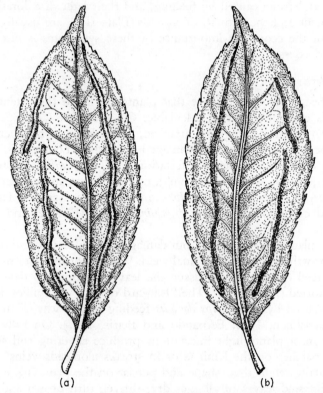

(a)   (b)

Fig. 79. Characteristic sand-papery lines on (a) the lower and (b) the upper surface of mature tea leaves, caused by the feeding of *Scirtothrips dorsalis* when the leaves were in bud (after Dev, 1964, and photo by R. Fordham).

but they also provide thrips with more attractive or nutritious food and therefore encourage heavier infestations (Mumford and Hey, 1930; Plank and Winters, 1949; Silva, 1964; Fennah, 1965). For the same reason tall spindly plants are often more heavily infested than short stocky ones. Other factors that affect the susceptibility of plants to attack by thrips are mentioned on pp. 44–46.

The incidence of *Diarthrothrips coffeae* on coffee in Kenya illustrates the dependence of outbreaks on dry weather. Between about 1950 and 1966 this insect caused little damage because mulching and overhead irrigation prevented infested plants from drying out, but during the rainy season of 1966, when only about two-thirds of the usual amount of rain fell, thrips became temporarily abundant and caused serious defoliation (Evans, 1967).

The extent of injury and loss of yield is easiest to assess when leaves themselves are harvested, as with tobacco and tea. On dark Virginian tobacco grown in South Africa up to 75% of the crop may be attacked by *Scirtothrips aurantii*, and the damaged leaves represent a complete loss because they are thin and papery and unfit for curing (Strofberg, 1948). Thrips are unimportant on American flue-cured crops but the white patches that they cause on cured cigar leaves seriously depreciate the value of wrapper tobacco (Akehurst, 1968). Infestations on tea lessen yields because the growth of the attacked shoot is arrested and the leaves become hard and brittle and finally fall (Fig. 79; Dev, 1964; Mound, 1968). In East Africa, sometimes almost all shoots on tea bushes may be infested with *Scirtothrips* spp. or *Heliothrips haemorrhoidalis* at the end of the dry season, so most leaves plucked when rapid growth is resumed at the beginning of the wet season are distorted. The damaged leaves are nevertheless processed, but may produce poorer quality tea than healthy leaves (Fordham, *in litt.*). When drought is severe in Kenya the leaves are not worth plucking and the whole crop may be lost (Benjamin, 1968a), and heavy infestations slow the growth of nursery stock (Benjamin, 1968b). Cabbages bronzed by *Thrips tabaci* feeding on the leaves, and with thrips present in the heads, are unmarketable, and in some fields in Iowa, more than a third of the cabbages may be spoiled (Wolfenbarger and Hibbs, 1958).

The importance of leaf injury to crops whose fruits or seeds are harvested depends on many different cultural and climatic factors, and is more difficult to assess accurately. In field experiments on cotton seedlings in Louisiana, feeding by *Frankliniella fusca* destroyed at least half the total leaf area and retarded growth so much that plants were 20% shorter than normal by the time they were six weeks old (Newsom *et al.*, 1953). Laboratory measurements made on the height and weight of seedlings attacked by *F. fusca* and *F. tritici* confirmed that they were 6–30% shorter and 11–36% lighter (Table 30) (Hightower, 1958). Despite these serious setbacks in early growth the plants have remarkable powers of recovery, and field experiments in Alabama and Arkansas showed that even when thrips were controlled on cotton there was no improvement in earliness or total yield of the crop (Lincoln and Leigh, 1957; Watson, 1965). The recovery of the seedlings probably depends on the type of soil, occurring most readily on rich, moist, alluvial deposits in river valleys. By contrast, on dry sandy hills, where plants probably experience water stress, thrips control increased yield by 167 lb of seed cotton/acre (184 kg/ha) at first picking and 266 lb/acre (283 kg/ha) altogether (Newsom *et al.*, 1953). Mature cotton plants in Texas also had smaller and fewer bolls per plant when attacked by thrips, and 358 lb/acre (394 kg/ha) of seed cotton were lost (Fletcher and Gaines, 1939).

Root and bulb crops yield less when their leaves are damaged at a critical stage of growth. Fields of onions, where *T. tabaci* was controlled by insecticides applied every five days, yielded 409 sacks of bulbs/acre compared with 363 sacks from untreated fields (Richardson, 1953). In

TABLE 30

The effects of laboratory infestations of *Frankliniella tritici* and *F. fusca* on the height and weight of cotton seedlings. All plants in each of 3 replicated groups containing 10 seedlings were exposed to 10 adult thrips from 2–14 days after emergence (after Hightower, 1958)

Replicate	1		2		3	
		% loss		% loss		% loss
Mean height (mm)—crown to base of cotyledon						
No thrips	12·3	—	13·8	—	12·9	—
Frankliniella tritici	11·5	7	12·7	8	11·0*	15
Frankliniella fusca	8·8*	28	9·7*	30	9·4*	27
Mean fresh weight (g)						
No thrips	0·77	—	0·63	—	0·63	—
Frankliniella tritici	0·60	22	0·56	11	0·33*	48
Frankliniella fusca	0·33*	57	0·35*	44	0·23*	63

* Significantly less ($P = 0.05$) than control.

parts of Japan the green tips of onion leaves are a popular vegetable, and feeding scars on the blades lower their market value as well as retarding growth (Sakimura, 1937a).

Defoliation of tree crops may indirectly affect the yield of fruit, because thrips move to it when preferred leaves wither. This occurs on pears in California, where there are records of such extensive defoliation that thrips spoiled 15% of the crop (Bailey, 1938a). The effects of defoliation in one year may also lessen yields in the next because the early loss of leaves depletes the accumulation of food reserves and also exposes new growth to sun-scalding in the hottest part of the summer.

Larch trees grown for timber need a vigorous leading shoot to produce a straight trunk. Infestation by larch thrips produces stunted, hook-shaped needles and fissures on the axis of young shoots, especially in the upper crowns of trees between 11 and 20 years old. The tips of leading shoots are sometimes killed, the tree axis is destroyed and it develops a shrubby form unsuitable for quality timber (Fig. 80) (Kratochvil and Farsky, 1942; Zenther Møller, 1966). In Germany up to 20% of the top shoots may be destroyed (Ewald and Burst, 1959); in Denmark 20–35% of all shoots in one stand were destroyed in 1961 and 1962, and damage

FIG. 80. The appearance of a young larch stem 3 years after attack by larch thrips. The original terminal shoot remains as a dead apex and the main stem is twisted by eccentric growth (after Vité, 1955).

may be even more severe (Zenther Møller, 1966). As with leaf damage to herbaceous crops, cool wet weather enables trees to withstand attack. The larch thrips also provides a good illustration of the incidence of damage in relation to dispersal (see Fig. 57); the risk of attack diminishes with distance from spruce trees on which the thrips overwinter, so in forests where hardwood species predominate serious injury is rare.

Defoliation of ornamental shrubs grown for the attractive appearance of their foliage or for shade spoils them. In parts of Connecticut and Illinois, the privet thrips is a serious perennial pest of *Ligustrum ovalifolium* as well as sometimes attacking *Tilia*, *Syringa* and *Alnus*. It sometimes damages privet so badly in late summer that the leaves fall instead of remaining on the shrubs through autumn and early winter (Plate XIIb) (Stannard, 1968; Schread, 1969). Similarly, the leaves of many varieties of garden crotons can be spoiled by *Heliothrips haemorrhoidalis*.

Floret-, grain- and seed-feeders

The direct damage and loss of yield caused by thrips infestation is easier to assess in seed and fruit crops than in those whose leaves or roots are harvested, and more reliable information has been collected for cereal and grass seed crops than for most others in this group. Infested florets of cereals usually appear white or silvery and many fail completely to develop grains, or produce thin shrivelled ones (Plate XIIIa). Experimental wheat plants grown in cages and infested with *Limothrips cerealium* developed no grain at the top or bottom of the head, and intermediate grains were also damaged. Field counts established that the amount of damage depended on the number of thrips present on the ears (Table 31)

TABLE 31

Proportion of damaged wheat florets caused by different levels of infestation by *Limothrips cerealium* (data from Sharga, 1933a)

	Light infestation	Heavy infestation
Mean number of adult thrips per ear	1·2	12·1
Mean number of immature thrips per ear	1·1	11·5
% damaged florets	1·9	17·9

(Sharga, 1933a). In Irkutsk, U.S.S.R., an infestation of 1 to 5 specimens of *Haplothrips aculcatus* or *H. tritici* per grain decreased yield by 5% (Rubtzov, 1935). More precisely Pavlov (1937) estimated that each specimen of *Haplothrips* present on a grain lessened the latter's weight by 1 mg. On oats in Canada, *L. cerealium* caused a 10% loss of grain (Seamans, 1928) and *Anaphothrips obscurus* a 36% loss (Hewitt, 1914). In 1964, *Frankliniella tenuicornis* destroyed so much barley in one district in Finland that it did not pay to harvest (Koppa, 1970). In Finnish rye crops *Limothrips denticornis* may decrease yields by 5–10% (Vappula, 1965) and Brummer (1939) found *H. aculeatus* responsible for 2·6–9·1% of all empty spikelets; serious losses of rye have also occurred in Bohemia (Stranack, 1912).

The actual loss of revenue attributable to *L. denticornis* infesting barley in North Dakota, has been computed. Yields from plots in 10 fields where the thrips population averaged 2 adults and 30 larvae per plant, were compared over a 3-year period with yields from plots in the same field where thrips were killed with organo-phosphorus insecticide. The mean loss of grain was 160 lb/acre (176 kg/ha) representing a loss of $4·75 U.S./acre at 1966 prices (Post and McBride, 1966).

By contrast, results of other experiments on wheat (Jablonowski, 1926;

Körting, 1930) and on oats (Rademacher, 1936; Sheals, 1950) suggested that blindness in growing plants was caused more by physiological disorders than by thrips, and that the appearance of damaged ears was misleading because the undamaged grains in them grew larger and plumper than usual, thereby compensating for some of the destroyed or shrunken grains. As with non-graminaceous crops, the ability of cereals to withstand attack and compensate for damaged florets probably depends on the soil, season and perhaps variety of plant. Thus, in some seasons and localities thrips damage can be serious, whereas in others their effect on yield is negligible.

In many parts of the world species of *Chirothrips* feed on the florets

TABLE 32

The calculated effects of infestations of *Chirothrips pallidicornis* on seed yields of cocksfoot in New Zealand (after Doull, 1956)

	Sprayed (DDT)	Unsprayed
Number of florets dissected	1,000	1,000
Number of female thrips	12	239
Presumed number of male thrips	12	239
Presumed number of seeds destroyed	24	478
% of seed destroyed	2·4	47·8

of useful grasses, and in grass seed crops they often destroy a much greater proportion of the seeds than cereal thrips destroy in grain crops. In New Zealand *C. pallidicornis* destroyed 30% of cocksfoot seed when the population density reached about 20 thrips per inflorescence. Females of this species usually lay one egg per floret, placing it near the top of the developing ovule. By the time the larva has reached the pupal stage the seed is destroyed. Feeding larvae do not move from floret to floret, so each thrips present in the inflorescence represents one ruined seed. The females overwinter in the dead florets, but males leave them in late summer and die. Thus, assuming that equal numbers of each sex develop, the number of seeds destroyed is simply twice the number of thrips present in heads examined during the winter months. When thrips on growing crops were controlled with DDT sprays the amount of good seed more than doubled (Doull, 1956; Table 32).

This work was extended during four summers by Morrison (1961), who derived regression equations for each year for % good seed on % of infested florets. A single equation was inappropriate for all four seasons because the amount of seed produced in each differed, but the average regression coefficient was 0·76, meaning that for every 1% increase in infestation there was a corresponding 0·76% loss of yield. The relationship between the level of infestation and good seed in one district was also a reliable indication of infestation and losses in other districts in the same year.

In Arizona, *Chirothrips mexicanus* and *C. falsus* decreased yields of Bermuda grass seed from between 350 and 450 to as little as 30 to 50 lb/acre (440 decreasing to 44 kg/ha), representing a loss to farmers of about $300 U.S./acre at prevailing prices (Roney, 1949), and in Texas, *C. mexicanus* caused an eight-fold decrease in yield of Rhodes grass seed (Riherd, 1954). In Sweden, *C. manicatus* and *C. hamatus* together destroyed 11 to 19% of timothy seed (Johansson, 1946); in Finland *C. hamatus* destroyed 25 to 37% of meadow foxtail seed (Hukkinen, 1936), and in Germany seven common species of thrips contributed to the loss of 15% of meadow foxtail and 10% of ryegrass seed (Wetzel, 1964). Indeed, one of the reasons for the decline in popularity of meadow foxtail grass in seed mixtures in the past three decades has been the difficulty of obtaining enough seed to establish the grass in a sward. In Oregon, 16% of the seed in a field of "Emerald" bentgrass (*Agrostis stolonifera*) was destroyed by *Anaphothrips obscurus* (Plate IIIb; Kamm, 1972).

Thrips also cause losses of seed in non-graminaceous crops, but generally these can tolerate much heavier infestations than grasses before measurable losses occur. For example, onion seed heads are not seriously damaged by *Frankliniella occidentalis* until the population reaches about 10 individuals per floret, a density rarely found on grass florets and equivalent to about 9,000 thrips per onion seed head (Carlson, 1964b). On safflowers grown for oil seed, sparser populations of the same species of thrips lowered the weight of individual seeds and their number per seed head, but even so the density of populations needed to cause serious losses was greater than is usual on grasses. The mean yield from caged safflower plants artificially infested with 50 adult thrips and left for 20 days was 20·3 seeds per head, each seed weighing 1·16 g, compared with 37·6 seeds each weighing 2·06 g from protected heads. Losses in the field are probably less than this; artificially disbudded plants compensate for the loss of flower heads by producing more seeds on the remaining good heads, and the response of the plants to thrips damage is perhaps similar (Carlson, 1964b). Later field experiments showed that 10 adult thrips with an average of 75 larvae per head did not decrease yield, but 20 adults with 150 larvae did (Carlson, 1966).

Measured and estimated losses in other seed crops include up to 20% destruction of peas by larvae of *Kakothrips robustus* in Holland (Plate XIIId; Franssen, 1960), 50% loss of carrot and beet seed by *Frankliniella* spp. in California (Bailey, 1938) and the spoiling of 50,000 lb (22,500 kg) of cardamom pods by *Ramakrishnothrips cardomomi* in Madras, India (Ramakrishna Ayyar and Kylasam, 1935).

A few species decrease seed production by damaging the flowers so that normal pollination by larger insects is prevented. A good example is *Odontothrips confusus* which feeds on lucerne flowers. These are pollinated by bees which "trip" the flowers to release the stamens and pistil from the protecting keel petals. When thrips feed on the keel it withers and

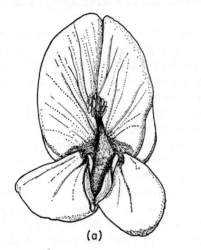 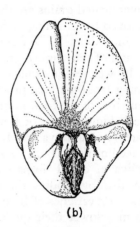

(a)   (b)

FIG. 81. Sterilization of lucerne flowers caused when *Odontothrips confusus* or *Thrips flavus* feed on the keel petals; (a) healthy flower visited by bees, with stamens released; (b) flower damaged by thrips with stamens trapped inside the shrivelled keel petals (original).

prevents the emergence of the pistil so the flowers remain unpollinated. Sometimes 30% of the flowers are sterilized in this way (Fig. 81; Bournier and Kochbav, 1963, 1965; Noll and Rohr, 1966). In India, *Taeniothrips lefroyi* congregates on flowers of broad beans and by feeding inside the corolla tube on the ovary and stamens sometimes completely prevents pod formation (Haq, 1961). By contrast there are other seed crops including pyrethrum (Bullock, 1962), flax (Henry and Tu, 1928) and onions (Carlson, 1964a) whose set may be increased by thrips on the flowers (see p. 57).

Seed-feeders and grain quality

As well as completely destroying some cereal and grass seeds, thrips may impair the germination of sown seed and spoil grain for milling by contaminating the flour with insect fragments. The high standards of cleanliness in flour imported into the U.S.A. demand that it be absolutely free from extraneous matter including insect fragments, rodent hairs and mites, although in practice up to 20 insect fragments per half-pound (227g) of flour are tolerated. A large proportion of insect fragments found in British flour exported to the U.S.A. are from thrips, mainly *Limothrips cerealium* in wheat, *L. denticornis* in barley and *Stenothrips graminium* in oats flour. They originate from adult insects that have lodged in the crease of grains (Plate XIV). Contamination occurs especially after a wet growing season, when thrips stick to the moist seeds and individuals that die while lying along the ventral side of a developing grain are enfolded by its swelling cheeks.

Contaminated grains are the same size as clean ones so are unlikely to be removed by screening or scouring. Assuming that an individual thrips breaks into about 20–24 fragments during milling, as few as 1–3 insects per pound of the original grain could produce an unacceptable sample of flour. In wet seasons millers could remove many of the fragments by excluding from the final product the "first break" flour, which accounts for only 6% of the total flour obtained (Kent, 1970).

The baking quality of flour milled from grains scarred by thrips is satisfactory (Nuorteva and Kanervo, 1952), and unlike the saliva of the grain-feeding shield bug, *Eurygaster integriceps* Put., there is no evidence that the saliva of cereal thrips contains a proteolytic enzyme to digest the gluten, thereby spoiling the flour for baking. However, attacked grains, especially of varieties of hard wheats, are susceptible to fungal infection which may lower their quality (Kanervo, 1950; Bournier and Bernaux, 1971).

Flower- and fruit-feeders

Feeding damage on the petals of blossoms or on developing fruitlets often lowers the amount and quality of fruit harvested. Many of the features of injury to fruits appear on oranges attacked by species of *Scirtothrips*. On fruitlets between 0·5 and 3 cm diameter the thrips feed around the calyx or under the sepals and produce a narrow ring of scar tissue. As the fruit grows the ring extends outwards and widens, the damaged skin turns scurvy and the fruit grows misshapen. Older fruits may be attacked by a later generation of thrips at the blossom end, where long, cracked scars develop, and while such fruits are marketable they are less valuable than unblemished oranges (Plate XV). In California, up to 80% of some crops have been spoilt (Quayle, 1938). In Southern Rhodesia, Hall (1930) compared the cost of controlling *Scirtothrips aurantii* with the ensuing returns, in terms of the number of cases of fruit produced which were suitable for export. In one year with an estimated export crop of 150,000 cases from 43 groves, the entire cost of a thrips control programme represented the cost of only 2,287 cases, a number which could easily be destroyed by thrips in a single grove. On young trees sometimes half the crop was destroyed. Modern orchard management has decreased the chances of such large losses, but the potential destructiveness of the thrips remains.

Poor initial set or distorted, scarred and cracked fruits similarly caused by different species feeding on blossoms or on young fruits, often occur on crops of pears, apples, prunes, peaches, nectarines, apricots, strawberries and grapes in different parts of the world (Fig. 82) (Horton, 1918;

FIG. 82. The development and spread of scar tissue caused by *Frankliniella occidentalis* on growing nectarines (after Bailey, 1938). Nectarines are especially susceptible to injury by this thrips.

Bailey, 1938; Quayle, 1938; Allman, 1948; Bournier and Blanche, 1956; Sharma and Bhalla, 1963; Bournier, 1970). Even the tough skin of coconuts is scarred when young nuts are attacked (Plate XIIIb).

A different kind of injury develops on the skin of Golden and Red Delicious apples which grow pitted where *Thrips madroni* have earlier laid eggs inside the blossoms (Swift and Madsen, 1956), and a combination of oviposition and feeding by *Frankliniella tritici* may destroy up to half the blossoms on sweet cherry and plum trees (Boyce, 1955).

Silver or white feeding blotches and streaks, and unsightly droplets of excrement on the petals of ornamental blooms, make them unmarketable. Irises and gladioli especially are disfigured by *Taeniothrips simplex* (Herr, 1934; Speyer, 1951), carnations by *Taen. dianthi* (Pelikan, 1951) and the flowers of many tropical shrubs fade and drop when other *Taeniothrips* spp. are present. Unsightly flowers and scarred leaves also occur on many exotic flowers grown in glasshouses (Pussard, 1946; Morison, 1957; Hussey *et al.*, 1969) (Plate IIIa).

Losses from virus diseases

The importance and incidence of diseases caused by thrips-transmitted tomato spotted wilt virus differ greatly between seasons, localities and crops. In a few areas heavy infections occur so often that some susceptible crops cannot be grown at all; in others, the poor quality and quantity of yields makes growing unprofitable.

Spotted wilt occurs on lettuces in many countries and causes serious damage especially in warm coastal areas. Infected plants are unmarketable; young plants turn yellow, droop and die; in older ones the leaves show marginal wilting, yellowing and necrotic spots. Growth is unequal and infected plants are vulnerable to rot-producing organisms (Grogan *et al.*, 1955). In South Africa and California the disease causes heavy losses of tomatoes and also seems worst in warm coastal areas, probably because there is no freezing weather to destroy infected winter crops, ornamentals and weeds, and possibly because populations of thrips in such equable climates are larger than where weather is more extreme. In a few Californian coastal districts spotted wilt is so prevalent that tomatoes cannot be grown profitably. Elsewhere in the state the disease is scattered and 10–25% of a crop may be infected early in the growing season, increasing later. Plants infected when young usually remain alive but grow stunted and rosetted, often showing severe mosaic symptoms; they bear no crop. When older bearing plants are infected, fruits develop circular patterns which spoil their appearance and marketability (Moore and Anderssen, 1939; MacGillivray *et al.*, 1950). A strain of the red currant tomato, *Lycopersicon pimpinellifolium*, is resistant o the disease in California (Smith and Gardner, 1951), although in South Africa a variety of this tomato showed 20% infection (Hean, 1940).

Species of tobacco are very susceptible to the virus and the damage caused by infection is usually economically more important than silvering caused by direct feeding. Symptoms are variable; the stem apex may bend over, and young leaves show concentric necrotic rings which often coalesce to produce necrotic patches. Plants are usually stunted and grow unevenly. Badly infected young plants often die, but some can recover partially and after initial infection successive leaves show progressively less severe symptoms (Smith, 1957; Akehurst, 1968). The virus causes tip chlorosis on *Nicotiana rusticana* in the Ukraine, where Razvyazkina (1953) found 5–12% of plants with symptoms at the beginning of the season increasing to 90–95% at the end. In Bulgarian tobacco-growing districts 70–80% of plants in some fields may be infected (Ivancheva-Gabrovska, 1959), and in South Africa kromneck disease is a constant threat to tobacco crops (van der Plank and Anderssen, 1945).

The percentage of infected plants in pineapple fields is rarely as great as in tobacco crops but in areas where the disease occurs, crown plantings often show 30–40% of diseased plants after one to six months. Plants infected early produce no fruit; the fruits on later infected ones are malformed and often rot. Even an infection as small as 3–5% of plants represents a loss of 700–800 pineapples per acre at the usual density of planting (Linford, 1943). Different planting materials vary in susceptibility to yellow spot; crowns are most susceptible, slips next and shoots least. The difference in susceptibility is probably dependent on the looseness of the leaf bases which determines the accessibility of the succulent parts of the plant to thrips vectors. The shoots have more tightly imbricated leaf bases and a tougher epidermis than either crowns or slips (Collins, 1960).

Cultural Control

Appropriate cultural methods coupled with clean management of farms and orchards can sometimes completely prevent damage by thrips and thrips-borne virus diseases, and often decrease it.

Irrigation

Because plants under water stress are so susceptible to direct injury (see p. 44) adequate irrigation in dry areas and in hot seasons alone decreases losses. Bullock (1963b) mentions a striking example in a field of pyrethrum, heavily and fairly uniformly infested with *Thrips nigropilosus*, in which one edge only received overhead irrigation. The watered plants were unharmed whereas the unwatered ones were seriously damaged. Similarly, mulching and overhead irrigation of coffee in Kenya has lessened trouble from thrips (Evans, 1967). In dry parts of the Lockyer Valley, Queensland, where *Thrips tabaci* used to damage onions, profitable crops can now be grown despite the presence of the thrips, by irrigating to ensure rapid growth. In fact, irrigation has so decreased the harmful effects of this pest that the benefits which might have been derived from the use of modern insecticides under older dry-farming conditions have been largely forestalled (Passlow, 1957). In California, infestation of Acala cotton by *Caliothrips fasciatus* invariably followed faulty irrigation; thrips were never found on plants that received enough water (Mumford and Hey, 1930).

Flooding can also be used to destroy species which spend part of their life-cycle in the soil. In experiments with bananas, 56% fewer adult *Frankliniella parvula* emerged from plots that had been flooded for 24 h than from dry ones (Harrison, 1963). When rains are delayed, the rice thrips, *Chloethrips oryzae*, may injure seedlings, but the pest can be destroyed in a few hours by flooding paddy fields, spraying a thin film of oil on the water, then knocking the thrips into it by dragging a rope over the plants (Grist, 1965). In the Sudan also, heavy watering of clay soils lessens the numbers of *Caliothrips impurus* on cotton, by sealing the pupae beneath the surface crust, and preventing adult emergence (Pearson, 1958). By contrast, *Thrips angusticeps* can survive flooding for several weeks, perhaps because before pupation larvae prepare a crude cell that retains air around them (Franssen and Huisman, 1958). Species that make stronger cells of earth and silk (see p. 27) are probably even more resistant to flooding.

Cultivation and burning

Soil-dwelling stages are vulnerable to mechanical cultivations. Ploughing cereal stubbles to a depth of about 25 cm destroys some species of cereal thrips, either because the adults that emerge from the buried immature stages cannot reach the surface, or because the soil at this depth is too moist for survival (Kurdjumov, 1913; Kolobova, 1926; Grivanov, 1939; Lyubenov, 1961). Harrowing the stubble before ploughing ensures that the straw is buried and no thrips left on the surface. Populations of *Haplothrips tritici* can be decreased by 30–60% by this treatment (see p. 169; Tansky, 1958a). The common practice in Europe and central Asia of burning stubble before ploughing to help control this thrips, kills larvae overwintering in dead stalks on the surface of the ground (Shurovenkov, 1961), but it is probably less effective than supposed because most larvae hibernate 10–11 cm below the surface and at this depth the soil never reaches lethal temperatures. Stubble burning is probably a more effective means of control for grass thrips, because many of the harmful species of *Chirothrips* overwinter in dead stubble and culms lying on the ground. Care should be taken when harvesting grass seed crops to leave as few heads as possible in the field, and to burn the chaff and straw after threshing.

An unconventional mechanical method of control that might usefully be modified for use on a field scale in low-growing crops with a high cash value, is the use of aluminium foil mulches. Foil wrapped around a stiff board and placed around the base of rose trees so that it extends 0·5–0·7 m beyond the canopy of the tree, deters thrips from landing on the blossoms. When placed in position several weeks before flowering, the method was superior to control with certain systemic insecticides applied to the soil (Ota and Smith, 1968). The mechanism of repellency is perhaps associated with a disturbance of the thrips' orientation prior to landing, caused by the presence of a bright, highly reflective surface beneath them instead of the usual duller, less reflective vegetation or earth.

Mulches beneath coffee perhaps help to decrease infestations of coffee thrips by providing less suitable pupation sites than bare soil.

Time of planting and harvesting

Well-established crops can withstand attack better than newly planted ones, so early planting is usually desirable. It is especially beneficial on light, dry soils where plants suffer from water shortage as the growing season progresses. In the central rainlands of the Sudan where upland strains of cotton may be grown with rainfall as the only source of water, July sowing used to be essential when insecticides were not used. Irrigated plants sown much later were often heavily infested but were able to withstand attack because of better water supplies (Pearson, 1958). Similarly, the key to successful onion production in areas with a hot, dry season is to sow early enough for plants to be almost mature before onion thrips increase in the dry weather. Most horticultural varieties of peas flower before agricultural types, so where *Kakothrips robustus* is abundant it is advisable to grow only the former, and to sow them early to avoid attack (Franssen, 1960).

In north Khazakhstan the slower-developing ears of spring-sown cereals are much more likely to be infested by *Haplothrips tritici* than ears of autumn-sown wheat (Tansky, 1958b). However, the benefits of early sowing depend on the climate, soil and perhaps variety of wheat, because in Bulgaria the same thrips, by contrast, lays more eggs on early varieties of winter wheat than on spring wheat (Lyubenov, 1961).

Careful timing of harvesting can also diminish loss of yield. Weekly plucking of tea leaves removes many thrips from bushes and lessens the chance of heavy infestations developing (Dev, 1964). Yields of some grass seed crops could perhaps be increased by grazing the grass early in the season to delay flowering so that plants have no attractive inflorescences exposed when *Chirothrips* migrate. Early harvesting of cereals may increase mortality in the second-generation larvae of *Limothrips* spp., but unless practised over a large area it would not noticeably lessen populations of adults in the following spring. Damage to lucerne seed crops by *Odontothrips confusus* is partly prevented by taking the seed from the first cut, before flowers are sterilized by heavy infestations of thrips later in the summer (Bournier and Kochbav, 1965).

Plant spacing

In some crops susceptible to virus diseases the proportion of virus-infected plants can be decreased by increasing the density of planting.

This effect of spacing applies especially if there is little subsequent spread of the virus within the crop after the initial infection, as when tomato spotted wilt virus is carried into tobacco crops by *Frankliniella schultzei* in South Africa. Van der Plank and Anderssen (1945) calculated that if the incidence of the disease were 50% with single plant spacing, it would be decreased to 9% by planting two plants per hill and to 1% by planting three, without altering the spacing between hills. This proved a satisfactory method in practice, and was particularly useful in preventing loss from infection during the first month or two after transplanting when there is little danger of overcrowding and stands do not need thinning. Similarly, Linford (1943) found that pineapples planted 45 cm apart in rows were 57·6% more heavily infected with pineapple yellow spot than plants 30 cm apart, whereas the number of plants infected per acre was unaffected by the density of planting.

On a much larger scale, the distance between larch trees in a forest and their arrangement in relation to other species of tree affects the incidence of damage by the larch thrips. In Germany, damage was lessened by planting larch of mixed ages in small groups in shaded sites, and by scattering the groups sparsely through stands of other trees (Vietinghoff-Riesch, 1958).

Previous cropping and neighbouring crops

Species that survive in soil or on remnants of previous crops can be controlled by crop rotation. In north-west Europe, *Thrips angusticeps* attacks several crops including flax, brassicas and peas, and a brachypterous generation hibernates in the soil. When susceptible crops are grown in consecutive years serious damage often results, but it can be prevented completely by following a three-course rotation, e.g. red clover or potatoes, oats, flax. Peas should not be grown after flax or brassicas, especially mustard (Gough, 1955; Franssen and Huisman, 1958). In warmer areas, *T. tabaci* survives in plant remains left in fields, and new crops should not be planted on or near the sites of old ones (see p. 170; Horsfall and Fenton, 1922). In Turkey, Ileri (1947) stressed the importance of destroying plant remains in tobacco fields and of ploughing fields immediately one crop was removed to prevent infestation of the next. In eastern Europe and Asia populations of *Haplothrips tritici* on cereals are smaller on crops grown after perennial crops than after other cereals (Tansky, 1958a). The size of populations persisting in stubbles and the effect of cultivation on their distribution and mortality is discussed on pp. 169, 177.

For species with a wide host range, the probability of large infestations developing on a young crop can sometimes be predicted by examining populations on older, different crops in fields nearby. Race (1965) noticed that in years when populations of *Frankliniella occidentalis* were sparse in onion, lucerne and lettuce fields before cotton was sown, cotton seedlings were likewise lightly infested when they eventually appeared. This type of information could be used to decide when insecticidal treatments of seeds or seed beds are necessary.

Isolation of crops susceptible to virus infection from infected crops lessens the chance of early or heavy infection. Safe distances probably differ from year to year depending on the direction of wind during the thrips' migratory period and on the species present. Linford (1943) believed that pineapple yellow spot was carried to pineapple fields by *T. tabaci* from uncultivated fields several hundred metres away. In California, several lettuce crops are often grown successively and provide a recurring source of thrips infected with tomato spotted wilt virus. Probably these infective thrips are disturbed when the lettuces are harvested and they then spread the virus to tomatoes and other nearby plants (Sakimura, 1961). Vector species are also stimulated to fly when infected plants wither and dry.

Control of weeds and alternative hosts

Many harmful thrips have a wide host range including wild plants, and they often survive on weeds growing in and around fields when there is no suitable crop present. *Caliothrips fasciatus* alone breeds on at least 28 species of crop plants and 48 wild or ornamental plants (Bailey, 1938), and *T. tabaci* has some hundreds of hosts, so the control of weeds in most crops clearly helps to lessen infestations. Wild grasses in grass-seed producing areas provide a permanent and uncontrollable source of *Chirothrips* spp. but in fields, host grasses and volunteer plants on which thrips can hibernate should be destroyed.

Weed control is of great importance in checking the spread of tomato spotted wilt virus which can survive in more than a hundred crop plants, ornamental flowers and wild plants (Smith, 1957). In crops, the first plants to be infected are often near the edges of fields, partly because airborne vectors accumulate there (see p. 158), but also because they acquire the virus from weeds growing along the margins. In the Ukraine where tomato spotted wilt virus causes tip chlorosis in tobacco, the disease first appears each spring near clumps of *Sisymbrium* and other weeds on which the overwintered thrips congregate (Razvyazkina, 1953). Ornamental plants that are alternative hosts for virus or vector may also encourage the spread of disease. Tomato spotted wilt virus was introduced into Rhodesian tobacco-growing areas in dahlia tubers commonly grown in town gardens. Quarantine restrictions were placed on nurseries, and in response to appeals and house-to-house inspections, most gardeners destroyed all their dahlias. Despite this, the virus became established in several other garden plants and weeds and continues to infect tobacco crops (Hopkins, 1943). In California, spotted wilt infection of lettuce is usually greatest in fields near urban areas where perennial ornamentals such as calla lily, nasturtium, dahlia and chrysanthemum abound (Grogan et al., 1955).

A careful combination of cultural and chemical methods has controlled pineapple yellow spot completely in Hawaii, so much so that diseased plants are difficult to find. The pineapple plant is attractive to the thrips vectors only immediately after planting and when the succulent crowns appear on the young fruit, and even at these times the thrips prefer to feed on weeds, especially *Emilia*, growing around the fields. The key to successful control thus depends on avoiding any disturbance of thrips on the weed while the pineapples are in one of these vulnerable stages, thereby ensuring that the insects do not migrate to the crop, and at other times destroying the thrips and weeds with a mixture of insecticide and herbicide. Chemicals may be applied to the weeds when the pineapples are vulnerable, but only where the weeds are downwind of the fields, so that any disturbed thrips will be blown away from them (Sakimura, personal communication).

Barrier crops

Although rows of barrier plants and cover crops are often successful in decreasing or delaying the infection of seedlings by aphid-borne viruses (Broadbent, 1964), the method is not known to have been tried for controlling thrips-borne viruses. Korolikoff (1910) suggested that a band of rye or oats sown around winter cereals two weeks before they were drilled would decrease infestations on them; the infested trap crop could be removed later. Apart from being unnecessarily costly in time and land, it is unlikely to have been an effective control measure because in autumn most cereal thrips seek hibernation sites, not young cereals.

Insecticidal Control

As thrips are so responsive to cultural methods of control (see p. 239), the need to resort to insecticides usually implies that some aspect of crop or farm management has been neglected, except in dry seasons when plants might succumb to infestations that would normally be harmless. Nevertheless, insecticides are often necessary to protect crops grown by intensive agricultural methods from damage by thrips.

Individual thrips are easily killed by a wide range of contact and systemic insecticides, but populations on crops are sometimes more difficult to control than might be expected for insects susceptible to so many toxicants. The problems are caused largely by the great numbers that infest individual plants, the rapid increase of field populations caused by breeding and airborne immigration, and by the secretive feeding habits of some species. Small numbers of individuals living hidden in the crevices of flowers may require a different method of control from large populations feeding on the exposed surfaces of leaves, and species that spend part of their life-cycle in the soil may be vulnerable at times different from species that remain above ground. These difficulties can be overcome only by using appropriate insecticides, formulations, and methods of application timed to attack the pests at the most vulnerable stage of their life-cycle, when they are most accessible on host plants, and before crops have been seriously injured.

The effectiveness of insecticides against field populations of thrips may be assessed by manufacturers before marketing or, more often, afterwards by agricultural extension organizations. Thus most of the insecticides sold for application to field crops have been tested against thrips and many found useful against several species on a few crops. Many that were useful for a period have been superseded; many were applied at rates and frequencies far in excess of requirements, so much so that improvements in methods of formulation and application have perhaps contributed as much to the effectiveness of present control by chemicals as the greater diversity of insecticides now available.

Insecticides commonly used against thrips, with proven formulations and dosages, are listed in Appendix 5.

Materials and formulations

Because thrips feed on the contents of superficial plant cells, stomach poisons on leaf surfaces are ineffective against them, and sap-translocated insecticides perhaps act more slowly than against insects that feed directly on the phloem vessels. Persistent, contact insecticides are usually unsatisfactory for controlling young and adults protected in leaf sheaths or flowers. They may be applied to control exposed leaf- or fruit-feeders but are safer and more useful for destroying stages in the soil. Non-persistent systemic insecticides are effective against thrips living in inaccessible places on plants and against sluggish species living on the underside of leaves where sprays may not reach. They are ineffective, and it is wasteful to use them, against soil-dwelling forms.

Between 1945 and 1960, persistent chlorinated hydrocarbons, mainly BHC, DDT, aldrin, dieldrin and toxaphene, were generously applied to growing crops as sprays and dusts to kill thrips living in exposed positions on plants. Predictably, they were most successful against leaf-feeders on plants with a fairly open form of growth including coffee (Anon., 1963), cotton (Lincoln and Leigh, 1957), tea (Rao, 1970), young onions (Richardson, 1957), tomatoes (MacGillivray et al., 1950) and flax (Franssen and

Huisman, 1958). Species feeding in crevices were less affected; Riherd (1954) obtained good control of *Chirothrips* spp. feeding in grass heads in a nursery, but heavy doses of insecticide were necessary (4 dressings of 10% DDT plus 2·5% dieldrin dust applied weekly at 30 lb/acre) and would probably have been uneconomical on a field scale.

For several reasons the use of chlorinated hydrocarbons applied as foliar sprays or dusts to growing crops declined. Signs of resistance to these compounds appeared in the mid-1950's probably encouraged by the heavy dosages applied. Of 18 insecticides tested against *T. tabaci* in Texas, in 1955, thrips were resistant to dieldrin, heptachlor and toxaphene, and aldrin and endrin gave no control when used alone (Richardson and Wene, 1956). Some chlorinated hydrocarbons also proved repellent to *Heliothrips haemorrhoidalis*, especially chlordane, and at strong doses DDT and heptachlor, and the toxicity of these chemicals was inversely related to their repellency (Rogoff, 1952). Where other pests occurred on a crop the persistence of the chlorinated hydrocarbons was sometimes a disadvantage because it interfered with the natural control of these pests by other insects (MacPhee, 1953; Tapley, 1960; Evans, 1967). To obtain effective control of thrips on rapidly growing crops, especially when new migrants were arriving daily, the insecticides needed applying every few days to maintain a toxic dose on the young, developing leaves (Richardson, 1953) and at high pressures (30 p.s.i.) to penetrate the crevices or leaf bases.

When formulated as dusts or granules some chlorinated hydrocarbons are more effective used as soil insecticides against immature and adult thrips living in the ground than against thrips on the plant. Granulated aldrin controlled infestations of *Frankliniella fusca* and *F. tritici* on peanuts (Dogger, 1956) and BHC dust incorporated into the upper 10 cm of soil in seed beds before sowing, followed by monthly dustings with DDT, controlled *T. tabaci* and the spread of tomato spotted wilt in tobacco crops (Ivancheva-Gabrovska, 1959)

The contact plant derivatives, nicotine, pyrethrum and derris are satisfactory for short-term experimental control.

Semi-persistent systemic insecticides, mostly organo-phosphorus derivatives or carbamates, have gradually replaced the chlorinated hydrocarbons and can be used to control thrips on most growing crops. Of the early organo-phosphates, parathion applied as sprays or dusts probably gave the best results over a wide spectrum of pests. Sprayed at 400 g a.i./hectare (0·36 lb/ac) it controlled *Thrips angusticeps* in flax better than chlorthion, malathion or diazinon applied at the same rates (Franssen and Huisman, 1958). Routine sprays of parathion were used for many years to control pests in citrus, including thrips (Hodgson, 1970) and were commonly used against species attacking cotton, groundnuts and onions. Damage to grass seed crops by *Chirothrips* spp. was prevented by 2% parathion dust applied at 15–18 lb/acre (16–20 kg/ha), and in two counties of Arizona, 60,000 lb (27,500 kg) was used in one year for this purpose alone (Roney, 1949). Other systemic insecticides sometimes applied as dusts include ethion and carbophenothion used against *Frankliniella vaccinii* on blueberries; applied when the plants are 2–3 cm tall they do less harm to bees than dieldrin (Boulanger and Abdalla, 1966). Trichlorophen is also dusted on young wheat to kill *Haplothrips tritici* (Lyubenov, 1961).

An important advance in the use of systemic insecticides against thrips was their formulation as seed dressings and granules for application to the soil at sowing or planting time to kill thrips attacking the developing seedlings. Demeton and phorate seed dressings protect young cotton plants for 4 to 6 weeks after planting. Applied in this way these compounds are slightly phytotoxic and retard germination, seedling emergence and plant growth, but eventual yield is unaffected (Hanna, 1958; Mistric and Spyhalski, 1959). Onion seeds treated with phorate before sowing produced seedlings immune to attack by *T. tabaci* for about a month (Shazli and Zazou, 1959). Granules of phorate or dimethoate sprinkled on mature onion plants intended for seed production lodged in the angle of the leaves and protected the plants from *Frankliniella occidentalis* for 4–21 days; longer protection, lasting for a month, was obtained when phorate granules were applied in a trench 8–10 cm deep and 10–15 cm to one side of the plant row (Hale and Shorey, 1965). Granules of aldicarb applied in the furrow at planting gave excellent control of *F. fusca* on cotton without the harmful effects on young plants caused by phorate (Pfrimmer, 1966).

Many systemic insecticides, including parathion, carbophenothion, methyl carbophenothion, methyl demeton, ethion, malathion, dicrotophos and carbaryl, are effective when applied as low-volume sprays. Malathion applied at 15-day intervals controls cocoa thrips (Silva, 1964); carbaryl is effective against thrips on cotton and leguminous forage crops: dicrotophos is preferable to parathion and fenitrothion for controlling coffee thrips because it is moderately persistent, killing adult immigrants and adults emerging from pupae in the soil, as well as thrips already on the leaves (Evans, 1967). The effectiveness of these compounds against coffee thrips perhaps results from their ability to penetrate leaf tissue and destroy the eggs, a valuable property which might be exploited more against other leaf-breeding species.

Application

No special modifications to the usual field insecticide dusting and spraying machinery are necessary for applications of materials against thrips.

Contact insecticides are more effective when applied at high pressures to ensure that the toxicant penetrates the crowns, leaf bases and partly opened shoots of plants. Sprays of systemic insecticides are usually applied at low volumes from hand- or powered-machines.

The cost of control can often be decreased by applying insecticides with routine herbicidal sprays; it may be uneconomic to control infestations if a special application is necessary. For example, demeton, applied with 2,4-D herbicide to cereals in spring 2–3 days after adult *Haplothrips tritici* emerged from the soil, prevented damage to young plants (Lyubenov, 1961), but it is doubtful if spraying against the thrips alone would be profitable. Similarly it is rarely worthwhile applying insecticides to control thrips on crops which recover quickly from damage, unless other pests are present.

The timing of insecticidal applications is often critical, and correct doses applied at the wrong time are ineffective. In Iowa, *Thrips tabaci* flourished on cabbages sprayed first with DDT until head formation, then with parathion. This was because the DDT did not kill the thrips; some insects penetrated the loosely developing heart leaves, and after a compact head formed the thrips were protected from the later parathion sprays, despite the systemic properties of this insecticide (Wolfenbarger and Hibbs, 1958).

Attempts to control infections of tomato spotted wilt virus with insecticides applied to growing crops are usually unsatisfactory, probably because there is little spread of thrips within the crop after the first migrants arrive. An infective thrips requires only 5–15 minutes feeding to transmit virus to a healthy plant (Razvyazkina, 1953; Sakimura, 1960, 1963; see p. 55), so unless arriving vectors are killed before they can feed, insecticidal treatment is useless. The timing of spraying is therefore rarely precise enough to prevent infection. For most crops, cultural methods (see p. 241), perhaps with insecticidal treatment of local sources of virus and vectors, remain the most reliable means of controlling thrips-transmitted virus diseases (MacGillivray *et al.*, 1950; Grogan *et al.*, 1955). In tobacco nurseries heavy applications of soil insecticides to seed beds before sowing killed pupae of *Thrips tabaci*; similar treatment of fields at transplanting followed by monthly applications of parathion or DDT to the young plants, restricted infection to 1–2% of the crop (Ivancheva-Gabrovska, 1959). Unfortunately, similar heavy doses of insecticide applied to rapidly growing food crops such as lettuce and tomatoes might contaminate the produce.

Aerial application of insecticides may be profitable in a few densely spaced or tall crops where the wheels of tractors and sprayers would damage the plants. Flax crops sprayed from the air with either parathion, dieldrin or heptachlor in a total volume of 30–40 litres/hectare to kill

T. angusticeps showed an increase in profitability of $146 U.S./hectare (Franssen and Kerssen, 1962). Post and McBride (1966) claimed that barley growers could increase profits by $2.25 U.S./acre ($5.56/ha) by controlling *Limothrips denticornis* with systemic insecticides applied by aircraft.

Systemic insecticides can be applied to larch trees through the trunks to control *Taeniothrips laricivorus* attacking the young shoots. Heavy infestations on 14-year-old trees were controlled by applying methyl demeton on a band of cellulose wadding tied round the scraped bark of the trunks, and covered with a plastic sheet to prevent evaporation. A 50% emulsion concentrate was used, diluted to contain 5% toxicant, and 250 ml of the diluent applied to each tree in July. In September the crowns and leading shoots of treated trees were healthy and showed little damage, whereas leading shoots on untreated trees were scarred and stunted (Martignoni and Zemp, 1956). The treatment was less effective the greater the girth of the tree, but was unaffected by tree height. Vité (1955, 1957) injected methyl demeton into the sap of trees by hammering either a hollow aluminium cartridge or a hollow knife with openings in the cutting edge, into the trunk to a depth of 2–3 cm, and pouring insecticide into the cavity of the tools. The knife method was effective without damaging the cambium; the cartridge caused local injury. Later, Vité (1961) showed that a 10% solution of methyl demeton sprayed directly on to the trunks of young trees before the summer generation of larch thrips hatched, killed the young nymphs when they emerged, but it was ineffective on old trees, because the chemical took too long to reach the vulnerable shoots. The trunk-spraying method might be useful against heavy infestations of thrips attacking some fruit trees, providing the insecticide used had disappeared by harvest. In apple and stone fruit orchards special control of thrips is rarely necessary because spray programmes directed against other sucking insects also usually kill thrips.

Control in glasshouses

Smokes, aerosols and dusts are preferable to sprays in glasshouses, because they are less likely to injure delicate foliage and flowers, or to encourage fungal infections. Smokes of DDT and BHC applied at 10–14-day intervals are cheap and effective except for plants such as cucurbits and carnations, which are susceptible to injury by chlorinated hydrocarbons. Aerosols of malathion or diazinon applied at similar intervals are equally effective and less likely to injure established plants but are slightly more expensive. Thrips on seedlings or very tender young plants, including cucumbers, are best killed by parathion smokes. Aerosols of dichlorvos or naled vaporized from heating pipes or by other means are also satisfactory if used often enough. As with field crops, the regular use of insecticides to control other glasshouse pests also diminishes thrips populations. For example, thrips are rarely a nuisance when diazinon is used to control red spider mite. To protect flowers from infestation, very frequent applications of smokes or sprays of nicotine or pyrethrum dusts may be required, especially where thrips can enter the glasshouse through ventilators. To try to prevent this Karlin *et al.* (1957) fitted screens impregnated with either heptachlor, malathion or dieldrin to ventilators in glasshouses containing roses, and blooms in the dieldrin-protected glasshouse remained uninfested.

In some cucumber houses the predatory mite, *Phytoseiulus persimilis* A.-H., is used to control red spider mite. When thrips occur in these houses a non-acaricidal treatment such as a soil drench of BHC should be used, and organo-phosphorus compounds avoided (Hussey *et al.*, 1969).

Gladiolus thrips can be removed from gladiolus corms before they are stored over winter with DDT or BHC dust, or by fumigation with powdered naphthalene in an airtight container (Speyer, 1951). Lily and gladiolus thrips on lily bulbs can also be killed with BHC dusts, by fumigation with HCN (Kuwayama, 1962) or by immersing the bulbs for an hour in water at 44°C.

Biological Control

Often large proportions of field populations of thrips are killed by predators and parasites (see Chapter 4, and Appendix 3) but there have been few deliberate attempts to introduce or encourage natural enemies to control pest species. Chrysopid larvae are amongst the most voracious predators of thrips but there are no records known of attempts to introduce these or other predators into infested crops. However, there have been several introductions of hymenopterous parasites. *Dasyscapus parvipennis* was introduced from the Gold Coast to the West Indies between 1933 and 1937, where it established in Trinidad, Jamaica and Puerto Rico (Adamson, 1936; Callan, 1943). After releases on cacao plants in one partly open glasshouse, almost all cacao thrips were parasitized, but out of doors only 20–30% of thrips were attacked and the parasites did not destroy enough to prevent their damaging the trees. During the 1960's it became very localized and uncommon in some years.

In the early 1930's *Thripoctenus russelli* was imported into Hawaii from California and *T. brui* from Japan to parasitize *Thrips tabaci* (Sakimura, 1937d). Between 1936 and 1942, *Dasyscapus parvipennis* was sent from Trinidad to the U.S.A., Canada, Hawaii, Bermuda and Grenada to try to control different species of thrips. In the early 1960's it was again sent from Trinidad to California for release in avocado orchards to attack *Heliothrips haemorrhoidalis*. The egg parasite *Megaphragma mymaripennis*

Timb. already attacked 22–75% of the greenhouse thrips in California, but had little effect on numbers. It is not known whether *Dasyscapus* was able to adapt to this different host or was any more successful in controlling it than *Megaphragma* (McMurtry and Johnson, 1963). In the early 1960's outbreaks of *Gynaikothrips* occurred in many parts of Brazil, and Bennett (1965) suggested transferring parasites, including *Tetrastichus thripophonus* Wtstn, from the south to the more recently colonized northern areas where the natural enemies were fewer. None of these attempts at biological control have succeeded alone in producing satisfactory control, but when established they probably help by contributing another mortality factor to the thrips' environment. They may be more successful in glasshouses where pests are protected from the most important short-term mortality factor, the weather, and natural enemies are being sought to control *Thrips tabaci* in European cucumber houses. Moreover, the less frequently crops are sprayed and the less persistent the insecticides used (see p. 246) the greater the opportunity for populations of endemic natural enemies to exert some control on pest populations.

One encouraging success with the biological control of a thrips has been achieved in Hawaii. In 1964 the cuban laurel thrips, *Gynaikothrips ficorum*, appeared there, and rapidly became so abundant and widespread that many of the banyan trees (*Ficus* sp.), used widely for shade, were completely defoliated. About 10,000 individuals of the anthocorid bug, *Montandoniola morguesi*, were introduced from the Philippines to two of the Hawaiian islands, and now this bug controls the thrips satisfactorily. Outbreaks of the pest still occur, but within about two weeks of an increase in their numbers the number of bugs also increases, often to 5 or 6 per leaf, and they quickly destroy the thrips, allowing the damaged leaves to recover. Since release the bug has also spread on its own to three of the other Hawaiian islands (Plate XVIa, b) (Davis, personal communication; DeBach, 1964).

Deep ploughing of cereal stubble probably encourages the destruction of overwintering thrips by the fungus *Beauveria bassiana* (Kurdjumov, 1913; Grivanov, 1939; Lyubenov, 1961). Attempts have also been made to control cacao thrips by spraying leaves with a suspension of spores of *Beauvaria globulifera*. Many of the thrips placed on sprayed cacao plants in a humid atmosphere developed fungal infections, but the experiments were not completed (Nowell, 1917). Control by fungi out of doors would probably be too dependent on weather to be reliable, and in glasshouses the conditions needed to stimulate the resting spores of many species to germinate are still unknown.

12 Beneficial thrips

About a tenth of all known species of thrips live on crop or ornamental plants, and many cause economically important injury (see Chapter 11). However, there is a small proportion of beneficial species, including some that prey on harmful thrips and other arthropod pests, and a few phytophagous species that have been exploited to control weeds. Compared with the chapter on "Thrips as crop pests", this final chapter is short, but nevertheless important because it shows how thrips can contribute to long-term, cheap, and environmentally safe methods of pest and weed control.

EFFECTS OF PREDATORY THRIPS ON POPULATIONS OF INSECT PESTS

The feeding habits of predatory thrips, and the range of arthropods attacked are described in Chapter 4. Their effect on the size of host populations depends largely on the density and reproductive potential of predator and prey. Predatory thrips are rarely as abundant as plant-feeders (see p. 75). For example, the ratio of predatory *Franklinothrips* larvae to cacao thrips larvae recorded in Trinidad over two years was about 1:260 (Callan, 1943). In these proportions *Franklinothrips* had little effect on populations of this pest. Rarely are more than two larvae of *Leptothrips mali* found on the largest apple leaves infested with eriophyid mites, and females probably lay only a few eggs each. The effect of *Leptothrips* on populations of scales is also small compared with the mortality caused by parasites, and it has little impact on field populations of the lepidopterous twig-borer *Anarsia lineatella* Zell. (Bailey, 1940a), even though Jones (1935) found 18% of the borers destroyed by the thrips in a small plot experiment. In many apple orchards in California *L. mali* appears to control populations of the silver mite, *Caliptrimerus baileyi*, by late summer (Bailey, 1940), but other factors may also contribute to their disappearance. Similarly, *Scolothrips sexmaculatus* is rarely abundant and females probably produce only 4–5 eggs each. This thrips requires 17–37 days to develop, so populations build up too slowly to exert any important control of *Tetranychus pacificus* McG., a spider mite which completes its development in only 10–14 days, each female laying 50–100 eggs (Lamiman, 1935). In Canadian peach orchards *Haplothrips faurei* probably has a greater effect on populations of a different spider mite, *Panonychus ulmi*, than either *L. mali* or *S. sexmaculatus* in the U.S.A., largely because it can destroy over 90% of the mites' winter eggs in autumn. Adult thrips emerging from hibernation in early spring also feed on the surviving winter eggs, but the mite populations developing later in the season increase too rapidly to be affected by this relatively trivial spring predation. In spring and summer, populations of *H. faurei* increase much more slowly than those of the mite because the thrips' life-cycle lasts twice as long (MacPhee, 1953), and their eventual density rarely exceeds about 1·6 individuals per branch (Putman, 1942). In unsprayed orchards where populations of mites remain small, thrips are usually too scarce to affect greatly the abundance of mites in summer. However, most larval and adult thrips captured from such orchards were shown by paper chromatography to have eaten mites, and Putman (1942) noticed that the numbers of *H. faurei* increased with rising density of prey.

In a survey spread over six years at nine different sites in Florida, Selhime *et al.* (1963) also showed a positive correlation ($P = 0·05$) between numbers of *Aleurodothrips fasciapennis* and purple scales [*Lepidosaphes beckii* (Newm.)] at four sites, correlation coefficients approaching significance at three more, and at one site the numbers of this thrips and Florida red scale (*Chrysomphalus ficus*) were also correlated. The maximum density of thrips recorded was 0·28 per leaf, but the usual density was only about a tenth of this. However, only small decreases in populations of scales occurred where the thrips were common, even though it bred faster than its prey (Muma, 1955).

In Nova Scotia, *H. faurei* attacks eggs of the eye-spotted bud-moth (*Spilonota ocellana*) and codling moth (*Cydia pomonella*), but this prey is unlikely to be common enough to allow population of thrips to increase rapidly. However, after dense populations of thrips have become established on mites, egg predation by the thrips may help to control these moths. As many as 69% of eggs of *S. ocellana* have been reported destroyed by predators, mostly thrips. Because the thrips must first feed on mites before becoming abundant enough to destroy many moth eggs, they act as a density-independent mortality factor on the moth populations (MacPhee, 1953).

The evidence available suggests that most predatory thrips are unlikely to be key factors (Morris, 1959; Varley and Gradwell, 1960) in limiting populations of mites and insects because they breed too slowly and are less fecund than their hosts. A possible exception is *Sericothrips variabalis*, which can cause rapid decreases in populations of red spider on cotton in Arkansas (Lincoln *et al.*, 1953). No deliberate introductions of predatory thrips to control agricultural pests are known, but they often establish naturally in perennial crops where ploughing and cultivation do not destroy their hibernation sites, and in these habitats probably help to control some injurious insects and mites. In Canada the potential value of *H. faurei* as a predator on orchard pests has been recognized and attempts made to formulate spray programmes harmless to it. This thrips is almost exterminated from orchards by DDT, parathion and BHC, and sulphur is also harmful, but the insecticides, cryolite, nicotine sulphate and summer oil, and the fungicides, Phygon and Tag (10% phenyl mercury acetate) have a less drastic effect; ferbam and copper mixtures have no effect on thrips but control phytophagus mites satisfactorily (MacPhee, 1953). Similarly *Aleurodothrips fasciapennis* is destroyed by parathion and acid lead arsenate but zineb, sulphur and basic copper sulphate do not affect it seriously (Selhime *et al.*, 1963).

THRIPS AND WEED CONTROL

The potential value of thrips for weed control is probably greater than for insect control, because phytophagous species breed and develop so much more quickly than predators, and reach much greater densities. Many species have a limited host range, which is an advantage in weed control programmes, but others can adapt to and rapidly exploit new habitats and food plants. For example, *Taeniothrips orionis* transferred from native plants to introduced crops in Alaska (Washburn, 1958), *Limothrips denticornis* became a common pest on barley in Dakota within 7–10 years of first being recorded there (Post and McBride, 1966), and *Organothrips bianchi* can even change its habits and possibly undergo physiological adaptations to survive in an aquatic habitat (see p. 61; Titschack, 1969). Depending on whether they establish solely on a weed or spread to useful plants, such adaptability may either be an asset or a serious disadvantage when introducing thrips to areas away from their original home, so species considered for weed-control programmes must be thoroughly screened.

The introduction of *Liothrips urichi* to Fiji from Trinidad to control the weed Koster's curse (*Clidemia hirta*) (Plate XVIc) has provided lasting benefits (Simmonds, 1933). The weed was probably introduced into Fiji from British Guiana (Guyana) before 1890, and by 1919 it covered thousands of hectares of agricultural land to the exclusion of nearly all other growth. Its rapid spread was probably caused by birds which feed upon the berries, especially the Indian mynah, assisted by doves and pigeons. Cultivation destroys it, but this was impractical and uneconomic in established plantations and permanent pastures.

Research on *Clidemia* in Trinidad and British Guiana revealed several insects that feed on the plant, especially *Liothrips urichi*, which is specific to *C. hirta* and does not even attack other species in the genus. This thrips was shipped to Fiji in 1930 on living plants; 20,000 adult specimens survived and were removed, and the remaining plant material was fumigated and dumped into the sea to destroy the thrips' natural enemies unavoidably carried with it.

The first releases were made in Fiji in March 1930, but the following season was very dry and establishment and spread were slow during the next ten months. Only a few thrips were found within 300–400 m of the liberation sites. However, after heavy rain followed by a long, moist period, the insect throve and spread rapidly across several kilometres of sea or jungle, so that eighteen months later most of the separate colonies established had joined up. The longest recorded distance travelled from release points after two years was 48 km by land or nearly 32 km over the ocean.

The first effect of this introduction was that over wide areas where *Clidemia* had been growing luxuriantly, the terminal shoots were attacked and growth arrested. This enabled rival plants to outgrow and choke it, producing a further check on its spread. In Trinidad, *Liothrips* prefers unshaded sites, but in Fiji, where populations of the insect are far denser, it has penetrated many kilometres into the jungle, colonizing and retarding growth in clusters of weed in areas only moderately lit.

Of particular interest is the fact that control of the weed has been established over wide areas not by killing the plant directly, but by so inhibiting its growth that it is unable to compete with the surrounding vegetation. In open pastures where some of the taller plant competition is absent, the thrips alone is unable to retard the mature plants enough and they need cutting as well.

In 1953 just over half a million *Liothrips urichi* were released in Hawaii (Weber, 1954) where *Clidemia* flourishes much better in shade than on exposed ground. The thrips is now widespread through the islands and *Clidemia* is still confined largely to the forested areas. Perhaps this is partly because the thrips has prevented its establishment in more exposed places (Davis, personal communication).

A more recent attempt at weed control using a thrips has been the importation of *Amynothrips andersoni* from Argentina into the U.S.A. for release on alligatorweed (*Alternanthera philoxeroides*). This aquatic plant, introduced into Florida about 1894, forms mats of interwoven stems that float on the surface of water, blocking drainage canals, impeding navigation and crowding out fish and other wild-life. By 1970 it covered about 67,000 acres (27,000 ha) extending through most of the southern States. Chemical herbicides provide only temporary relief, and heavy, expensive doses are needed because the nodes of the plant interfere with the translocation of herbicides from the leaves to underwater roots.

In its native South American habitat alligatorweed grows far less luxuriantly than in the U.S.A. One of the five or so insects that retard its growth in South America is the alligatorweed thrips which is widespread in Brazil, Argentina and Paraguay (Vogt, 1960, 1961). It attacks the developing terminal leaves and produces scarred lesions along their margins. These dry and curl to provide a protective crevice in which the thrips continue to feed even after the injured leaves have matured. The general effect is a stunting of leaves and retarded growth. Heavy infestations also spread to the young flower-stalks in the terminal flower-clusters. The thrips prefers alligatorweed growing adjacent to water and in terrestrial situations rather than in truly aquatic stands.

Adults of *Amynothrips andersoni* were received in California from Buenos Aires in 1966 and reared through several generations before release in Florida, South Carolina and Georgia. It remains to be seen whether it alone, or in conjunction with an introduced flea-beetle (*Agasicles hygrophila* Selman and Vogt) and a phycitid stem-borer (*Vogtia malloi* Pastrana), solves the problem of alligatorweed in the U.S.A. So far, the flea-beetle seems to have had the greatest effect, but it is hoped the thrips and the moth will complement its action (Maddox *et al.*, 1971), although the stunting of the growing tips by the thrips may have the undesirable effect of encouraging the weed to branch.

PLATE I. (a) The aerial roots of a banyan tree (arrowed) providing an unusual habitat for several fungus-feeding thrips. (b) The fleshy flower of the cannonball tree (*Couroupita guianensis*), a typically attractive site for flower-feeding thrips, providing them with abundant food and crevices in which to shelter. This flower contained about a hundred *Thrips hawaiiensis*, visible as black specks (photos: T. Lewis).

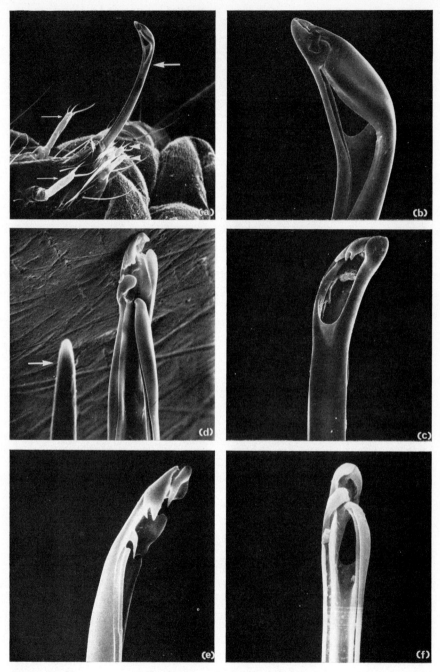

PLATE II. Mouthparts of fungal spore-feeding thrips photographed by scanning electron microscope. (a)–(c) *Dinothrips sumatrensis*, II instar larva: (a) mouthcone from right-hand side showing maxillary palps (small white arrows), maxillary stylets (large

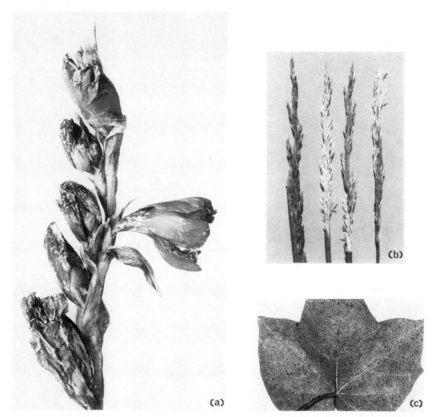

PLATE III. Examples of feeding damage: (a) withered gladiolus flowers caused by *Taeniothrips simplex* (photo: M.A.F.F., U.K.), (b) "silvertop" on inflorescences of bent grass caused by *Anaphothrips obscurus*—normal head on left (photo: J. A. Kamm, 1971), (c) silvering and scarring of cotton leaf caused by *Caliothrips impurus*; the black fragments are faeces and mould (photo: H. Schmutterer, 1969).

white arrow) and labial palps (arrow heads), × 225, (b) detail of stylet apices, "tongue" of right maxillary stylet disengaged from "groove" of left maxillary stylet; feeding aperture on opposite side, × 1500, (c) feeding aperture at apex of maxillary stylets, × 900. (d) *D. sumatrensis*, adult: left maxillary stylet, showing "grooves" and complex structure of apex, and left mandible (arrowed), × 1225. (e) and (f) *Megathrips lativentris*: apices of left and right maxillary stylets, × 2000 (photos: B. R. Pitkin).

PLATE IV. Thrips galls from India. (a) *Schedothrips tumidus* on *Ventilago* sp. (b) *Aneurothrips* sp. on *Cordia* sp. (c) *Liothrips ramakrishnai* on *Schefflera* sp.—with *Liothrips associatus* as an inquiline. (d) *Dixothrips onerosus* on *Terminalia chebula*— with *Corycidothrips inquilinus* as an inquiline (photos: T. N. Ananthakrishnan).

PLATE V. Systemic symptoms of tomato spotted wilt virus on (a) tomato, (b) pineapple, (c) tobacco, (d) pepper (photos: K. Sakimura).

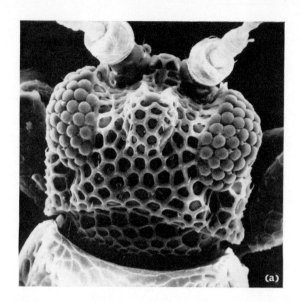

PLATE VI. (a) Head of *Bhattithrips frontalis* photographed by scanning electron microscope to emphasize details of surface sculpture (photo: L. A. Mound, 1970e). (b) Privet
thrips (*Dendrothrips ornatus*) on a leaf from which they leap when disturbed (a Shell
photograph).

PLATE VII. Examples of polymorphism in males of tropical fungus-feeding Tubulifera (mounted specimens). (a), (b) *Kleothrips gigans*, showing large, toothed forelegs (arrowed) in the giant form (a) compared with the relatively slender, scarcely armed legs of the supressed (gynaecoid) form (b). (c) Giant form of *Dinothrips sumatrensis* with armed forelegs and thoracic extensions (arrowed). (d) *Veerabahuthrips bambusae* with grotesquely modified forelegs (photos: T. N. Ananthakrishnan).

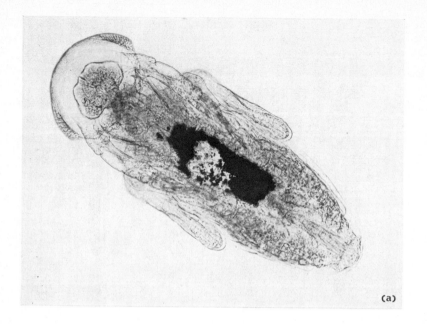

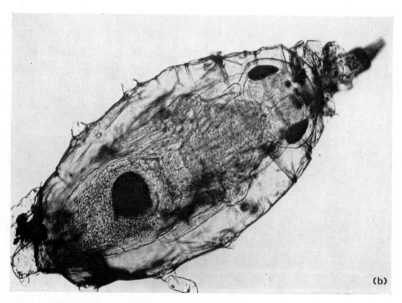

PLATE VIII. (a) Ventral view of a pupa of internal parasite *Tetrastichus gentilei*. (b) A pupa inside the abdomen of a *Gynaikothrips ficorum* II stage larva, facing its hind end (photos: A. Bournier, 1967: INRA)

PLATE IX. (a) A small 23 cm suction trap, a water trap and a sticky trap placed to sample thrips flying at crop level in a wheat field. (b) A series of suction traps (arrowed) placed in a wheat field to measure the aerial density of thrips to leeward of a windbreak of tall trees (photos: T. Lewis).

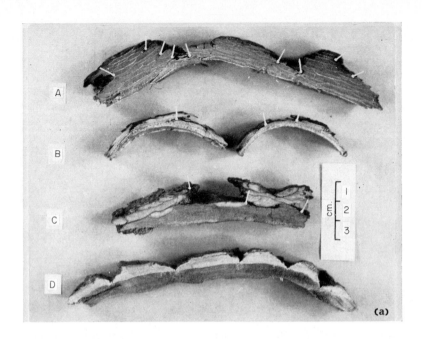

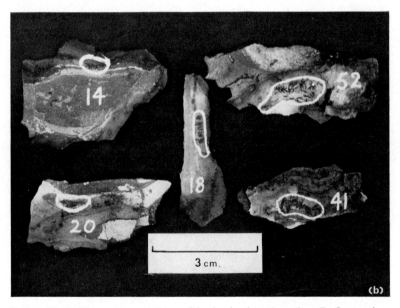

PLATE X. (a) Transverse sections of bark. A: Austrian pine; B: Scots pine; C: horse-chestnut; D: Lombardy poplar. The pointers indicate crevices suitable for overwintering thrips. (b) Aggregations of dead thrips (ringed) found between bark scales on Austrian pine. The figures give the number in each aggregation (photos: T. Lewis).

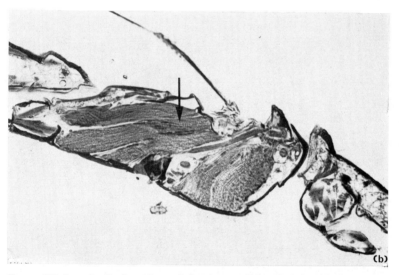

PLATE XI. Longitudinal sections of the thorax of female *Limothrips cerealium:* (a) individual caught in flight in September. (b) Individual collected from hibernation in December, showing that the flight muscles (arrowed) remain intact during the winter in England (photos: T. Lewis).

PLATE XII. (a) Young onion plant killed by *Thrips tabaci* (photo: H. Schmutterer). (b) Healthy privet twig compared with one heavily infested with *Dendrothrips ornatus* (a Shell photograph).

PLATE XIII. (a) Shrivelled tip ("whitehead") of a wheat ear caused by *Limothrips cerealium* (a Shell photograph). (b) Green coconut scarred by an unidentified thrips (a Shell photograph). (c) Dried brown patches on a cashew leaf caused by *Selenothrips rubrocinctus* feeding alongside the veins. (d) Scarred and silvered pea pod caused by *Kakothrips robustus* (photos: T. Lewis). (e) "Leaf-curl" of chillies caused by *Scirtothrips dorsalis* (photo: T. N. Ananthakrishnan, 1971).

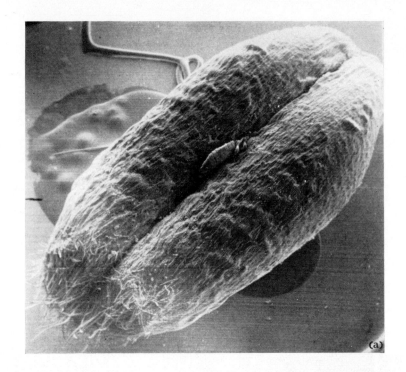

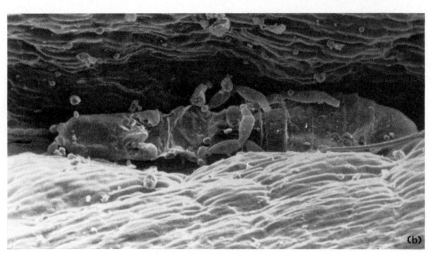

PLATE XIV. Dead corn thrips (*Limothrips cerealium*) trapped between the cheeks of wheat grains, photographed by scanning electron microscope. (a) Head and thorax trapped, (b) close-up of a thrips wholly enfolded by the cheeks (photos: Cambridge Instrument Co., Milling, and N. L. Kent, 1970.)

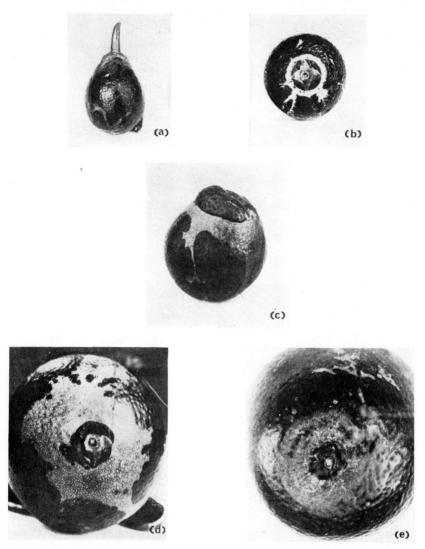

PLATE XV. Damage to oranges caused by *Scirtothrips aurantii* in Rhodesia. (a), (b) Initial scarring round the calyx, (c) deformed fruitlet, (d) spread of scarred tissue on half-grown fruit, (e) mature fruit on which damage becomes less unsightly (photos: W. J. Hall, 1930).

PLATE XVI. (a) Galls on a banyan tree (*Ficus* sp.) caused by *Gynaikothrips ficorum* in Hawaii. (b) A few days later than stage (a) some galled leaves drop, but by this time populations of the introduced predator, *Montandoniola moraguesi*, have usually killed most of the thrips, thus preventing serious damage and allowing many leaves to recover (arrowed) (photos: T. Lewis). (c) Koster's Curse (*Clidemia hirta*), the weed so successfully controlled by *Liothrips urichi* in Fiji (photo: C. J. Davis).

REFERENCES

(Journal titles given in square brackets indicate English translation)

ADAMSON, A. M. (1936). Progress report on the introduction of a parasite of the cacao thrips from the Gold Coast to Trinidad, B.W.I. *Trop. Agric. Trin.* **13**, 62–63.

AHLBERG, O. (1924). Über die Biologie und Entwicklung des *Euchaetothrips ingens* Priesner. *Ent. Tidskr.* **41**, 1–6.

AHLBERG, O. (1926). Svensk insektfauna. 6. Tripsar Thysanoptera. *Ent. For. Rekv.* **29**, 1–62.

AKEHURST, B. C. (1968). *Tobacco.* 551 pp. Longmans, London.

ALLMAN, S. L. (1948). Skin blemish of nectarines caused by Plague thrips. *Agric. Gaz. N.S.W.* **59**, 423–426.

ANANTHAKRISHNAN, T. N. (1951). Variations in the mouth parts of some Thripidae (Thysanoptera–Terebrantia). *Indian J. Ent.* **12**, 237–239.

ANANTHAKRISHNAN, T. N. (1955). Miscellaneous note. 28. Notes on *Thrips palmi* Karny, attacking *Sesamum indicum. J. Bombay nat. Hist. Soc.* **52**, 951–952.

ANANTHAKRISHNAN, T. N. (1956). A survey of our present knowledge of Indian Thysanoptera. *Agra Univ. J. Res.* **4**, 35–40.

ANANTHAKRISHNAN, T. N. (1967). On the multiple effects of œdymerism in *Nesothrips falcatus* Ananthakrishnan. *Curr. Sci. Bangalore* **36**, 610–611.

ANANTHAKRISHNAN, T. N. (1968). Patterns of structural diversity in the males of some phlaeophilous Tubulifera (Thysanoptera). *Annls Soc. ent. Fr. (N.S.)* **4**, 413–418.

ANANTHAKRISHNAN, T. N. (1969a). *Indian Thysanoptera.* CSIR Zool. Monograph No. 1. 171 pp. New Dehli.

ANANTHAKRISHNAN, T. N. (1969b). *Kleothrips gigans* Schmutz—A study in intraspecific diversity. *Bull. Ent. Loyola Coll.* **10**, 130–133.

ANANTHAKRISHNAN, T. N. (1971). Thrips (Thysanoptera) in agriculture, horticulture and forestry—diagnosis, bionomics and control. *J. scient. ind. Res.* **30**, 113–140.

ANANTHAKRISHNAN, T. N. and JAGADISH, A. (1969). Studies on Indian gall thrips (Thysanoptera). *Marcellia* **36**, 7–104.

ANDRE, F. (1941). Two new species of *Chirothrips* Haliday with notes on *Chirothrips frontalis* Williams (Thysanoptera: Thripidae). *Ann. ent. Soc. Am.* **34**, 451–457.

ANDREWARTHA, H. G. (1934). Thrips investigation. 5. On the effect of soil moisture on the viability of the pupal stages of *Thrips imaginis* Bagnall. *J. Coun. scient. ind. Res. Aust.* **7**, 239–244.

ANDREWARTHA, H. G. (1935). Thrips investigation. 7. On the effect of temperature and food on egg production and length of adult life of *Thrips imaginis* Bagnall. *J. Coun. scient. ind. Res. Aust.* **8**, 281–288.

ANDREWARTHA, H. G. and BIRCH, L. C. (1954). *The distribution and abundance of animals.* 782 pp. University of Chicago Press, Chicago.

ANDREWARTHA, H. V. (1936). The influence of temperature on the rate of development of the immature stages of *Thrips imaginis* Bagnall and *Haplothrips victorienensis* Bagnall. *J. Coun. scient. ind. Res. Aust.* **9**, 57–64.

K

ANNAND, P. N. (1926). Thysanoptera and the pollination of flowers. *Am. Nat.* **60**, 177–182.

ANON. (1963). *An atlas of coffee pests and diseases.* 146 pp. Coffee Board, Kenya, Nairobi.

ARNOLD, A. J. (1971). Particle (insect) counter. *Rep. Rothamsted exp. Stn for 1970*, Pt 1. 175–176.

AUCAMP, J. L. and RYKE, P. A. J. (1964). A preliminary report on a grease film extraction method for soil microarthropods. *Pedibiologia* **4**, 77–79.

BAGNALL, R. S. (1915). On a collection of Thysanoptera from the West Indies with descriptions of new genera and species. *J. Linn. Soc. Zool.* **32**, 495–507.

BAGNALL, R. S. (1928). On some Samoan and Tongan Thysanoptera with special reference to *Ficus* gall-causers and their inquilines. *Insects of Samoa* **7**, 55–76.

BAILEY, S. F. (1932). A method employed in rearing thrips. *J. econ. Ent.* **25**, 1194–1196.

BAILEY, S. F. (1933). The biology of the bean thrips. *Hilgardia* **7**, 467–522.

BAILEY, S. F. (1934a). Factors influencing pear thrips abundance and effectiveness of cultural control. *J. econ. Ent.* **27**, 879–884.

BAILEY, S. F. (1934b). A winter study of the onion thrips in California. *Mon. Bull. Calif. Dep. Agric.* **23**, 149–152.

BAILEY, S. F. (1935). Thrips as vectors of plant diseases. *J. econ. Ent.* **28**, 856–863.

BAILEY, S. F. (1936). Thrips attacking Man. *Can. Ent.* **68**, 95–98.

BAILEY, S. F. (1937). The bean thrips. *Mon. Bull. Calif. Dep. Agric.* **609**, 1–36.

BAILEY, S. F. (1938). Thrips of economic importance in California. *Circ. Calif. agric. Exp. Stn* **346**.

BAILEY, S. F. (1939). The six-spotted thrips, *Scolothrips sexmaculatus* Perg. *J. econ. Ent.* **32**, 43–47.

BAILEY, S. F. (1940a). The black hunter, *Leptothrips mali* Fitch. *J. econ. Ent.* **33**, 539–544.

BAILEY, S. F. (1940b). Cocoon-spinning Thysanoptera. *Pan-Pacif. Ent.* **16**, 77–79.

BAILEY, S. F. (1941). Breeding vegetables for resistance to insect attack. *J. econ. Ent.* **34**, 352–358.

BAILEY, S. F. (1944). The pear thrips in California. *Univ. Calif. Bull.* **687**.

BAILEY, S. F. (1948). Grain and grass infesting thrips. *J. econ. Ent.* **41**, 701–705.

BAILEY, S. F. (1957). The thrips of California. Part 1. Suborder Terebrantia. *Bull. Calif. Insect Surv.* **4**, 143–220.

BAILEY, S. F. and CAMPOS, L. E. (1962). The Thysanoptera of Chile. *Publ. Inst. Estud. Entomol. Univ. Chile* **4**, 19–26.

BALD, J. G. and SAMUEL, G. (1931). Investigations on "spotted wilt" of tomatoes. II. *Bull. Commonw. Sci. Ind. Res. Org.* **54**.

BARNES, H. F. (1930). A new thrips-eating gall midge, *Thripsobremia liothrips*. *Bull. ent. Res.* **21**, 331–332.

BARNES, H. F. (1948). *Gall midges of economic importance. (IV) Galls of ornamental plants and shrubs.* 165 pp. Crosby Lockwood and Son, London.

BAWDEN, F. C. (1964). *Plant viruses and virus diseases.* 361 pp. Ronald Press Co., New York.

BECKHAM, C. M. (1969). Colour preference and flight habits of thrips associated with cotton. *J. econ. Ent.* **62**, 591–592.

BEDFORD, H. W. (1921). The cotton thrips (*Heliothrips indicus* Bagn.) in the Sudan. *Bull. Wellcome Trop. Res. Labs, Entom. Sect.* **18**.

BENJAMIN, D. M. (1968a). Insects and mites on tea in Africa and adjacent islands. *E. Afr. agric. For. J.* **33**, 345–357.

BENJAMIN, D. M. (1968b). Economically important insects and mites on tea in East Africa. *E. Afr. agric. For. J.* **34**, 1–16.

BENNETT, F. D. (1965). Observations on the natural enemies of *Gynaikothrips ficorum* Marchal, in Brazil. *Tech. Bull. Commonw. Inst. biol. Control* **5**, 117–125.

BERLAND, L. (1935). Premier résultats de mes recherches en avion sur la faune et la flore atmosphériques. *Annls Soc. ent. Fr.* **104**, 73–96.

BERLÈSE, A. (1905). Apparecchio per raccogliere presto ed in gran numero piccoli artropodi. *Redia* **2**, 85–89.

BERRY, R. E. and TAYLOR, L. R. (1968). High altitude migration of aphids in maritime and continental climates. *J. Anim. Ecol.* **37**, 713–722.

BILLES, D. J. (1941). Pollination of *Theobroma cacao* L. in Trinidad, B.W.I. *Trop. Agric. Trin.* **18**, 151–156.

BOLD, T. J. (1869). Great abundance of thrips. *Entomologists mon. Mag.* **6**, 171.

BONDAR, G. (1924). O "mosaico" provocado pelo Thysanoptero *Euthrips manihoti* sp. n. *Chacaras Quint.* **30**, 216–218.

BONNEMAISON, L. and BOURNIER, A. (1964). Les thrips du lin, *Thrips angusticeps* Uzel et *Thrips linarius* Uzel (Thysanoptères). *Annals Épiphyt.* **15**, 97–169.

BORDEN, A. D. (1915). The mouthparts of the Thysanoptera and the relation of thrips to the non-setting of certain fruits and seeds. *J. econ. Ent.* **8**, 354–360.

BOULANGER, L. W. and ABDALLA, D. A. (1966). Control of blueberry thrips in Maine. *J. econ. Ent.* **59**, 1007–1008.

BOURNIER, A. (1956a). Contribution à l'étude de la parthénogenèse des thysanoptères et de sa cytologie. *Arch. Zool. exp. gén.* **93**, 219–317.

BOURNIER, A. (1956b). Un nouveau cas de parthénogenèse arrhénotoque. *Liothrips oleae* Costa (Thys. Tubulif.). *Arch. Zool exp. gén.* **93**, 135–141.

BOURNIER, A. (1957a). Un deuxieme cas d'ovoviviparité chez les Thysanoptères *Caudothrips buffai* Karny (Tubulifère, Megathripidae). *C.r. hebd. Séanc. Acad. Sci., Paris* **244**, 506–508.

BOURNIER, A. (1957b). *Drepanothrips reuteri* Uzel. Le thrips de la vigne. *Annals Éc. natn. Agric. Montpellier* **30**, 145–157.

BOURNIER, A. (1961). Remarques au sujet du brachyptérisme chez certaines espèces de Thysanoptères. *Bull. Soc. ent. Fr.* **66**, 188–191.

BOURNIER, A. (1966). L'embryogenèse de *Caudothrips buffai* Karny. (Thys. Tubulifera). *Ann. Soc. ent Fr.* (*N.S.*) **2**, 415–435.

BOURNIER, A. (1967). Un intéressant parasite de Thysanoptères: *Tetrastichus gentilei* (Hym. Chalcididae). *Ann. Soc. ent. Fr.* (*N.S.*) **3**, 173–179.

BOURNIER, A. (1970). Principaux types de dégats de Thysanoptères sur les plantes cultivées. *Ann. Zool. Ecol. Anim.* **2**, 237–259.

BOURNIER, A. and BERNAUX, P. (1971). *Haplothrips tritici* Kurdj. et *Limothrips cerealium* Hal. agents de la moucheture des blés durs. *Ann. Zool. Econ. Anim.* **3**, 247–259.

BOURNIER, A. and BLACHE, M. (1956). Thysanoptères nuisibles aux pêchers. *Rev. Zool. Agric. appl.* **1**, 7 pp.

BOURNIER, A. and KOCHBAV, A. (1963). Action stérilisante d'un Thysanoptère sur les fleurs de luzerne. *Bull. Soc. ent. Fr.* **68**, 28–30.

BOURNIER, A. and KOCHBAV, A. (1965). *Odontothrips confusus* Pr. nuisible à la luzerne. *Annls Épiphyt.* **16**, 53–69.

BOYCE, H. R. (1955). Note on injury to tree fruits by *Frankliniella tritici* (Fitch). *Can. Ent.* **87**, 238–239.

BOYCE, K. E. and MILLER, L. A. (1954). Overwintering habitats of the onion thrips, *Thrips tabaci* Lind. (Thysanoptera: Thripidae) in south western Ontario. *Rep. ent. Soc. Ont.* **84**, 82–86.

BROADBENT, L. (1964). Control of plant virus diseases. In Corbett, M. K. and Sisler, H. D. (eds) (1964). *Plant virology.* 527 pp. Univ. Fla. Press, Gainesville. 330–364.

BRUMMER, V. (1939). Beiträge zum Problem der durch Thysanopteren verursachten Schartigkeit des Roggens. *Acta agric. fenn.* **11**, 127–145.

BUCHANAN, D. (1932). A bacterial disease of beans transmitted by *Heliothrips femoralis* Reut. *J. econ. Ent.* **25**, 49–53.

BUFFA, P. (1907). Trentuna specie di Tisanotteri italiani. *Atti Soc. Tosc. Sci. Nat. Mem.* **23**. Pisa. 60 pp.

BUFFA, P. (1911). Studi untorno al ciolo partenogenetico dell' *Heliothrips haemorrhoidalis* Bouché. *Redia* **7**, 71–109.

BUHL, C. (1937). Beitrage zur Kenntnis der Biologie, wirtschaftliche Bedeutung und Bekämpfung von *Kakothrips robustus* Uz. *Z. angew. Ent.* **23**, 65–113.

BULLOCK, J. A. (1962). Fertilisation of the pyrethrum flower. *Pyrethrum Post.* **6**, 39.

BULLOCK, J. A. (1963a). Extraction of Thysanoptera from samples of foliage. *J. econ. Ent.* **56**, 612.

BULLOCK, J. A. (1963b). Thysanoptera associated with pyrethrum and the control of *Thrips tabaci* Lind. *Trop. Agric. Trin.* **40**, 329–335.

BULLOCK, J. A. (1964). A note on the soil fauna of a pyrethrum field. *E. Afr. agric. For. J.* **30**, 8–10.

BULLOCK, J. A. (1965) The assessment of populations of *Thrips nigropilosus* Uzel on pyrethrum. *Ann. appl. Biol.* **55**, 1–12.

BURRILL, A. C. (1918). New economic pests of red clover. *J. econ. Ent.* **11**, 423.

CALDIS, P. D. (1927). Etiology and transmission of endosepsis (internal rot) of the fruit of the fig. *Hilgardia* **2**, 289–328.

CALLAN, E. McC. (1943). Natural enemies of the cacao-thrips. *Bull. ent. Res.* **34**, 313–321.

CALLAN, E. McC. (1947). Technique for rearing thrips in the laboratory. *Nature, Lond.* **160**, 432.

CALLAN, E. McC. (1951). Biology of *Dinurothrips hookeri* (Hood) (Thysanoptera, Thripidae). *Revta Ent., Rio de J.* **22**, 357–362.

CAMERON, A. E. and TREHERNE, R. C. (1918). The pear thrips (*Taeniothrips inconsequens* Uzel) and its control in British Columbia. *Bull. Dep. Agric. Can. ent. Brch* **15**.

CARLSON, E. C. (1964a). Effect of flower thrips on onion seed plants and a study of their control. *J. econ. Ent.* **57**, 735–741.

CARLSON, E. C. (1964b). Damage to safflower plants by thrips and lygus bugs and a study of their control. *J. econ. Ent.* **57**, 140–145.

CARLSON, E. C. (1966). Further studies of damage to safflower plants by thrips and lygus bugs. *J. econ. Ent.* **59**, 138–141.

CARTER, W. (1939). Populations of *Thrips tabaci*, with special reference to virus transmission. *J. Anim. Ecol.* **8**, 261–276.

CEDERHOLM, L. (1963). Ecological studies on Thysanoptera. *Opusc. ent. suppl.* **22**, 215 pp.

CHARLES, V. K. (1941). A preliminary check list of the entomogenous fungi of North America. *Tech. Bull. U.S. Dep. Agric.* **21**, 707–785.

CHITTENDEN, F. J. (ed.) (1951). *The Royal Horticultural Society Dictionary of gardening* **3**, 1712 pp. Clarendon Press, Oxford.

CLARK, L. R., GEIER, P. W., HUGHES, R. D. and MORRIS, R. F. (1967). *The ecology of insect populations in theory and practice.* 232 pp. Methuen, London.

COLLINS, J. (1960). *The Pineapple; botany, cultivation and utilization.* 249 pp. Leonard Hill, London.

COMEGYS, G. R. and SCHMITT, J. B. (1966). A list of the Thysanoptera or Thrips of New Jersey. *Jl N.Y. ent. Soc.* **73**, 195–222.

COON, B. F. and RINICKS, H. B. (1962). Cereal aphid capture in yellow baffle trap. *J. econ. Ent.* **55**, 407–408.

COTT, H. E. (1956). Systematics of the suborder Tubulifera (Thysanoptera) in California. *Univ. Calif. Publs. Ent.* **13**, 216 pp.

COTTERELL, G. S. (1927). A new parasite of *Heliothrips rubrocincta*. *Bull. Dep. Agric. Gold Cst* **12**, 47–48.

CRAWFORD, D. C. (1909). Some Thysanoptera of Mexico and the South. I. *Pomona Coll. J. Ent.* **1**, 109–119.

CRAWFORD, D. C. (1910). Thysanoptera of Mexico and the South. II. *Pomona Coll. J. Ent.* **2**, 153–170.

CRAWFORD, J. C. (1940). *Heliothrips haemorrhoidalis* male. *Proc. ent. Soc. Wash.* **42**, 90.

CRICHTON, M. I. (1960). A study of captures of Trichoptera in a light-trap near Reading, Berkshire. *Trans. R. ent. Soc. Lond.* **112**, 319–344.

CUMBER, R. A. (1958). The insect complex of sown pastures in the North Island. I. The general structure revealed by summer sweep-sampling. *N.Z. Jl agric. Res.* **1**, 719–749.

CUMBER, R. A. (1959). The insect complex of sown pastures in the North Island. VII. The Thysanoptera as revealed by summer sweep-sampling. *N.Z. Jl agric. Res.* **2**, 1123–1130.

CURTIS, J. (1860). *Farm insects*. 528 pp. Blackie and Son, Glasgow.

CURZI, M. (1932). I tripidi come causa della "malattia del pennacchio" del pesco. *Boll. Staz. Patol. veg. Roma (n.s.)* **12**, 238–243.

DAVEY, A. E. and SMITH, R. F. (1933). The epidemiology of fig spoilage. *Hilgardia* **7**, 523–551.

DAVIDSON, J. (1936). The apple thrips (*Thrips imaginis* Bagnall) in South Australia. *J. Dep. Agric. S. Aust.* **39**, 930–939.

DAVIDSON, J. (1944). On the relationship between temperature and rate of development of insects at constant temperatures. *J. Anim. Ecol.* **13**, 26–38.

DAVIDSON, J. and ANDREWARTHA, H. G. (1948a). Annual trends in a natural population of *Thrips imaginis* (Thysanoptera). *J. Anim. Ecol.* **17**, 193–199.

DAVIDSON, J. and ANDREWARTHA, H. G. (1948b). The influence of rainfall, evaporation and atmospheric temperature on fluctuations in size of a natural population of *Thrips imaginis* (Thysanoptera). *J. Anim. Ecol.* **17**, 200–222.

DAVIES, R. G. (1958). Observations on the morphology of the head and mouthparts in the Thysanoptera. *Proc. R. ent. Soc. Lond.* (*A*) **33**, 97–106.

DAVIES, R. G. (1961). The postembryonic development of the female reproductive system in *Limothrips cerealium* Haliday (Thysanoptera: Thripidae). *Proc. zool. Soc. Lond.* **136**, 411–437.

DAY, M. F. and IRZYKIEWICZ, H. (1954). Physiological studies on thrips in relation to transmission of tomato spotted wilt virus. *Aust. J. biol. Sci.* **7**, 274–281.

DEBACH, P. (1964) (ed). *Biological control of insect pests and weeds*. 844 pp. Chapman and Hall, London.

DEL CANIZO, J. (1944). Las agallas foliares de los *Ficus* ornamentales. *Boll. Patol. veg. Ent. agric.* **13**, 323–334.

DEL GUERCIO, G. (1911). Il *Tetrastichus gentilei* Del Guercio nei suoi rapporti col Fleotripide dell' olivo. *Atti Accad. Georgof. Firenze* **8**, 222–227.

DEMPSTER, J. P. (1960). A quantitative study of the predators on the eggs and larvae of broom beetle, *Phytodecta olivacea* Forester, using the precipitin test. *J. Anim. Ecol.* **29**, 149–167.

DEMPSTER, J. P. (1961). A sampler for estimating populations of active insects upon vegetation. *J. Anim. Ecol.* **30**, 425–427.

DERBENEVA, N. N. (1959). I. Life-cycle of *Haplothrips yuccae* Sav. (Thysanoptera: Phlaeothripidae). (in Russian) *Ént. Obozr.* **36**, 64–81.

DE SANTIS, L. (1971). Nota sobra "*Symphyothrips concordiensis*" (Thysanoptera). *Revta Fac. Agron. Univ. nac. La Plata* **47**, 45–48.

DEV, H. N. (1964). Preliminary studies on the biology of the Assam thrips *Scirtothrips dorsalis* Hood on tea. *Indian J. Ent.* **26**, 184–194.

DIETRICK, E. J., SCHLINGER, E. I. and VAN DEN BOSCH, R. (1959). A new method for sampling arthropods using a suction collecting machine and modified Berlèse funnel separator. *J. econ. Ent.* **52**, 1085–1091.

DIGBY, P. S. B. (1955). Factors affecting the temperature excess of insects in sunshine. *J. exp. Biol.* **32**, 279–298.

DOCTERS VAN LEEUWEN, W. M. (1956). The aetiology of some thrips galls found on leaves of Malaysian *Schefflera* species. *Acta bot. neerl.* **5**, 80–89.

DOEKSEN, J. (1941). Bijdrage tot de Vergelijkende morphologie der Thysanoptera. *Meded. LandbHoogesch. Wageningen* **45**, 114 pp.

DOESBURG, P. H. VAN (1964). *Termatophylidae opaca* Carvalho, a predator of thrips (*Selenothrips rubrocinctus* Giard). *Ent. Ber. Amst.* **24**, 248–253.

DOGGER, J. R. (1956). Thrips control on peanuts with granulated insecticides. *J. econ. Ent.* **49**, 632–635.

DOULL, K. M. (1956). Thrips infesting cocksfoot in New Zealand. II. The biology and economic importance of the cocksfoot thrips *Chirothrips manicatus* Haliday. *N.Z. Jl Sci. Technol.* (A) **38**, 56–65.

DOWNES, J. A. (1964). Arctic insects and their environment. *Can. Ent.* **96**, 279–307.

DOWNEY, J. C. (1965). Thrips utilize exudations of Lycaenidae. *Ent. News* **76**, 25–27.

EDWARDS, C. A. (1967). Relationships between weights, volumes and numbers of soil animals. In *Progress in Soil Biology*, GRAF, O. and SATCHELL, J. E. (eds.). 656 pp. Verlag. Friedr. Veiweg & Sohn, Braunschweig. 585–594.

EDWARDS, C. A. and FLETCHER, K. E. (1970). The assessment of terrestrial invertebrate populations. *Methods of study in soil ecology. Proc. UNESCO and IBP Symp. Paris.* 57–66.

EDWARDS, C. A. and HEATH, G. W. (1963). Improved methods of extracting arthropods from soil. *Rep. Rothamsted exp. Stn for 1962* **57**, 158.

EDWARDS, D. K. (1960a). Effects of artificially produced atmospheric electrical fields upon the activity of some adult Diptera. *Can. J. Zool.* **38**, 899–912.

EDWARDS, D. K. (1960b). Effects of experimentally altered unipolar air-ion density upon the amount of activity of the blowfly, *Calliphora vicina* R.D. *Can. J. Zool.* **38**, 1079–1091.

EECKE, VAN R. (1931). Fauna van Nederland, aflevering V. *Tysanoptera* (Q VI). 154 pp. Leiden.

ELBADRY, E. A. and TAWFIK, M. S. F. (1966). Life-cycle of the mite *Adactylidium* sp. (Acarina: Pyemotidae), a predator of thrips eggs in the United Arab Republic. *Ann. ent. Soc. Am.* **59**, 458–461.

ELLIOTT, C. and POOS, F. W. (1940). Seasonal development, insect vectors and host range of bacterial wilt of sweet corn. *J. agric. Res.* **60**, 645–686.

EMDEN, H. F. VAN (1967). An increase in the longevity of adult *Aphis fabae* fed artificially through parafilm membranes on liquids under pressure. *Entomologia exp. appl.* **10**, 166–170.

EVANS, D. E. (1967). Insecticide field trials against the coffee thrips (*Diarthrothrips coffeae* Williams) in Kenya. *Turrialba* **17**, 376–380.

EVANS, J. W. (1932). The bionomics and economic importance of *Thrips imaginis* Bagnall with special reference to the effect on apple production in Australia. *Pamph. Coun. scient. ind. Res. Aust.* **30**.

EVANS, J. W. (1933a). A simple method of collecting thrips and other insects from blossom. *Bull. ent. Res.* **24**, 349–350.

EVANS, J. W. (1933b). Thrips investigation. I. The seasonal fluctuations in numbers of *Thrips imaginis* Bagnall and associated blossom thrips. *J. Coun. scient. ind. Res. Aust.* **6**, 145–159.

EWALD, G. and BURST, R. (1959). Untersuchungen über Lärchenblasenfuss und Lärchenminiermotte. *Ang. Forst-u. Jagdztg.* **130**, 173–181.

FAULKNER, L. R. (1954). Economic thrips of Southern New Mexico. *Bull. New Mex. agric. Expt Stn* **387**.

FENNAH, R. G. (1955). The epidemiology of cacao-thrips on cacao in Trinidad. *Rep. Cacao Res.* **24**, 7–26.

FENNAH, R. G. (1963). Nutritional factors associated with seasonal population increases of cacao thrips, *Selenothrips rubrocinctus* (Giard) (Thysanoptera), on cashew, *Anacardium occidentale. Bull. ent. Res.* **53**, 681–713.

FENNAH, R. G. (1965). The influence of environmental stress on the cacao tree in predetermining the feeding sites of cacao thrips, *Selenothrips rubrocinctus* (Giard), on leaves and pods. *Bull. ent. Res.* **56**, 333–349.

FERNANDO, H. E. and PEIRIS, J. W. L. (1957). Investigations on the chilli leaf curl complex and its control. *Trop. Agric. Mag. Ceylon agric. Soc.* **113**, 305–323.

FERRIÈRE, C. (1958). Un nouveau parasite de thrips en Europe centrale (Hym. Euloph.). *Mitt. schweiz. ent. Ges.* **31**, 320–324.

FLETCHER, R. K. and GAINES, J. C. (1939). The effect of thrips injury on production in cotton. *J. econ. Ent.* **32**, 78–80.

FRANSSEN, C. J. H. (1960). Biology and control of the pea thrips. *Versl. Landbouwk. Onderz. RijkslandbProefstn* **66**, 36 pp.

FRANSSEN, C. J. H. and HUISMAN, P. (1958). The biology and control of *Thrips angusticeps* Uzel. *Versl. Landbouwk. Onderez. RijkslandbProefstn* **64**, 104 pp.

FRANSSEN, C. J. H. and KERSSEN, M. C. (1962). Aerial control of thrips on flax in the Netherlands. *Agric. Aviat.* **4**, 50–54.

FRANSSEN, C. J. H. and MANTEL, W. P. (1961). De Door Tripse Veroorzaakte Beschadigingen in Het Vlasgewas en Het Voorkomen Daarvan. *Tijdschr. Plziekt* **67**, 39–51.

FRANSSEN, C. J. H. and MANTEL, W. P. (1962). Lijst van in Nederland aangetroffen Thysanoptera met beknopte aantekeningen over kun levenswijze en kun betekensis voor onze cultuurgewassen. *Tijdschr. Ent.* **105**, 97–133.

FRANSSEN, C. J. H. and MANTEL, W. P. (1965). Thrips in cereal crops (biology, economic importance and control). 1. Biology. *Versl. Landbouwk. Onderz. RijkslandbProefstn* **662**, 97 pp.

FREEMAN, J. A. (1945). Studies on the distribution of insects by aerial currents. *J. Anim. Ecol.* **14**, 128–154.

FRITZSCHE, R. (1958). Zur Kenntnis der Raubinsekten van *Tetranychus urticae* Koch (Thysanoptera; Heteroptera). *Beitr. Ent.* **8**, 716–724.

FROGGATT, W. W. (1906). Thrips or black fly (Thysanoptera). *Agric. Gaz. N.S.W.* **17**, 1005–1011.

FROGGATT, W. W. (1927). The bubble leaf-gall thrips (*Kladothrips* sp.). In *Forest insects and timber borers.* 107 pp. Sydney.

GAWAAD, A. A. A. and EL SHAZLI, A. Y. (1970). Studies on *Thrips tabaci* Lindeman. VI. New stage in the life cycle. *Z. angew. Ent.* **66**, 395–398.

GHABN, A. A. E. (1931). Zur Biologie und Bekämpfung eines neuen Nelkenschädlings aus der Gruppe Thysanopteren in Aegypten (quoted by PRIESNER, 1964a).

GHABN, A. A. E. (1948). Contribution to the knowledge of the biology of *Thrips tabaci* Lind. in Egypt (Thysanoptera). *Bull. Soc. Fouad I Ent.* **32**, 123–174.

GLICK, P. A. (1939). The distribution of insects, spiders and mites in the air. *Tech. Bull. U.S. Dep. Agric.* **673**.

GLICK, P. A. (1960). Collecting insects by airplane, with special reference to dispersal of the potato leafhopper. *Tech. Bull. U.S. Dep. Agric.* **1222**.

GOUGH, H. C. (1955). *Thrips angusticeps* Uzel attacking peas. *Plant Pathology* **4**, 53.

GRANOVSKY, A. A. and LEVINE, M. N. (1932). The dissemination of cereal rust spores in the greenhouse by terrestrial invertebrates. *Phytopathology* **22**, 9–10.

GRAY, H. and TRELOAR, A. (1933). On the enumeration of insect populations by the method of net collection. *Ecology* **14**, 356–367.

GREGORY, P. H. (1951). Deposition of air-borne *Lycopodium* spores on cylinders. *Ann. appl Biol.* **38**, 357–376.

GRESSITT, J. L., LEECH, R. E. and O'BRIEN, C. W. (1960). Trapping of airborne insects in the Antarctic area. *Pacif. Insects* **2**, 245–250.

GRESSITT, J. L., LEECH, R. E., LEECH, T. S., SEDLACEK, J. and WISE, K. A. J. (1961). Trapping of airborne insects in the Antarctic area (Part 2). *Pacif. Insects* **3**, 559–562.

GRINFEL'D, E. K. (1959). The feeding of thrips (Thysanoptera) on pollen of flowers and the origin of asymmetry in their mouthparts. (in Russian) *Ént. Obozr.* **38**, 798–804.

GRIST, D. H. (1965). *Rice.* 4th Edn. 548 pp. Longmans, London.

GRIVANOV, K. P. (1939). Deep ploughing instead of burning the stubble for the control of the wheat thrips (*Haplothrips tritici* Kurd.). (in Russian) *Social. Grain Fmg* **6**, 182–188 [*Rev. appl. Ent.* (A) **28**, 69–70].

GROBBELAAR, J. H., MORRISON, G. J., BAART, E. E. and MORAN, V. C. (1967). A versatile, highly sensitive activity recorder for insects. *J. Insect Physiol.* **13**, 1843–1848.

GROGAN, R. G., SNYDER, W. C. and BARDIN, R. (1955). Diseases of lettuce. *Circ. Calif. agric. Exp. Stn Ext. Serv.* **448**.

GROMADSKA, M. (1954). Thysanopteran flower-fauna of sand-dune biotope. (in Polish) *Ekol. pol.* **2**, 93–137.

HAARLØV, N. (1947). A new modification of the Tullgren apparatus. *J. Anim. Ecol.* **16**, 115–121.

HAGERUP, O. (1950). Thrips pollination in *Calluna*. *Biol. Meddr* **18**, 1–16.

HALE, R. L. and SHOREY, H. H. (1965). Systemic insecticides for the control of western flower thrips on bulb onions. *J. econ. Ent.* **58**, 793–794.

HALL, W. J. (1930). The South African citrus thrips in Southern Rhodesia. *Rep. Mazoe Citrus exp. Stn* **1**, 5–55.

HAMILTON, W. D. (1967). Extraordinary sex ratios. *Science, N.Y.* **156**, 447–488.

HAMILTON, W. J. (1930). Notes on the food of the American toad. *Copeia* **18**, 45.

HANNA, R. L. (1958). Insecticidal seed treatments for cotton. *J. econ. Ent.* **51**, 160–163.

HANSEN, H. N. (1929). Thrips as carriers of fig-decaying organisms. *Science, N.Y.* **69**, 356–357.

HANSEN, H. N. and DAVEY, A. E. (1932). Transmission of smut and molds in figs. *Phytopathology* **22**, 247–252.

HAQ, A. (1961). A note on the broad bean thrips, *Taeniothrips lefroyi* Bagnall. *Indian J. Ent.* **22**, 132–133.

HARDING, J. A. (1961). Effect of migration, temperature, and precipitation on thrips infestations in South Texas. *J. econ. Ent.* **54**, 77–79.

HARDING, J. P. (1949). The use of probability paper for the graphical analysis of polymodal frequency distributions. *J. mar. biol. Ass. U.K.* **28**, 141–153.

HARDY, A. C. and MILNE, P. S. (1938). Studies in the distribution of insects by aerial currents. Experiments in aerial tow-netting from kites. *J. Anim. Ecol.* **7**, 199–229.

HARDY, G. H. (1916). A new gall-making thrips. *Pap. Proc. R. Soc. Tasm.* **1915**, 102.

HARRIS, H. M., DRAKE, C. J. and TATE, H. D. (1936). Observations on the onion thrips. *Iowa Coll. J. Sci.* **10**, 155–172.

HARRISON, J. O. (1963). Notes on the biology of the banana flower thrips *Frankliniella parvula*, in the Dominican Republic. *Ann. ent. Soc. Am.* **56**, 664–666.

HARTWIG, E. K. (1967). Termitophilous Thysanoptera from South Africa. *J. Ent. Soc. S. Africa* **29**, 44–47.

HEALEY, V. (1964). The density and distribution of two species of *Aptinothrips* (Thysanoptera) in the grass of a woodland. *Entomologist* **97**, 258–263.

HEAN, A. F. (1940). Kromneck in South Africa. Its host range and distribution. *Fmg S. Afr.* **15**, 388–390.

HEIKINHEIMO, O. and RAATIKAINEN, M. (1962). Comparison of suction and netting methods in population investigations concerning the fauna of grass leys and cereal fields, particularly in those concerning the leafhopper, *Calligypona pellucida* (F.). *Valt. Maatalouskoet. Julk.* **191**, 31 pp.

HEMING, B. S. (1970). Postembryonic development of the male reproductive system in *Frankliniella fusca* (Thripidae) and *Haplothrips verbasci* (Phlaeothripidae) (Thysanoptera). *Misc. Publs. ent. Soc. Am.* **7**, 237–272.

HEMING, B. S. (1971). Functional morphology of the thysanopteran pretarsus. *Can. J. Zool.* **49**, 91–108.

HEMING, B. S. (1972). Functional morphology of the pretarsus in larval Thysanoptera. *Can. J. Zool.* **50**, 751–766.

HENDERSON, C. F. and McBURNIE, H. V. (1943). Sampling technique for determining populations of the citrus red mite and its predators. *Circ. Bur. Ent. U.S. Dep. Agric.* **671**.

HENNEBERRY, T. J., TAYLOR, E. A. and SMITH, F. F. (1961). Foliage and soil treatments for the control of flower thrips in outdoor roses. *J. econ. Ent.* **54**, 233–235.

HENRY, A. W. and TU, C. (1928). Natural crossing in flax. *J. Am. Soc. Agron.* **20**, 1183–1192.

HERR, E. A. (1934). The gladiolus thrips, *Taeniothrips gladioli* M. & S. *Bull. Ohio. agric. Exp. Stn* **537**.

HERTING, B. and SIMMONDS, F. J. (1971). *A catalogue of parasites and predators of terrestrial arthropods. Section A, Host or prey/enemy.* 1. *Arachnida to Heteroptera.* 129 pp. Commnw. Agric. Bur., Farnham Royal.

HEWITT, C. G. (1914). Sterility in oats caused by thrips. *J. econ. Ent.* **7**, 211.

HIGHTOWER, B. G. (1958). Laboratory study on the effect of thrips infestation on the height and weight of seedling cotton. *J. econ. Ent.* **51**, 115–116.

HIGHTOWER, B. G. and MARTIN, D. F. (1956). Ecological studies of thrips found on cotton in central Texas. *J. econ. Ent.* **49**, 423–424.

HINDS, W. E. (1902). Contribution to a monograph of the insects of the order Thysanoptera inhabiting North America. *Proc. U.S. natn. Mus.* **26**, 79–247.

HODGSON, C. J. (1970). Pests of citrus and their control. *Pest Artic. News Summ. (PANS)* **16**, 647–666.

HODSON, W. E. H. (1935). The lily thrips (*Liothrips vaneeckei* Priesner). *Bull. ent. Res.* **26**, 469–474.

HOERNER, J. L. (1947). A separator for onion thrips. *J. econ. Ent.* **40**, 755.

HOLTMANN, H. (1962). Untersuchungen zur Biologie der Getreide-Thysanopteren. Teil I. *Z. angew. Ent.* **51**, 1–41.

HOLTMANN, H. (1963). Untersuchungen zur Biologie der Getreide-Thysanopteren. Teil II. *Z. angew. Ent.* **51**, 285–299.

HOLZAPFEL, E. P. and HARRELL, J. C. (1968). Transoceanic dispersal studies of insects. *Pacif. Insects* **10**, 115–153.

HOOD, J. D. (1914). *Frankliniella insularis. Proc. ent. Soc. Wash.* **16**, 29.

HOOD, J. D. (1927). A blood-sucking thrips. *Entomologist* **60**, 201.

HOOD, J. D. (1934). New Thysanoptera from Panama. *Jl N.Y. ent. Soc.* **41**, 407–434.

HOOD, J. D. (1935). Eleven new Thripidae (Thysanoptera) from Panama. *Jl N.Y. ent. Soc.* **43**, 143–170.

HOOD, J. D. (1939). *Organothrips bianchii*, a new Hawaiian thrips from taro. *Proc. Hawaii ent. Soc.* **10**, 423–427.

HOOD, J. D. (1940). *Microscopical whole mounts of insects.* 31 pp. Cornell University. 2nd Mimeographed Edn.

HOOD, J. D. (1950). Thrips that "talk". *Proc. ent. Soc. Wash.* **52**, 42–43.

HOOD, J. D. (1955). Brazilian Thysanoptera VI. *Revta bras. Ent.* **4**, 51–160.

HOPKINS, J. C. F. (1943). Mycological notes. 16. The campaign against the Kromneck virus. *Rhodesia agric. J.* **40**, 47–49.

HORRIDGE, G. A. (1956). The flight of very small insects. *Nature, Lond.* **178**, 1334–1335.

HORSFALL, J. L. and FENTON, F. A. (1922). Onion thrips in Iowa. *Iowa Agric. Expt. Stn Res. Bull.* **205**, 65–66.

HORTON, J. R. (1918). The citrus thrips. *Bull. U.S. Dep. Agric.* **616**.

HOSNY, M. M. (1964). Testing the validity of a simple method for estimating thrips infestation on cotton-seedlings in the field. *Agric. Res. Rev., Cairo* **42**, 136–140.

HOUARD, C. (1924). Les collections cécidologiques du Laboratoire d'Entomologie du Museum d'Histoire Naturelle de Paris: Galles de Nouvelle-Calédonie. *Marcellia* **21**, 59–93.

HOWARD, N. O. (1923). The relation of an undescribed species of *Pestallozzia* to a disease of *Cinnamomum camphora* Nees and Eberm. *Phytopathology* **13**, 47–48.

HOWLETT, T. M. (1914). A trap for thrips. *J. econ. Biol.* **9**, 21–23.

HUFFAKER, C. B. and MESSENGER, P. S. (1964). The concept and significance of natural control. In *Biological control of insect pests and weeds.* (P. Debach, ed.). 844 pp. Chapman and Hall, London. 74–117.

HUGHES, R. D. (1963). Population dynamics of the cabbage aphid, *Brevicoryne brassicae* (L.). *J. Anim. Ecol.* **32**, 393–424.

HUKKINEN, Y. (1936). Investigations on the seed pests of the meadow foxtail grass, *Alopecurus pratensis*. I. *Chirothrips hamatus*. (in Finnish) *Valt. Maatalouskoet. Julk.* **81**, 1–132 [*Rev. appl. Ent. (A)*. **25**, 306].

HUKKINEN, Y. and SYRJANEN, V. (1921). V. Contribution to knowledge of the Thysanoptera of Finland. *Ann. ent. fenn.* **41**, 115–128.

HURD, P. D. and LINDQUIST, E. E. (1958). Analysis of soil invertebrate samples from Barrow, Alaska. *Arctic Inst. N. America, Washington Final Rpt Projects ONR-173 and ONR-193*.

HURST, G. W. (1964). Effects of weather conditions on thrips activity. *Agric. Meteorol.* **1**, 130–141.

HUSSEY, N. W., READ, W. H. and HESLING, J. J. (1969). *The pests of protected cultivation.* 404 pp. Arnold, London.

ILERI, M. (1947). The bionomics, mode of infection and control of the tobacco thrips. (in Turkish) *Zir. Derg.* **8**, 19–57 [*Rev. appl. Ent. (A)*. **38**, 142–143].

ILLINGWORTH, J. F. (1931). Yellow spot of pineapples in Hawaii. *Phytopathology* **21**, 865–880.

IVANCHEVA-GABROVSKA, T. (1959). Tomato spotted wilt (*Lycopersicum* virus 3 Smith) on tobacco in Bulgaria. (in Bulgarian) *Nauchi Trud. nauchno-izsled. Inst. Zasht. Rast.* **2**, 5–32 [*Rev. appl. Ent. (A)*. **49**, 339–340].

JABLONOWSKI, J. (1926). Zur Klärung der Thripschädenfrage. *Z. angew. Ent.* **12**, 223–242.

JACCARD, P. (1912). The distribution of the flora in the alpine zone. *New Phytol.* **11**, 37–50.

JACOT-GUILLARMOD, C. F. (1970). Catalogue of the Thysanoptera of the world (Pt 1). *Ann. Cape Prov. Mus. (Nat. Hist.)* **7**, 1–216.

JACOT-GUILLARMOD, C. F. (1971). Catalogue of the Thysanoptera of the world (Pt 2). *Ann. Cape Prov. Mus.* (*Nat. Hist.*) **7**, 217–515.

JAGOTA, U. K. (1961). The life-history of *Microcephalothrips abdominalis* (Crawford). *Bull. Ent. Loyola Coll.* **2**, 12–20.

JARY, S. G. (1934). A note on injury caused by two species of Thysanoptera. *J. S-east agric. Coll. Wye.* **34**, 63–64.

JENSEN, M. (1954). *Shelter effects.* 264 pp. Danish Technical Press, Copenhagen.

JOHANSSON, E. (1946). Studier och försök rörande de pa gras och sadesslag levande tripsarnas biologi och skadegörelse. II. Tripsarnas frekvers och spridning i jämförelse med andra sugande insekters samt deras fröskadegörande betydelse. *Meddn St. VaxtskAmst* **46**, 1–59.

JOHN, O. (1923). Fakultative Viviparitat bei Thysanopteren. *Ent. Mitt.* **12**, 227–232.

JOHNSON, C. G. (1950). A suction trap for small airborne insects which automatically segregates the catch into successive hourly samples. *Ann. appl. Biol.* **37**, 80–91.

JOHNSON, C. G. (1952). The changing numbers of *Aphis fabae* Scop., flying at crop level, in relation to current weather and to the population on the crop. *Ann. appl. Biol.* **39**, 525–547.

JOHNSON, C. G. (1957). The distribution of insects in the air and the empirical relation of density to height. *J. Anim. Ecol.* **26**, 479–494.

JOHNSON, C. G. (1969). *Migration and dispersal of insects by flight.* 763 pp. Methuen, London.

JOHNSON, C. G., SOUTHWOOD, T. R. E. and ENTWISTLE, H. M. (1957). A new method of extracting arthropods and molluscs from grassland and herbage with a suction apparatus. *Bull. ent. Res.* **48**, 211–218.

JOHNSON, C. G., TAYLOR, L. R. and SOUTHWOOD, T. R. E. (1962). High altitude migration of *Oscinella frit* L. (Diptera: Chloropidae). *J. Anim. Ecol.* **31**, 373–383.

JOHNSON, E. C. (1911). Floret sterility of wheats in the Southwest. *Phytopathology* **1**, 18–27.

JOHNSTON, H. B. (1925). *Heliothrips indicus* (Bagnall) injurious to man in the Sudan. *Entomologists mon, Mag.* **61**, 132–133.

JONES, H. A., BAILEY, S. F. and EMSWELLER, S. L. (1934). Thrips resistance in the onion. *Hilgardia* **8**, 215–232.

JONES, L. S. (1935). Observations of the habits and seasonal life history of *Anarsia lineatella* in California. *J. econ. Ent.* **28**, 1002–1011.

JONES, T. (1954). The external morphology of *Chirothrips hamatus* (Trybom) (Thysanoptera). *Trans. R. ent. Soc. Lond.* **105**, 163–187.

JORDAN, K. (1888). Anatomie und Biologie der Physopoda. *Z. wiss. Zool.* **47**, 541–620.

KAMM, J. A. (1971). Silvertop of bluegrass and bentgrass produced by *Anaphothrips obscurus. J. econ. Ent.* **64**, 1385–1387.

KAMM, J. A. (1972). Environmental influence on reproduction, diapause and morph determination of *Anaphothrips obscurus* (Thysanoptera: Thripidae). *Environ. Ent.* **1**, 16–19.

KANERVO, V. (1950). *Frankliniella tenuicornis* Uzel (Thysanoptera) als intrafloraler Schädling der Gerste. *Proc. VIII Int. Congr. Ent.* 647–653.

KARLIN, E. J., NAEGELE, J. A. and JOHNSON, G. V. (1957). Screening greenhouses with insecticide-impregnated cloth for thrips control. *J. econ. Ent.* **50**, 55–58.

KARNY, H. H. (1911). Über Thripsgallen und Gallenthrips. *Zentbl. Bakt. ParasitKde abt. II* **30**, 556–572.

KARNY, H. H. (1922). A remarkable new gall-thrips from Australia. *Proc. Linn. Soc. N.S.W.* **47**, 266–274.

KARNY, H. H. and DOCTERS VAN LEEUWEN, W. J. (1914). Beitrage zur Kenntnis des

Gallen von Java. II. Über die javanischen Thysanopterocecidien und deren Bewohner. *Z. wiss. InsektBiol.* **10,** 201–208, 288–296, 355–369.

KELER, S. (1936). A catalogue of the Polish Thysanoptera. (in Polish) *Pr. Wydz. Chorób Szkod. Rósl. pánst. Inst. nauk Gospod. wiejsk. Bydgoszczy* **15,** 81–149.

KELLY, E. O. G. (1915). A new wheat thrips. *J. agric. Res.* **4,** 219–223.

KELLY, R. and MAYNE, R. J. B. (1934). *The Australian Thrips.* 81 pp. Australian Medical Publishing Co., Sydney.

KEMPSON, D., LLOYD, M. and GHELARDI, R. (1963). A new extractor for woodland litter. *Pedobiologia* **3,** 1–21.

KENNEDY, J. S. and BOOTH, C. O. (1956). Reflex and instinct. *Discovery, Lond.* **17,** 311–312.

KENNEDY, J. S. and BOOTH, C. O. (1963a). Free flight of aphids in the laboratory. *J. exp. Biol.* **40,** 67–85.

KENNEDY, J. S. and BOOTH, C. O. (1963b). Co-ordination of successive activities in an aphid. The effect of flight on the settling responses. *J. exp. Biol.* **40,** 351–369.

KENNEDY, J. S., BOOTH, C. O. and KERSHAW, W. J. S. (1961). Host finding by aphids in the field. III. Visual attraction. *Ann. appl. Biol.* **49,** 1–21.

KENT, N. L. (1970). Thrips in home-grown wheat. *Milling* **152,** 22–26.

KIRKPATRICK, T. W. (1957). *Insect life in the tropics.* 310 pp. Longmans, London.

KITTEL, R. (1958). Untersuchungen über den Einfluss atmosphärischelectrischer Felder auf die Schwarmbildung bei Thysanopteren. *Verh. dt. zooh. Ges.* **67,** 177–181.

KLEE, O. (1958). Über die Biologie und Saugtäkigkeit des Thysanopteren *Taeniothrips laricivorus* Krat. und das Larchenwipfelsterben. *Waldhygiene* **2,** 166–181.

KLOCKE, F. (1926). Bieträge zur Anatomie und Histologie der Thysanopteren. *Z. wiss. Zool.* **128,** 1–36.

KLOFT, W. and EHRHARDT, P. (1959). Zur Frage der Speichelinjektion beim Saugakt von *Thrips tabaci* Lind. (Thysanoptera, Terebrantia). *Naturwissenschaften* **46,** 586–587.

KNECHTEL, W. K. (1923). Thysanoptere din Romania. (in Rumanian) *Bul. Minist. Agric. Ind.* **2–4,** 235 pp.

KNECHTEL, W. K. (1956). Oekologisch-phaenologische Forschungen über Thysanopteren. *Proc. XI Int. Congr. Ent.* **2,** 689–695.

KNECHTEL, W. K. (1960). Phaenologische Forschungen über Thysanopteren (Die Thysanopteren der Dobrogeasteppe). *Proc. XI Int. Congr. Ent.* **1,** 489–493.

KNOWLTON, G. F. (1938). Lizards in insect control. *Ohio J. Sci.* **38,** 235–238.

KOLOBOVA, A. N. (1926). *Stenothrips graminum* Uzel. (in Russian) *Trudy poltav. sel-khoz. opyt. Sta.* **49,** 1–26 [*Rev. appl. Ent.* (*A*). **14,** 606–607].

KONTKANEN, P. (1950). Quantitative and seasonal studies on the leafhopper fauna of the field stratum in open areas in North Karelia. *Ann. zool. Soc. zool.—bot. fenn. Vanamo* **13,** 1–91.

KOPPA, P. (1967). The composition of the thrips species in cereals in Finland. *Ann. agric. fenn.* **6,** 30–45.

KOPPA, P. (1969a). The sex index of some species of thrips living on cereal plants. *Ann. ent. fenn.* **35,** 65–72.

KOPPA, P. (1969b). Studies on the hibernation of certain species of thrips living on cereal plants. *Ann. agric. fenn.* **8,** 1–8.

KOPPA, P. (1970). Studies on the thrips (Thysanoptera) species most commonly occurring on cereals in Finland. *Ann. agric. fenn.* **9,** 191–265.

KOROLIKOFF, S. M. (1910). Tripsi jivoustchi na nacikhstakakh. *Izr. mosk. sel'.-khoz. Inst.* **16,** 192–204.

KÖRTING, A. (1930). Beitrag zur Kenntnis der Lebensgewohnheiten und der phytopathogen Bedeutung einiger an Getreide lebender Thysanopteren. *Z. angew. Ent.* **16,** 451–512.

KÖRTING, A. (1931). Beobachtung über die Fritfliege und einiger Getreidethysanopteren. *Z. angew. Ent.* **18**, 154–160.

KRATOCHVIL, J. and FARSKY, O. (1942). Das Absterben der diesjährigen terminalen Larchentriebe. *Z. angew. Ent.* **29**, 177–218.

KREUTZBERG, V. E. (1940). A new virus disease of *Pistacia vera* L. *Dokl. Acad. Sci. U.S.S.R.* **27**, 614–617.

KROMBEIN, K. V. (1958). Miscellaneous prey records of solitary wasps. III (Hymenoptera, Aculeata). *Proc. biol. Soc. Wash.* **71**, 21–26.

KUENEN, D. J. (1958). Some sources of misunderstanding in the theories of regulation of animal numbers. *Archs. néerl. Zool.* **13**, (*Suppl.* 1), 335–341.

KURDJUMOV, N. V. (1913). The more important insects injurious to grain crops in Middle and South Russia. (in Russian) *Studies Poltava agric. Expt. stn* **17**, *Dept. agric. Ent.* **6**, 119 pp. [*Rev. appl. Ent.* (*A*). **2**, 170–173].

KUROSAWA, M. (1968). Thysanoptera of Japan. *Insecta Matsum. Suppl.* **4**, 95 pp.

KUTTER, H. (1936). Über einen neuen Endoparasiten (*Thripoctenus:* Chalcididae) des Erbsenblasenfusses (*Kakothrips robustus* Uzel) seine Lebensweise und Entwicklung. *Mitt. schweiz. ent. Ges.* **16**, 640–652.

KUTTER, H. and WINTERHALTER, W. (1933). Untersuchungen über die Erbsenschädlinge im St. Gallischen Rheintale während der Jahre 1931 und 1932. *Landw. Jbr.* **47**, 275–338.

KUWAYAMA, S. (1962). On the lily bulb thrips *Liothrips vaneeckei* Pr. in Japan. (in Japanese) *Res. Bull. Hokkaido natn. agric. Exp. Stn* **78**, 90–98 [*Rev. appl. Ent.* (*A*). **52**, 90].

LACK, D. (1966). *Population studies of birds.* 341 pp. Clarendon Press, Oxford.

LALL, B. S. and SINGH, L. M. (1968). Biology and control of the onion thrips in India. *J. econ. Ent.* **61**, 676–679.

LAMB, H. H. (1965). Britain's changing climate. In *The biological significance of climatic changes in Britain* (Johnson, C. G. and Smith, L. P., eds) 222 pp. Academic Press, London and New York. 2–34.

LAMIMAN, J. F. (1935). The Pacific mite, *Tetranychus pacificus* McG. in California. *J. econ. Ent.* **28**, 900–903.

LAUGHLIN, R. (1970). The gum tree thrips, *Isoneurothrips australis* Bagnall. The population of a single tree. *Entomologia exp. appl.* **13**, 247–259.

LE PELLEY, R. H. (1942). A method of sampling thrips populations. *Bull. ent. Res.* **33**, 147–148.

LEWIS, T. (1955). Two interesting British records of Thysanoptera. *J. Soc. Br. Ent.* **5**, 110–113.

LEWIS, T. (1958). The distribution and dispersal of thysanopteran populations on Gramineae. Ph.D. Thesis, London University.

LEWIS, T. (1959a). The annual cycle of *Limothrips cerealium* Haliday (Thysanoptera) and its distribution in a wheat field. *Entomologia exp. appl.* **2**, 187–203.

LEWIS, T. (1959b). A comparison of water traps, cylindrical sticky traps and suction traps for sampling thysanopteran populations at different levels. *Entomologia exp. appl.* **2**, 204–215.

LEWIS, T. (1960). A method for collecting Thysanoptera from Gramineae. *Entomologist* **93**, 27–28.

LEWIS, T. (1961). Records of Thysanoptera at Silwood Park, with notes on their biology. *Proc. R. ent. Soc. Lond.* (*A*) **36**, 89–95.

LEWIS, T. (1962). The effects of temperature and relative humidity on mortality in *Limothrips cerealium* Haliday (Thysanoptera) overwintering in bark. *Ann. appl. Biol.* **50**, 313–326.

LEWIS, T. (1963). The effect of weather on emergence and take off of overwintering *Limothrips cerealium* Haliday (Thysanoptera). *Ann. appl. Biol.* **53**, 489–502.

LEWIS, T. (1964). The weather and mass flights of Thysanoptera. *Ann. appl. Biol.* **53,** 165–170.

LEWIS, T. (1965a). The species, aerial density and sexual maturity of Thysanoptera caught in mass flights. *Ann. appl. Biol.* **55,** 219–225.

LEWIS, T. (1965b). The effects of an artificial windbreak on the aerial distribution of flying insects. *Ann. appl. Biol.* **55,** 503–517.

LEWIS, T. (1969). The distribution of flying insects near a low hedgerow. *J. appl. Ecol.* **6,** 443–452.

LEWIS, T. (1970). Patterns of distribution of insects near a windbreak of tall trees. *Ann. appl. Biol.* **65,** 213–220.

LEWIS, T. and DIBLEY, G. C. (1970). Air movement near windbreaks and a hypothesis of the mechanism of the accumulation of airborne insects. *Ann. appl. Biol.* **66,** 477–484.

LEWIS, T. and NAVAS, E. (1962). Thysanopteran populations overwintering in hedge bottoms, grass litter and bark. *Ann. appl. Biol.* **50,** 299–311.

LEWIS, T. and SIDDORN, J. W. (1972). Measurement of the physical environment. In *Aphid Technology.* (van Emden, H. F., ed.), 344 pp. Academic Press, London and New York. 235–273.

LEWIS, T. and STEPHENSON, J. W. (1966). Permeability of artificial windbreaks and the distribution of flying insects. *Ann. appl. Biol.* **58,** 355–363.

LEWIS, T. and TAYLOR, L. R. (1964). Diurnal periodicity of flight by insects. *Trans. R. ent. Soc. Lond.* **116,** 393–479.

LEWIS, T. and TAYLOR, L. R. (1967). *Introduction to experimental ecology.* 401 pp. Academic Press, London and New York.

LINCOLN, C. and LEIGH, T. F. (1957). Timing insecticide applications for cotton insect control. *Bull. Ark. agric. Exp. Stn* **588.**

LINCOLN, C., WILLIAMS, F. J. and BARNES, G. (1953). Importance of thrips in red spider control. *J. econ. Ent.* **46,** 899–900.

LINFORD, M. B. (1943). Influence of plant populations upon incidence of pineapple yellow spot. *Phytopathology* **33,** 408–410.

LOAN, C. and HOLDAWAY, F. G. (1955). Biology of the red clover thrips, *Haplothrips niger* (Osborn). *Can. Ent.* **87,** 210–219.

LYSAGHT, A. M. (1936). A note on the adult female of *Anguillulina aptini* (Sharga), a nematode parasitizing *Aptinothrips rufus* Gmelin. *Parasitology* **28,** 290–292.

LYSAGHT, A. M. (1937). An ecological study of a thrips (*Aptinothrips rufus*) and its nematode parasite (*Anguillulina aptini*). *J. Anim. Ecol.* **6,** 169–192.

LYUBENOV, YA. (1961). A contribution to the bionomics of the wheat thrips (*Haplothrips tritici* Kurd.) in Bulgaria and possibilities of reducing the injury done by it. (in Bulgarian) *Izr. bsent. nauchnoizsled. Inst. Zasht. Rast. Sof.* **1,** 205–238 [*Rev. appl. Ent.* (*A*). **50,** 348].

MACFADYEN, A. (1961). Improved funnel-type extractors for soil arthropods. *J. Anim. Ecol.* **30,** 171–184.

MACFADYEN, A. (1962). Control of humidity in three funnel-type extractors for soil arthropods. In *Progress in soil zoology.* (Murphy, P. W. ed.) 398 pp. Butterworths, London. 158–168.

MACGILL, E. I. (1927). The biology of Thysanoptera with reference to the cotton plant. 2. The relation between temperature and life-cycle in a saturated atmosphere. *Ann. appl. Biol.* **14,** 501–512.

MACGILL, E. I. (1937). The biology of Thysanoptera with reference to the cotton plant. 8. The relation between variations in temperature and the life-cycle. *Ann. appl. Biol.* **24,** 95–109.

MACGILL, E. I. (1939). A gamasid mite (*Typhlodromus thripsi* n. sp.), a predator of *Thrips tabaci* Lind. *Ann. appl. Biol.* **26,** 309–317.

MACGILLIVRAY, J. H., MICHELBACHER, A. E. and SCOTT, C. E. (1950). Tomato production in California. *Circ. Calif. agric. Exp. Stn Ext. Serv.* **167**.

MACKIE, W. W. and SMITH, F. L. (1935). Evidence of field hybridization in beans. *J. Am. Soc. Agron.* **27**, 903–909.

MACPHEE, A. W. (1953). The influence of spray programs on the fauna of apple orchards in Nova Scotia. V. The predacious thrips *Haplothrips faurei* Hood. *Can. Ent.* **85**, 33–40.

MADDOCK, D. R. (1949). Thrips infesting tomatoes and their occurrence on hosts. *J. econ. Ent.* **42**, 146.

MADDOX, D. M., ANDRES, L. A., HENNESSEY, R. D., BLACKBURN, R. D. and SPENCER, N. R. (1971). Use of insects to control alligatorweed, an invader of aquatic ecosystems in the United States. *Bioscience* **21**, 985–991.

MAKSYMOV, J. K. (1965). Die Überwinterung des Lärchenblasenfusses *Taeniothrips laricivorus* Kratochvil und Farsky. *Mitt. schweiz. Anst. forstl. VersWes.* **41**, 1–17.

MALLMANN, R. J. DE (1959). Observations sur le thigmotactisme de *Limothrips cerealium* (Thysan.). *Bull. Soc. ent. Fr.* **64**, 151–157.

MALLMANN, R. J. DE (1964). Contribution a l'étude de la thigmotaxie chez les insectes. *Ann. Soc. ent. Fr.* **133**, 1–141.

MALTBAEK, J. (1932). Frynsevinger eller blaere fodder (Thysanoptera). *Danmarks Fauna* **37**, 146 pp. Kobenhaven.

MALTBAEK, J. (1938). Thysanoptera. In *The Zoology of Iceland* **3**, 6 pp. Lerin and Munksgaard, Copenhagen.

MANI, M. S. (1964). Ecology of plant galls. *Monographiae biol.* **12**, 434 pp.

MARSHAM, T. (1796). Observations on the insects that infested the corn in the year 1795. In a letter to the Rev. Samuel Goodenough, LL.D., F.R.S., Tr.L.S. *Trans. Linn. Soc. Lond.* **3**, 242–251.

MARTIGNONI, M. E. and ZEMP, H. (1956). Versuch zur Bekämpfung des Lärchenblasenfusses *Taeniothrips laricivorus* Kratochvil and Farsky (Thripidae, Thysanoptera) mit einem systemischen Insektizid. *Mitt. schweiz. Zent. Anst. forstl. Versuchsw.* **32**, 1–22.

MAW, M. G. (1962). Some biological effects of atmospheric electricity. *Proc. ent. Soc. Ont.* **92**, 33–37.

MAW, M. G. (1963a). Behaviour of insects in electrostatic fields. *Proc. ent. Soc. Manitoba* **18**, 30–36.

MAW, M. G. (1963b). Physics in entomology: sound and electricity in insect behaviour. *Proc. N. cent. Brch Am. Ass. econ. Ent.* **18**, 6–10.

McGREGOR, E. A. (1926). A device for determining the relative degree of insect occurrence. *Pan. Pacif. Ent.* **3**, 29–33.

McMURTRY, J. A. (1961). Current research on biological control of avocado insect and mite pests. *Yb. Calif. Avocado Soc.* **45**, 104–106.

McMURTRY, J. A. and JOHNSON, H. G. (1963). Progress report on the introductions of a thrips parasite from the West Indies. *Yb. Calif. Avocado Soc.* **47**, 48–51.

MELIS, A. (1935). Tisanotter italiani. Studio anatomo-morphologico e biologico de Liothripidae dell'olivo (*Liothrips oleae* Costa). *Redia* **21**, 1–188.

MELIS, A. (1959). I. Tisanotteri italini. *Redia* **44** (append), 1–184.

MELIS, A. (1960). II. Tisanotteri italiani. *Redia* **45** (append.), 185–329.

MELIS, A. (1961). III. Tisanotteri italiani. *Redia* **46** (append.), 331–530.

MICKOLEIT, E. (1963). Untersuchungen zur Kopfmorphologie der Thysanopteren. *Zool. Jb. (Anat.)* **81**, 101–150.

MISTRIC, W. J. and SPYHALSKI, E. J. (1959). Response of cotton and cotton pests to thimet seed-treatment. *J. econ. Ent.* **52**, 807–811.

MITTLER, T. E. and DADD, R. H. (1962). Artificial feeding and rearing of the aphid,

Myzus persicae (Sulzer) on a completely defined synthetic diet. *Nature, Lond.* **195**, 404.

MOERICKE, V. (1950). Über das Farbsehen der Pfirschblattlaus (*Myzodes persicae* Sulz.). *Z. Tierpsychol.* **7**, 265–274.

MOFFITT, H. R. (1964). A color preference of the western flower thrips, *Frankliniella occidentalis. J. econ. Ent.* **57**, 604–605.

MOLL, E. O. (1963). Life history of the small-mouthed salamander, *Ambystoma texanum* (Malthes) in Illinois. M.S. Thesis, Univ. of Urbana.

MOLLER-RACKE, I. (1952). Farbensinn und Farbenblindheit bei Insekten. *Zool. Jb. Zool. u. Physiol.* **63**, 237–274.

MOORE, E. S. and ANDERSSEN, E. E. (1939). Notes on plant virus diseases in S. Africa. I. The Kromnek disease of tobacco and tomato. *Scient. Bull. Dep. Agric. S. Afr.* **182**.

MORGAN, A. C. (1925). A new genus, a new subgenus and seven new species of Thysanoptera from Porto Rico. *Florida Ent.* **9**, 1–9.

MORGAN, A. C. and CRUMB, S. E. (1928). Notes on the chemotactic responses of certain insects. *J. econ. Ent.* **21**, 913–920.

MORISON, G. D. (1943). Notes on Thysanoptera found on flax (*Linum usitatissimum* L.) in the British Isles. *Ann. appl. Biol.* **30**, 251–259.

MORISON, G. D. (1947–1949). Thysanoptera of the London Area. *Lond. Nat.* (Suppl.) **59**, 1–36; 1948, *Ibid.* 37–75; 1949, *Ibid.* 77–131.

MORISON, G. D. (1957). A review of British glasshouse Thysanoptera. *Trans. R. ent. Soc. Lond.* **109**, 467–534.

MORISON, G. D. (1958). The Thysanoptera of Kew Gardens. *Kew Bull.* **2**, 295–301.

MORRIS, R. F. (1959). Single-factor analysis in population dynamics. *Ecology* **40**, 580–588.

MORRISON, L. (1961). Thrips infestation of cocksfoot seed crops. *N.Z. Jl agric. Res.* **4**, 246–252.

MOULTON, D. (1927). New gall-forming Thysanoptera of Australia. *Proc. Linn. Soc. N.S.W.* **52**, 153–160.

MOULTON, D. (1932). The Thysanoptera of South America. *Revta Ent., Rio de J.* **2**, 464–465.

MOULTON, D. (1944). Thysanoptera of Fiji. *Occ. Pop. Bernice P. Bishop Mus.* **17**, 268–311.

MOUND, L. A. (1967a). A new genus and species of Thysanoptera predatory on mites in Iraq. *Bull. ent. Res.* **57**, 315–319.

MOUND, L. A. (1967b). The British species of the genus *Thrips* Thysanoptera. *Entomologist's Gaz.* **18**, 13–22.

MOUND, L. A. (1968a). A review of R. S. Bagnall's Thysanoptera collections. *Bull. Br. Mus. nat. Hist.* (Ent.) Suppl. **11**, 181 pp.

MOUND, L. A. (1968b). A taxonomic revision of the Australian Aeolothripidae (Thysanoptera). *Bull. Br. Mus. nat. Hist.* (Ent.) **20**, 41–74.

MOUND, L. A. (1970a). Thysanoptera from the Solomon Islands. *Bull. Br. Mus. nat. Hist.* (Ent.) **24**, 85–126.

MOUND, L. A. (1970b). Convoluted maxillary stylets and the systematics of some phlaeothripine Thysanoptera from *Casuarina* trees in Australia. *Aust. J. Zool.* **18**, 439–463.

MOUND, L. A. (1970c). Intra gall variation in *Brithothrips fuscus* Moulton with notes on other Thysanoptera induced galls on *Acacia* phyllodes in Australia. *Entomologists mon. Mag.* **105**, 159–162.

MOUND, L. A. (1970d). Sex intergrades in Thysanoptera. *Entomologists mon. Mag.* **105**, 186–189.

MOUND, L. A. (1970e). Studies on heliothripine Thysanoptera. *Proc. R. ent. Soc. Lond.* (B) **39**, 41–56.

MOUND, L. A. (1971a). Gall-forming thrips and allied species (Thysanoptera: Phlaeothripinae) from *Acacia* trees in Australia. *Bull. Br. Mus. nat. Hist.* (Ent.) **25**, 389–466.

MOUND, L. A. (1971b). The feeding apparatus of thrips. *Bull. ent. Res.* **60**, 547–548.

MOUND, L. A. (1973). Thysanoptera. In *Viruses and Invertebrates.* (Gibbs, A. J., ed.) North Holland Publishing Co. [in press].

MOUND, L. A. and PITKIN, B. R. (1972). Microscopic whole mounts of thrips (Thysanoptera). *Entomologist's Gaz.* **23**, 121–125.

MUESBECK, C. F., KROMBEIN, K. V. and TOWNES, H. K., (1951). Hymenoptera of America north of Mexico, synoptic catalog. *Agriculture Monogr.* **2**, 1420 pp.

MÜLLER, K. (1927). Beiträge zur Biologie, Anatomie, Histologie und inneren Metamorphose der Thripslarven. *Z. wiss. Zool.* **130**, 251–303.

MUMA, M. H. (1955). Factors contributing to the natural control of citrus insects and mites in Florida. *J. econ. Ent.* **48**, 432–438.

MUMFORD, E. P. and HEY, D. H. (1930). The water balance of plants as a factor in their resistance to insect pests. *Nature, Lond.* **125**, 411–412.

MUNGER, F. (1942). Notes on the biology of the citrus thrips. *J. econ. Ent.* **35**, 455.

MURPHY, P. W. (1962). Extraction methods for soil animals. I. Dynamic methods with particular reference to funnel processes. In *Progress in Soil Zoology.* (Murphy, P. W., ed.) 398 pp. Butterworths, London. 75–114.

MYERS, J. G. (1935). Notes on cacao-beetle and cacao-thrips. *Trop. Agric. Trin.* **12**, 22.

NARAYANAN, E. (1970). Survey of natural enemies of *Thrips tabaci* (for the U.K.). *Ann. Rept for 1969. Commnw. Inst. Biol. Control. Comm. Agric. Bur.* 54.

NEWSOM, L. D., ROUSSEL, J. S. and SMITH, C. E. (1953). The tobacco thrips, its seasonal history and status as a cotton pest. *Bull. La. agric. Exp. Stn.* **474**.

NICHOLSON, A. J. (1958). Dynamics of insect populations. *A. Rev. Ent.* **3**, 107–136.

NOLL, J. and ROHR, C. (1966). Thysanopterenschäden an Luzerneblüten. *NachrBl. dt. PflSchutzdienst, Berl.* **20**, 47–52.

NOLTE, H. W. (1951). Beiträge zur Morphologie und Biologie des Lärchenblasenfusses (*Taeniothrips laricivorus* Krat.). *Beitr. Ent.* **1**, 110–139.

NOTLEY, F. B. (1948). The *Leucoptera* leaf miners of coffee on Kilimanjaro. I. *Bull. ent. Res.* **39**, 399–416.

NOWELL, W. (1916). The fungus on cacao thrips. *Agric. News, Bridgetown* **15**, 430.

NOWELL, W. (1917). Preliminary trials with the cacao thrips fungus. *Agric. News, Bridgetown,* **16**, 94.

NUORTEVA, P. and KANERVO, V. (1952). Einwirkung von Thysanopteranschädigung auf die Backfähigkeit des Weizens. *Ann. ent. fenn.* **18**, 137–147.

OBRTEL, R. (1963). Subterranean phase of metamorphosis in *Odontothrips loti* Hal. (Thysanoptera, Thripidae). *Zool. Listy.* **12**, 139–148.

OBRTEL, R. (1965). The economic importance of the birdsfoot trefoil thrips (*Odontothrips loti* Hal.). *Ochr. Rost.* **2**, 57–64.

OETTINGEN, H. VON (1942). Die Thysanopteren des norddeutschen Graslands. *Ent. Beih. Berl. Dahlem,* **9**, 79–141.

OETTINGEN, H. VON (1954). Beiträge zur Thysanopterenfauna Schwedens. *Ent. Tidskr.* **75**, 134–150.

OLIVER, D. R. (1963). Entomological studies in the Lake Hazen area, Ellesmere Island, including lists of species of Arachnida, Collembola and insects. *Arctic* **16**, 175–180.

OLIVER, D. R., CORBETT, P. S. and DOWNES, J. A. (1964). Studies on arctic insects: the Lake Hazen project. *Can. Ent.* **96**, 138–139.

O'NEILL, K. (1960). Identification of the newly introduced phlaeothripid *Haplothrips clarisetis* Priesner (Thysanoptera). *Ann. ent. Soc. Am.* **53**, 507–510.

OSBORN, H. (1888). The food habits of the Thripidae. *Ins. Life* **1**, 137–142.

OTA, A. K. (1968). Comparison of three methods of extracting the flower thrips from rose flowers. *J. econ. Ent.* **61**, 1754–1755.

OTA, A. K. and SMITH, F. F. (1968). Aluminium foil—thrips repellent. *Am. Rose A.* **53**, 135–139.

PAESCHKE, W. (1927). Experimentelle Untersuchungen zum Rauhigkeits- und Stabilitätsproblem in der bodennahen Luftschicht. *Beitr. Phys. freien Atmos.* **24**, 163–189.

PASSLOW, T. (1957). Control of *Thrips tabaci* Lind. in onion crops in the Lockyer Valley. *Qd J. agric. Sci.* **14**, 53–72.

PATEL, G. N. and PATEL, G. A. (1953). Bionomics of the wheat thrips (*Anaphothrips flavicinctus* Karny) in the Bombay State. *Indian J. Ent.* **15**, 251–261.

PATTERSON, J. P. (1943). Drsophilidae of the south west. *Univ. Tex. Publs* **4313**, 17–216.

PAVLOV, I. F. (1937). Noxiousness of *Haplothrips tritici* Kurd. (in Russian) *Rev. appl. Ent.* (*A*). **25**, 143–144.

PEARSON, E. O. (1958). *The insect pests of cotton in tropical Africa.* 355 pp. Commonwealth Agricultural Bureau, London.

PELIKAN, J. (1951). On carnation thrips *Taeniothrips dianthi* Pr. (in Czech) *Ent. Listy* **14**, 5–38.

PELIKAN, J. (1952). A list of Czechoslovak Thysanoptera. *Zool. Ent. Listy* **15**, 185–195.

PERTTUNEN, V. (1959). Effect of temperature on the light reactions of *Blastophagus piniperda* L. (Col. Scolytidae). *Ann. ent. fenn.* **25**, 65–71.

PERTTUNEN, V. (1960). Seasonal variation in the light reactions of *Blastophagus piniperda* L. (Col., Scolytidae) at different temperatures. *Ann. ent. fenn.* **26**, 86–92.

PESSON, P. (1951). Ordre des Thysanoptera. In *Traité de Zoologie* **10**, 1948 pp. (Grassé, P. P., ed.) Masson, Paris. 1805–1869.

PETERSON, A. (1915). Morphological studies of the head and mouthparts of the Thysanoptera. *Ann. ent. Soc. Am.* **8**, 20–66.

PFRIMMER, T. R. (1966). Systemic insecticides for cotton insect control in 1965. *J. econ. Ent.* **59**, 1113–1118.

PITKIN, B. R. & MOUND, L. A. (1973). Check list of West African Thysanoptera. *Bull. Inst. franç. Afr. noire* (in press).

PITTMAN, H. A. (1927). Spotted wilt of tomatoes. Preliminary note concerning the transmission of the "spotted wilt" of tomatoes by an insect vector (*Thrips tabaci* Lind.). *J. Coun. scient. ind. Res. Aust.* **1**, 74–77.

PLANK, VAN DER J. E. and ANDERSSEN, E. E. (1945). Kromnek disease of tobacco: a mathematical solution to a problem of disease. *Sci. Bull. Dep. Agric. For. Un. S. Afr.* **240**.

PLANK, H. K. and WINTERS, H. F. (1949). Insect and other animal pests of *Cinchona* and their control in Puerto Rico. *Bull. fed.* (*agric.*) *Exp. Sta. P.R.* **45**.

POOS, F. W. and ELLIOT, C. (1936). Certain insect vectors of *Aplanobacter stewarti*. *J. agric. Res.* **52**, 585–608.

POST, R. L. (1947). Special apparatus for collecting insects. *Bi-m. Bull. N. Dak. agric. Exp. Stn* **9**, 78–80.

POST, R. L. and COLBERG, W. J. (1958). Barley thrips in North Dakota, *Circ. N. Dak. agric. Coll. Extn. Serv.* **A292**.

POST, R. L. and McBRIDE, D. K. (1966). Barley thrips: biology and control. *Circ. N. Dak. Stat. Univ. Extn. Serv.* **A292**.

POST, R. L. and OLSON, E. (1960). Barley thrips—1960. *N. Dak. Seed. Journ.* **29**, 4.

POST, R. L. and THOMASSON, G. L. (1966). The relative abundance and over-wintering mortality of sod inhabiting thrips. *N. Dak. agric. Coll. Exp. Stn Ins. Publs* **6**.

POWELL, D. M. and LANDIS, B. J. (1965). A comparison of two sampling methods for estimating population trends of thrips and mites on potatoes. *J. econ. Ent.* **58,** 1141–1144.

PRIESNER, H. (1925). Die Winterquartiere der Thysanopteren. *Ent. Jb.* **34,** 152–162.

PRIESNER, H. (1926–1928). *Die Thysanopteren Europas.* 755 pp. Wagner-Verlag, Wien.

PRIESNER, H. (1929). Indomalayische Thysanopteren. I. *Treubia* **10,** 447–462.

PRIESNER, H. (1930). Indomalayische Thysanopteren. II. *Treubia* **11,** 357–371.

PRIESNER, H. (1932a). Indomalayische Thysanopteren. IV. *Konowia* **13,** 193–202; **14,** 58–67; 159–174; 241–255; 323–339.

PRIESNER, H. (1932b). Thysanopteren aus dem Belgischen Congo. *Rev. zool. Bot. Afr.* **22,** 192–221.

PRIESNER, H. (1932c). Thysanopteren aus dem Belgischen Congo. *Rev. zool. Bot. Afr.* **22,** 320–344.

PRIESNER, H. (1939). Zur Kenntnis der Gattung *Gynaikothrips* Zimmerman (Thysanoptera). *Mitt. Müncher Ent. Ges.* **29,** 475–487.

PRIESNER, H. (1949). Genera Thysanopterorum. *Bull. Soc. Fouad I. Ent.* **33,** 31–157.

PRIESNER, H. (1964a). A monograph of the Thysanoptera of the Egyptian deserts. *Publs Inst. Désert Égypte* (1964) **13,** 549 pp.

PRIESNER, H. (1964b). Ordnung Thysanoptera. *Bestimmungsbücher zur Bodenfauna Europas.* Lief **2,** 242 pp. Akademie-Verlag, Berlin.

PRINGLE, J. W. S. (1957). *Insect flight.* Monographs in Experimental Biology **9,** 132 pp. Cambridge University Press.

PUSSARD, R. (1946). Remarques sur deux thrips des cultures florales. *C.r. Acad. Agric. Fr.* **32,** 772–775.

PUSSARD-RADULESCO, E. (1930). Quelques observations biologiques sur *Parthenothrips dracaenae* Heeg. et *Aptinothrips rufus* Gmel. var. *connaticornis* Uzel. *Revue Path. veg. Ent. agric. Fr.* **17,** 24–28.

PUTMAN, W. L. (1942). Notes on the predacious thrips *Haplothrips subtilissimus* Hal. and *Aeolothrips melaleucus* Hal. *Can. Ent.* **74,** 37–43.

PUTMAN, W. L. (1965a). Paper chromatography to detect predation on mites. *Can. Ent.* **97,** 435–441.

PUTMAN, W. L. (1965b). The predacious thrips *Haplothrips faurei* Hood (Thysanoptera: Phaeothripidae) in Ontario peach orchards. *Can. Ent.* **97,** 1208–1221.

QUAYLE, H. J. (1938). *Insects of citrus and other tropical fruit.* 583 pp. Comstock, Ithaca, N.Y.

RACE, S. R. (1965). Predicting thrips populations on seedling cotton. *J. econ. Ent.* **58,** 1013–1014.

RADEMACHER, B. (1936). Flussigkeit, Blassenfüss Schaden und Fritfliegenbefall an Hafarispen. *Kranke Pfl.* **13,** 129.

RAINEY, R. C. (1958). Biometeorology and the displacement of airborne insects. I. *Int. bioclim. Congr.* (Vienna 1957), **3** (B), 7 pp.

RAJASEKHARA, K., CHATTERJI, S. and RAMDAS MENON, M. G. (1964). Biological notes on *Psallus* sp. (Miridae), a predator of *Taeniothrips nigricornis* Schmutz. *Indian J. Ent.* **26,** 62–66.

RAMACHANDRA RAO, Y. (1924). A gall-forming thrips on *Calycopteris floribunda: Austrothrips cochinchinensis. Agric. J. India* **19,** 435–437.

RAMAKRISHNA AYYAR, T. V. (1928). A contribution to our knowledge of the Thysanoptera of India. *Mem. Dep. Agric. India ent. Ser.* **10,** 215–316.

RAMAKRISHNA AYYAR, T. V. and KYLASAM, M. S. (1935). A new disease of cardomom (*Elatteria cardomomi*) apparently due to insect damage in South India. *Bull. ent. Res.* **26,** 359–361.

RAMBIER, A. (1958). Les tétranyques nuisibles a la vigne en France continentale. *Rev. Zool. agric.* **57,** 1–20.

RAO, G. N. (1970). Tea pests in Southern India and their control. *Pest Artic. News Summ.* (*PANS*) **16**, 667–672.

RAW, F. (1955). A flotation extraction process for soil micro-arthropods. In *Soil Zoology*. (Kevan, D. K. McE., ed.) 512 pp. Butterworths, London. 314–346.

RAW, F. (1962). Flotation methods for extracting soil arthropods. In *Progress in Soil Zoology*. (Murphy, P. W., ed.) 398 pp. Butterworths, London. 199–201.

RAZVYAZKINA, G. M. (1953). The importance of the tobacco thrips in the development of outbreaks of tip chlorosis of makhorka. (in Russian) *Dokl. vses. Akad. sel-khoz. Nauk* **18**, 27–31 [*Rev. appl. Ent.* (*A*). **42**, 146].

REYNE, A. (1921). De Cacaothrips (*Heliothrips rubrocinctus*, Giard). *Bull. Dep. Landb. Suriname* **44**.

REYNE, A. (1927). Untersuchungen über die Mundteile der Thysanopteren. *Zool. Jb.* **49**, 391–500.

RICHARDSON, B. H. (1953). Control of onion thrips in the winter garden area of Texas. *J. econ. Ent.* **46**, 92–95.

RICHARDSON, B. H. (1957). Control of onion thrips in the winter garden area of Texas. *J. econ. Ent.* **50**, 504–505.

RICHARDSON, B. H. and WENE, G. P. (1956). Control of onion thrips and its tolerance to certain chlorinated hydrocarbons. *J. econ. Ent.* **49**, 333–335.

RIHERD, P. T. (1954). Thrips as a limiting factor in grass seed production. *J. econ. Ent.* **47**, 709–719.

RISLER, H. (1957). Der Kopf von *Thrips physapus* L. (Thysanoptera, Terebrantia). *Zool. Jb.* **76**, 251–302.

RISLER, H. and KEMPTER, E. (1961). Die Haploidie der Männchen und die Endopolyploidie in einigen gewehen von *Haplothrips* (Thysanoptera). *Chromosoma* **12**, 351–361.

RIVNAY, E. (1935). Ecological studies of the greenhouse thrips, *Heliothrips haemorrhoidalis*, in Palestine. *Bull. ent. Res.* **26**, 267–278.

RIVNAY, E. (1938). Factors affecting the fluctuations in the population of *Toxoptera aurantii* Boy in Palestine. *Ann. appl. Biol.* **25**, 143–154.

ROGOFF, W. M. (1952). The repellency of chlordane, D.D.T. and other residual insecticides to greenhouse thrips. *J. econ. Ent.* **45**, 1065–1071.

RONEY, J. N. (1949). Bermuda seed grass insects in Arizona. *J. econ. Ent.* **42**, 555.

ROSS, W. A. (1918). Report on insects for the year; Division No. 7, Niagara District. *Rep. ent. Soc. Ont.* **48**, 29–30.

RUBTZOV, I. A. (1935). *Haplothrips tritici* Kurd. and the coefficient of its injury. (in Russian) *Pl. Prot. Leningr.* (*1935*), 41–46 [*Rev. appl. Ent.* (*A*). **23**, 565–566].

RUSSELL, H. M. (1912a). The bean thrips (*Heliothrips fasciatus* Pergande). *Bull. Bur. Ent. U.S. Dep. Agric.* **118**, 49 pp.

RUSSELL, H. M. (1912b). The red-banded thrips. *Bull. Bur. Ent. U.S. Dep Agric.* **99**, 17–29.

RUSSELL, H. M. (1912c). An internal parasite of Thysanoptera. *Tech. Ser. Bur. Ent. U.S.* **23**, 25–52.

SAKIMURA, K. (1932). Life history of *Thrips tabaci* on *Emilia sagitata* and its host plant range in Hawaii. *J. econ. Ent.* **25**, 884–891.

SAKIMURA, K. (1937a). The life and seasonal histories of *Thrips tabaci* Lind. in the vicinity of Tokyo, Japan. *Oyo Dobuts.-Zasshi* **9**, 1–24.

SAKIMURA, K. (1937b). On the bionomics of *Thripoctenus brui* Vuillet, a parasite of *Thrips tabaci* Lind. in Japan. (I) *Kontyû.* **11**, 370–390.

SAKIMURA, K. (1937c). On the bionomics of *Thripoctenus brui* Vuillet, a parasite of *Thrips tabaci* Lind. in Japan. (II) *Kontyû.* **11**, 410–424.

SAKIMURA, K. (1937d). Introduction of *Thripoctenus brui* Vuillet, a parasite of *Thrips tabaci* Lind. from Japan to Hawaii. *J. econ. Ent.* **30**, 799–802.

SAKIMURA, K. (1947). Thrips in relation to gall-forming and plant disease transmission: a review. *Proc. Hawaii. ent. Soc.* **13**, 59–96.

SAKIMURA, K. (1960). The present status of thrips-borne viruses. In *Biological transmission of disease agents*. (Maramorosch, K., ed.), 192 pp. Academic Press, New York and London. 30–40.

SAKIMURA, K. (1961). Techniques for handling thrips in transmission experiments with the tomato spotted wilt virus. *Pl. Dis. Reptr* **45**, 766–771.

SAKIMURA, K. (1963). *Frankliniella fusca*, an additional vector for the tomato spotted wilt virus, with notes on *Thrips tabaci*, another vector. *Phytopathology* **53**, 412–415.

SAKIMURA, K. (1969). A comment on the color forms of *Frankliniella schultzei* (Thysanoptera: Thripidae) in relation to transmission of the tomato-spotted wilt virus. *Pacif. Insects* **11**, 761–762.

SAKIMURA, K. and CARTER, W. (1934). The artificial feeding of Thysanoptera. *Ann. ent. Soc. Am.* **27**, 341.

SAKIMURA, K. and KRAUSS, N. L. H. (1944). Thrips from Maui and Molokai. *Proc. Hawaii. ent. Soc.* **12**, 113–122.

SAKIMURA, K. and KRAUSS, N. L. H. (1945). Collections of thrips from Kauai and Hawaii. *Proc. Hawaii. ent. Soc.* **12**, 319–331.

SALT, G. and HOLLICK, F. S. J. (1944). Studies of wireworm populations. 1. A census of wireworms in pasture. *Ann. appl. Biol.* **31**, 53–64.

SCHIMPER, A. F. W. (1903). *Plant-geography upon a physiological basis*. 840 pp. Clarendon Press, Oxford.

SCHMUTTERER, H. (1969). *Pests of crops in north east and central Africa*. 296 pp. Gustav Fischer Verlag, Stuttgart.

SCHREAD, J. C. (1969). Privet thrips. *Circ. Conn. agric. Exp. Stn* **230**.

SEAMANS, H. L. (1928). Insects of the season 1928 in Canada. *Rep. ent. Soc. Ont.* **59**, 30.

SĘCZKOWSKA, K. (1963). Thysanoptera of the reserve Stawska Góra near Chelm. (in Polish) *Annls Univ. Mariae Curie-Sklodowska* **18**, 135–142.

SELHIME, A. G., MUMA, M. H. and CLANCY, D. W. (1963). Biological, chemical and ecological studies on the predatory thrips *Aleurodothrips fasciapennis* in Florida citrus groves. *Ann. ent. Soc. Am.* **56**, 709–712.

SENEVET, G. (1922). Présence à Alger d'un insecte fort gênant pour l'homme. *Bull. Soc. Hist. nat. Afr. N.* **13**, 97–98.

SESHADRI, A. R. (1953). Observations on *Trichionthrips breviceps* (Bagnall), a little known predatory thrips from South India. *Indian J. agric. Sic.* **23**, 27–39.

SHARGA, U. S. (1932). A new nematode, *Tylenchus aptini* n. sp., parasite of Thysanoptera (Insecta: *Aptinothrips rufus* Gmelin). *Parasitology* **24**, 268–279.

SHARGA, U. S. (1933a). Biology and life-history of *Limothrips cerealium* Hal. and *Aptinothrips rufus* Gmelin feeding on Gramineae. *Ann. appl. Biol.* **20**, 308–326.

SHARGA, U. S. (1933b). On the internal anatomy of some Thysanoptera. *Trans. R. ent. Soc. Lond.* **81**, 185–204.

SHARMA, P. L. and BHALLA, O. P. (1963). Occurrence of the thrips damaging apple blossoms in Himachal Pradesh. *Indian J. Ent.* **25**, 85–86.

SHAW, H. B. (1914). Thrips as pollinators of beet flowers. *Bull. U.S. Dept. of Agric.* **104**.

SHAZLI, A. and ZAZOU, M. H. (1959) Control of onion thrips by surface and systemic insecticides. *Bull. Soc. ent. Égypte* **43**, 185–191.

SHEALS, J. G. (1950). Observations on blindness in oats. *Ann. appl. Biol.* **37**, 397–406.

SHIRCK, F. M. (1948). Collecting and counting onion thrips from samples of vegetation. *J. econ. Ent.* **41**, 121–123.

SHULL, A. F. (1911). Thysanoptera (and Orthoptera). In *A biological survey of the sand dune region on the south shore of Saginaw Bay, Michigan* (Ruthven, A. G. ed.). *Publs Mich. geol. biol. Surv. Biol. Ser.* 2 **4**, 177–216.

SHULL, A. F. (1914a). Biology of the Thysanoptera. I. Factors governing local distribution. *Am. Nat.* **48**, 161–176.

SHULL, A. F. (1914b). Biology of the Thysanoptera. II. Sex and the life-cycle. *Am. Nat.* **48**, 236–247.

SHULL, A. F. (1917). Sex determination in *Anthothrips verbasci. Genetics* **2**, 480–488.

SHUROVENKOV, B. G. (1961). Biological peculiarities of the larvae of *Haplothrips tritici* Kurd. under the conditions of Siberian Trans-Ural. *Zool. Zh.* **40**, 1568–1571.

SILVA, P. (1964). Tripes do cacaueiro—causador do "queima" da fôlha e da "ferrugem" du fruto. *Cacau Atual.* **1**, 9–10.

SIMBERLOFF, D. S. and WILSON, E. O. (1969). Experimental zoogeography of islands. The colonisation of empty islands. *Ecology* **50**, 278–296.

SIMBERLOFF, D. S. and WILSON, E. O. (1970). Experimental zoogeography of islands. A two-year record of colonization. *Ecology* **51**, 934–937.

SIMMONDS, H. W. (1933). The biological control of the weed *Clidemia hirta*, D. Don., in Fiji. *Bull. ent. Res.* **24**, 345–348.

SKUHRAVY, V., NOVÁK, K. and STARÝ, P. (1959). Entomofauna jetele (*Trifolium pratense* L.) a jeji vývoj. *Rozpr. čsl. Akad. Věd.* **69**, 1–83.

SLEESMAN, J. P. (1943). Variations in thrips populations on onions. *Bi-m. Bull. Ohio agric. Exp. Stn* **28**, 96–100.

SLEESMAN, J. P. (1946). D.D.T. for the control of onion thrips. *Bi-m. Bull. Ohio agric. Exp. Stn* **31**, 39–40.

SMITH, F. E. (1961). Density dependence in the Australian thrips. *Ecology* **42**, 403–407.

SMITH, F. F. and WEISS, F. (1942). Relationship of insects to the spread of azalea flower spot. *Tech. Bull. U.S. Dep. Agric.* **798**.

SMITH, K. G. V. (1955). Thrips on the stinkhorn fungus (*Phallus impudicus* Pers.). *J. Soc. Br. Ent.* **5**, 109.

SMITH, K. M. (1932). Studies on plant virus diseases. XI. Further experiments with a ringspot virus: its identification with spotted wilt of the tomato. *Ann. appl. Biol.* **19**, 305–330

SMITH, K. M. (1957). *A text book of plant virus diseases.* 2nd Edn. 652 pp. Churchill, London.

SMITH, P. G. and GARDNER, M. W. (1951). Resistance in tomato to the spotted-wilt virus. *Phytopathology* **41**, 257–260.

SOLOMON, M. E. (1951). Control of humidity with potassium hydroxide, sulphuric acid or other solutions. *Bull. ent. Res.* **42**, 543–554.

SOUTHWOOD, T. R. E. (1960). The flight activity of Heteroptera. *Trans. R. ent. Soc. Lond.* **112**, 173–220.

SOUTHWOOD, T. R. E. (1962). Migration of terrestrial arthropods in relation to habitat. *Biol. Rev.* **37**, 171–214.

SOUTHWOOD, T. R. E. (1966). *Ecological methods, with particular reference to the study of insect populations.* 391 pp. Methuen, London.

SOUTHWOOD, T. R. E. and PLEASANCE, H. J. (1962). A hand-operated suction apparatus for the extraction of arthropods from grassland and similar habitats, with notes on other models. *Bull. ent. Res.* **53**, 125–128.

SPEYER, E. R. (1938). 3. Animal pests. 1. Rose thrips (*Thrips fuscipennis*, Hal.). *Rep. exp. Res. Stn, Cheshunt* **23**, 64–65.

SPEYER, E. R. (1939). 3. Animal pests. 8. Thrips. *Rep. exp. Res. Stn, Cheshunt* **24**, 43.

SPEYER, E. R. (1951). Gladiolus thrips (*Taeniothrips simplex* Mor.) in England. *Proc. R. ent. Soc. Lond.* (B) **20**, 53–62.

SPEYER, E. R. and PARR, W. J. (1941). The external structure of some thysanopterous larvae. *Trans. R. ent. Soc. Lond.* **91**, 559–635.

SPILLER, D. (1951). Notes on Thysanoptera. *N.Z. Jl Sci. Technol.* (B) **33**, 142–143.

STANNARD, L. J. (1952). Peanut-winged thrips. *Ann. ent. Soc. Am.* **45**, 327–330.

STANNARD, L. J. (1957). The phylogeny and classification of the N. American genera of the suborder Tubulifera. *Illinois biol. Monogr.* **25**, 200 pp.

STANNARD, L. J. (1968). The thrips or Thysanoptera of Illinois. *Bull. Ill. St. nat. Hist. Surv.* **29**, 215–552.

STATHOPOULOS, D. G. (1964). Studies on the identification and bio-ecology of *Aphis* spp., *Thrips tabaci* Lind., *Bemisia tabaci* Genn., *Empoasca* sp. and *Tetranychus telarius* L., cotton pests. (in Greek) *Rep. Pl. Prot. agric. Res. Stn, Thessaloniki* **2**, 39–47.

STEELE, H. V. (1935). Thrips investigation: some common Thysanoptera in Australia. *Pamphl. sci. industr. Res. Aust.* **54**.

STEGWEE, D. (1964). Respiratory chain metabolism in the Colorado potato beetle. II. Respiration and oxidative phosphorylation in 'sarcosomes' from dispausing beetles. *J. Insect Physiol.* **10**, 97–102.

STEINHAUS, E. A. (1949). *Principles of insect pathology.* 757 pp. McGraw-Hill, New York.

STOKES, R. H. and ROBINSON, R. A. (1949). Standard solutions for humidity control at 25°C. *Industr. Engng Chem.* **41**, 2103.

STRADLING, D. J. (1968). Investigations on the natural enemies of the onion thrips (*Thrips tabaci* Lindeman). A. Work undertaken at the European station during 1968. *Interim Rept Commonw. Inst. biol. Control, Dec. 1968.* 1–8.

STRANACK, F. (1912). A contribution to the knowledge of the phytopathologic importance of grain thrips. *Dt. landw. GenossPr.* **39**, 771.

STRASSEN, R. ZUR (1960). Catalogue of the known species of South African Thysanoptera. *J. ent. Soc. S. Afr.* **23**, 321–367.

STRASSEN, R. ZUR (1967). Daten zur Thysanopteren-Faunistik des Rhein-Main-Gebietes (Ins., Thysanoptera). *Senckenberg. biol.* **48**, 83–116.

STRASSEN, R. ZUR (1968). Okologische und zoogeographisch Studien über die Fransenflügler-Fauna (Ins., Thysanoptera) des südlichen Marokko. *Abh. senckenb. naturforsch Ges.* **515**, 125 pp.

STROFBERG, F. J. (1948). New host of *Scirtothrips aurantii* (Faure). *J. ent. Soc. S. Afr.* **10**, 196–197.

STUCKENBERG, B. R. (1954). The immature stages of *Sphaerophoria quadrituberculata* Beazi (Diptera: Syrphidae), a predator of *Cercothrips afer* Priesner. *J. ent. Soc. S. Afr.* **17**, 58–61.

STULTZ, H. T. (1955). The influence of spray programs on the fauna of apple orchards in Nova Scotia. VIII. Natural enemies of the eye-spotted bud moth (*Spilonota ocellana*). *Can. Ent.* **87**, 79–85.

SUTTON, O. G. (1953). *Micrometeorology.* 333 pp. McGraw-Hill, New York.

SWEEZEY, O. H. (1945). Insects associated with orchids. *Proc Hawaii. ent. Soc.* **12**, 343–403.

SWIFT, J. E. and MADSEN, H. F. (1956). Thrips damage to apple. *J. econ. Ent.* **49**, 398–399.

TAKAHASHI, R. (1934). Association of different species of thrips in their galls (in Japanese). *Bot. Zool., Tokyo* **2**, 1827–1836.

TANI, N., INOUE, E. and IMAI, K. (1955). Some measurements of wind over the cultivated field. Pt 3. *J. agric. Met., Tokyo* **10**, 105–108.

TANSKY, V. I. (1958a). The biology of wheat thrips *Haplothrips tritici* Kurd. (Thysanoptera, Phlaeothripidae) in Northern Kazakhstan and the proposed cultural methods of its control. (in Russian) *Rev. d'ent U.R.S.S.* **37**, 785–797.

TANSKY, V. I. (1958b). Comparative infestation of varieties of spring wheat by the wheat thrips, *Haplothrips tritici* Kurd. and its injuriousness in Northern Kazakhstan. (in Russian) *Trudy vses. Inst. Zashch.* **11**, 7–25 [*Rev. appl. Ent. (A).* **49**, 288].

TANSKY, V. I. (1961). The formation of the thrips fauna (Thysanoptera) on wheat crops in the new soil of Northern Kazakhstan. (in Russian) *Ént. Obozr.* **40**, 785–793.

TANSKY, V. I. (1965). On some peculiarities of the Thysanoptera fauna as a component of steppe and wheat-field biocoenosis. (in Russian) *Trudy vses. ént. Obshch.* **50**, 67–72.

TAPLEY, R. G. (1960). Coffee leaf miner epidemics in relation to the use of persistent insecticides. *Ann. Rpt Coffee Res. Stn, Lyamungu, Tanganyika, 1960* 43–55.

TAYLOR, E. A. and SMITH, F. F. (1955). Three methods for extracting thrips and other insects from rose flowers. *J. econ. Ent.* **48**, 767–768.

TAYLOR, L. R. (1951). An improved suction trap for insects. *Ann. appl. Biol.* **38**, 582–591.

TAYLOR, L. R. (1955). The standardization of the air-flow in insect suction traps. *Ann. appl. Biol.* **43**, 390–408.

TAYLOR, L. R. (1957). Temperature relation of teneral development and behaviour in *Aphis fabae* Scop. *J. exp. Biol.* **34**, 189–208.

TAYLOR, L. R. (1958). Aphid dispersal and diurnal periodicity. *Proc. Linn. Soc. Lond.* **169**, 67–73.

TAYLOR, L. R. (1960). The distribution of insects at low levels in the air. *J. Anim. Ecol.* **29**, 45–63.

TAYLOR, L. R. (1961). Aggregation, variance and the mean. *Nature, Lond.* **189**, 732–735.

TAYLOR, L. R. (1962). The absolute efficiency of insect suction traps. *Ann. appl. Biol.* **50**, 405–421.

TAYLOR, L. R. (1963). Analysis of the effect of temperature on insects in flight. *J. Anim. Ecol.* **32**, 99–117.

TAYLOR, L. R. and PALMER, J. (1972). Aerial sampling. In *Aphid Technology* (van Emden, H. F., ed.) 344 pp. Academic Press, London and New York. 189–234.

TAYLOR, T. A. (1969). On population dynamics and flight activity of *Taeniothrips sjostedti* (Tryb.) (Thysanoptera: Thripidae) on cowpea. *Bull. ent. Soc., Nigeria* **2**, 60–71.

TAYLOR, T. H. C. (1935). The campaign against *Aspidiotus destructor* Sign. in Fiji. *Bull. ent. Res.* **26**, 1–102.

Theobald, F. V. (1926). Notes on some unusual insect pests on fruit. *J. Pomol.* **5**, 245–247.

THOMAS, H. E. and ARK, P. A. (1934). Fire blight of pears and related plants. *Bull. Calif. agric. Exp. Stn* **586**.

THOMPSON, W. R. (1950). *A catalogue of the parasites and predators of insect pests. Section 1. Pt 2. Parasites of the Neuroptera, Odonata, Orthoptera, Psocoptera, Siphonaptera and Thysanoptera.* 35 pp. Commonw. agric. Bur., Ottawa.

THOMPSON, W. R. and SIMMONDS, F. J. (1964). *A catalogue of the parasites and predators of insect pests. Section 3. Predator Host Catalogue.* 204 pp. Commonw. agric. Bur., Farnham Royal.

THOMPSON, W. R. and SIMMONDS, F. J. (1965). *A catalogue of the parasites and predators of insect pests. Section 4. Host predator catalogue.* 198 pp. Commonw. agric. Bur., Farnham Royal.

TITSCHACK, E. (1969). Der Tarothrips, ein neues Schadinsekt in Deutschland. *Sonderdr. Anz. Schädlingsk. Pflschutz* **42**, 1–6.

TOMINIC, A. (1950). Observations on the biology and control of the olive thrips (*Liothrips oleae* Costa). (in Serbo-Croat) *Zaranie Slas.* **2**, 73–85 [*Rev. appl. Ent. (A).* **41**, 245–246].

TRAVERSI, B. A. (1949). Estudio inicial sobre una enfermedad del Girasol (*Helianthus annuus* L.) en Argentina. *Rev. invest. agr., Buenos Aires* **3**, 345–351.

TREHERNE, R. C. (1923). Notes on *Frankliniella tritici* (Fitch). *Rep. ent. Soc. Ont.* **53**, 39–43.

TULLGREN, A. (1918). Ein sehr einfacher Ausleseapparat fur terricole Tierformen. *Z. angew. Ent.* **4**, 149–150.

TUTEL, B. (1963). Morphological, anatomical and physiological studies on the Malatya and Bursa varieties of tobacco in relation to their resistance to thrips. (in Turkish) *Tekel Enst. Raporl.* **9**, 37–74 [*Rev. appl. Ent.* (*A*). **52**, 467].

UICHANCO, L. B. (1919). A biological and systematic study of Philippine plant galls. *Philipp. J. Sci.* **14**, 527–554.

URICH, F. W. (1928). San Thomé cacao industry. *Trop. Agric. Trin.* **5**, 275–278.

UZEL, H. (1895). *Monographie der Ordnung Thysanoptera.* 472 pp. Königgrätz.

VAPPULA, N. A. (1965). Pests of cultivated plants in Finland. *Ann. agric. fenn.* **1**, Suppl. **1**, 1–239.

VARLEY, G. C. and GRADWELL, G. R. (1960). Key factors in population studies. *J. Anim. Ecol.* **29**, 399–401.

VERMA, S. K. (1966). Studies on the host preference of the onion thrips, *Thrips tabaci* Lindeman, to the varieties of onion. *Indian J. Ent.* **28**, 396–398.

VIEIRA, C. (1960). Sobre a hibridaçãe natural em *Phaseolus vulgaris. Revta Ceres* **11**, 103–107.

VIETINGHOFF-RIESCH, A. F. VON (1958). Untersuchungen über Verbreitung und Schadwirkung des Lärchenblasenfusses (*Taeniothrips laricivorus* Krat.) in den Randzonen seines Verbreitungegebietes in Norddeutschland, der Schweiz und Frankreich. *Z. angew. Ent.* **41**, 449–474.

VITÉ, J. P. (1955). Attempts at simultaneous control of thrips and larch casebearers: its fundamental principals and prospects. *Hofchen-Breife, Bayer Pflanzenschutz Nachrichten* **1**, 34–52.

VITÉ, J. P. (1956). Populationsstudien am Lärchenblasenfuss *Taeniothrips laricivorus* Krat. *Z. angew. Ent.* **38**, 417–488.

VITÉ, J. P. (1957). Uber einige versuche zur Bekämpfung des Lärchenblasenfusses durch Baumimpfung. *Schweiz Z. Forstwes.* **108**, 81–92.

VITÉ, J. P. (1961). On the prognosis and control of the larch thrips. (in German) *Allg. Forstztg* **16**, 202–204 [*Rev. appl. Ent.* (*A*). **51**, 242].

VOGT, G. B. (1960). Exploration for natural enemies of alligatorweed and related plants in South America. *U.S.D.A. Agr. Res. Serv., Entomol. Res. Div., Special Rpt* P1–4, 58 pp.

VOGT, G. B. (1961). Exploration for natural enemies of alligatorweed and related plants in South America. *U.S.D.A. Agr. Res. Serv., Entomol. Res. Div., Special Rpt* 50 pp.

VUILLET, A. (1914). Deux Thysanoptères nouveaux du Soudain Francais. *Insecta* **4**, 121–132.

WAHLGREN, E. (1945). Gallbildande Thysanoptera. *Opusc. ent.* **10**, 119–126.

WALSH, B. D. (1864). On the insects, coleopterous, hymenopterous and dipterous inhabiting the galls of certain species of willow. *Proc. Ent. Soc. Phila.* **3**, 611–612.

WALSH, B. D. (1867a). Notes by B. D. Walsh. *Pract. Ent. Philad.* **2**, 19.

WALSH, B. D. (1867b). The true thrips and the bogus thrips. *Pract. Ent. Philad.* **2**, 49–51.

WARD, L. K. (1966). The biology of Thysanoptera. Ph.D. thesis, London University.

WARDLE, R. A. and SIMPSON, R. (1927). The biology of Thysanoptera with reference to the cotton plant. 3. The relation between feeding habits and plant lesions. *Ann. appl. Biol.* **14**, 513–528.

WASHBURN, R. H. (1958). *Taeniothrips orionis* Leh., a thrips destructive to vegetables in Alaska. *J. econ. Ent.* **51**, 274.

WATSON, J. R. (1923). Synopsis and catalogue of the Thysanoptera of N. America. *Tech. Bull. Univ. Florida* **168**.

WATSON, T. F. (1965). Influence of thrips on cotton yields in Alabama. *J. econ. Ent.* **58**, 1118–1122.

WATTS, J. G. (1936). A study of the biology of the flower thrips *Frankliniella tritici* (Fitch) with special reference to cotton. *Bull. S. Carol. agric. Exp. Stn* **306**, 1–46.

WEBER, P. W. (1954). Recent liberations of beneficial insects.—III. *Proc. Hawaii. ent. Soc.* **15**, 369–370.

WEBLEY, D. (1957). A method of estimating the density of frit fly eggs in the field. *Pl. Path.* **6**, 49–51.

WEIS-FOGH, T. and JENSEN, M. (1956). Biology and physics of locust flight. 1. Basic principals in insect flight. A critical review. *Phil. Trans. R. Soc.* (B) **239**, 415–458.

WEITMEIER, H. (1956). Zur Oekologie der Thysanopteren Frankens. *Dt. ent. Z.* **3**, 285–330.

WELLINGTON, W. G. (1950). Effects of radiation on the temperatures of insectan habitats. *Scient. Agric.* **30**, 209–234.

WETZEL, T. (1963). Zur Frage der Überwinterung der Gräser-Thysanopteren. *Z. angew. Ent.* **51**, 429–441.

WETZEL, T. (1964). Untersuchungen zum Auftreten, zur Schadwirkung und zur Bekämpfung von Thysanopteren in Grassamenbeständen. *Beitr. Ent.* **14**, 427–500 [*Rev. appl. Ent.* (*A*). **54**, 101–102].

WHITING, P. W. (1945). The evolution of male haploidy. *Q. Rev. Biol.* **20**, 231–260.

WILDE, W. H. A. (1962). A note on colour preferences of some Homoptera and Thysanoptera in British Colombia. *Can. Ent.* **94**, 107.

WILLIAMS, C. B. (1915). The pea thrips (*Kakothrips robustus*). *Ann. appl. Biol.* **1**, 222–246.

WILLIAMS, C. B. (1921). A blood-sucking thrips. *Entomologist* **54**, 163–164.

WILLIAMS, C. B. (1944). The index of diversity as applied to ecological problems. *Nature, Lond.* **155**, 390–391.

WILLIAMS, C. B. (1949). Jaccard's generic coefficient and coefficient of floral community in relation to the logarithmic series. *Ann. Bot. N.S.* **13**, 53–58.

WILLIAMS, C. B. (1950). The application of the logarithmic series to the frequency of occurrence of plant species in quadrats. *J. Ecol.* **38**, 107–138.

WILLIAMS, C. B. (1964). *Patterns in the balance of nature.* 324 pp. Academic Press, London and New York.

WILLIAMS, L. T. (1916). A new species of *Thripoctenus* (Chalcidoidea). *Psyche* **43**, 54–61.

WILSON, E. O. and SIMBERLOFF, D. S. (1969). Experimental zoogeography of islands. Defaunation and monitoring techniques. *Ecology* **50**, 267–278.

WITTWER, S. H. and HASEMAN, L. (1945). Soil nitrogen and thrips injury on spinach. *J. econ. Ent.* **38**, 615–617.

WOGLUM, R. S. and LEWIS, H. C. (1936). Nitrogen trichloride as a fumigant. *J. econ. Ent.* **29**, 631–632.

WOLFENBARGER, D. and HIBBS, E. T. (1958). Onion thrips (*Thrips tabaci* Lind.) infesting cabbage. *J. econ. Ent.* **51**, 394–396.

YAKHONTOV, V. V. (1935). An ally of the cotton grower, the acariphagous thrips *Frankliniella vaccinii* Morgan. *Can. J. agric. Sci.* **36**, 510.

WOOD, G. W. (1960). Note on the occurrence of two species of thrips on low-bush blueberry in New Brunswick and Nova Scotia. *Can. Ent.* **92**, 757–758.

WYNIGER, R. (1962). *Pests of crops in warm climates and their control.* 555 pp. Verlag fur Recht und Gesellschaft, Basel.

YAKHONTOV, V. V. (1935). An ally of the cotton grower, the acariphagous thrips (*Scolothrips acariphagus* Yakh.). *Sots. Nauka Tekh.* **12**, 96–98.

YARWOOD, C. E. (1943). Association of thrips with powdery mildews. *Mycologia* **35**, 189–191.

YARWOOD, C. E. (1957). A brush-extraction method for transmission of viruses. *Phytopathology* **47**, 613–614.

YOSHIMOTO, C. M. and GRESSITT, J. L. (1961). Trapping of airborne insects on ships on the Pacific. Pt 4. *Pacif. Insects* **3**, 556–558.

YOSHIMOTO, C. M., GRESSITT, J. L. and MITCHELL, C. J. (1962). Trapping of airborne insects in the Pacific–Antarctic area. 1. *Pacif. Insects* **4**, 847–858.

ZANON, D. V. (1924). Nuova specie di "*Franklinothrips*" (Thysanoptera) rivenuta a Bengasi. *Atti Accad. pontif. Nuovi Lincei* **77**, 88–96.

ZAWIRSKA, I. (1963). A contribution to the bionomics of *T. linarius* Uzel. (in Polish) *Biul. Inst. Ochr. Rośl.* **19**, 1–10 [*Rev. appl. Ent.* (*A*). **52**, 429].

ZENTHER-MØLLER, O. (1965). *Taeniothrips laricivorus* (Krat.) (Thripidae, Thysanoptera) in Danish stands of *Larix decidua* (Mill.). *Oikos* **16**, 58–69.

ZENTHER-MØLLER, O. (1966). Investigations of the biology, geographical distribution and forestry importance of *Taeniothrips laricivorus* Krat. in Denmark 1961–1963. *Det. forst. Forsgsvæsen* **30**, 101–166.

ZIMMERMAN, E. C. (1948). *Insects of Hawaii* **2**, *Apterygota to Thysanoptera.* 425 pp. University of Hawaii Press, Honolulu.

Appendix 1

AN INTRODUCTORY GUIDE TO CATALOGUES AND DESCRIPTIVE LITERATURE OF SOME REGIONAL AND NATIONAL FAUNAS

Europe

General

Priesner, H. (1926–28). *Die Thysanopteren Europas*. 775 pp. Wagner-Verlag, Wien.
Priesner, H. (1964b). Ordnung Thysanoptera. *Bestimmungsbücher zur Bodenfauna Europas*. Lief 2, 242 pp. Akademie-Verlag, Berlin.

Austria

Weitmeier, H. (1956). Zur Oekologie der Thysanoptera Frankens. *Dt. ent. Z.* **3,** 285–330.

Czechoslovakia

Pelikan, J. (1952). A list of Czechoslovak Thysanoptera. *Zool. ent. Listy* **15,** 185–195.

Denmark

Maltbaek, J. (1932). Frynsevinger eller blaere fodder (Thysanoptera). *Danmarks Fauna* **37,** 146 pp. Kobenhaven.

Great Britain

Morison, G. D. (1947–1949). Thysanoptera of the London Area. *Lond. Nat.* (Suppl.) **59,** 1–36; 1948, *Ibid.* 37–75; 1949, *Ibid.* 77–131.

Holland

Eecke, van R. (1931). Fauna van Nederland, aflevering V *Tysanoptera* (Q VI). 154 pp. Leiden.
Franssen, C. J. H. and Mantel, W. P. (1962). Lijst van in Nederland aangetroffen Thysanoptera met beknopte aantekeningen over hun levenswijze en hun betekenis voor onze cultuurgewassen. *Tidjdschr. Ent.* **105,** 97–133.

Iceland

Maltbaek, J. (1938). *Thysanoptera*. The Zoology of Iceland **3,** 6 pp. Lerin and Munksgaard, Copenhagen.

Italy

Melis, A. (1959). I. Tisanotteri italiani. *Redia* **44** (append.) 1–184.
Melis, A. (1960). II. Tisanotteri italiani. *Ibid.* **45** (append.) 185–329.
Melis, A. (1961). III. Tisanotteri italiani. *Ibid.* **46** (append.) 331–530.

Poland

Keler, S. (1936) Tripsy (przylzence) Polski (A catalogue of the Polish Thysanoptera). *Pr. Wydz. Chorób Szkod. Rósl. pánst. Inst. nauk. Gospod. wiejsk. Bydgoszczy* **15**, 81–149.

Rumania

Knechtel, W. K. (1923) Thysanoptere din Romania. *Bul. Minist. Agric. Ind.* **2–4**, 235 pp.

Sweden

Ahlberg, O. (1926). Svensk insektfauna. 6. Tripsar. Thysanoptera. *Ent. For. Rekv.* **29**, 1–62.

North America

Genera

Stannard, L. J. (1957). The phylogeny and classification of the North American genera of the suborder Tubulifera (Thysanoptera). *Illinois biol. Monogr.* **25**, 200 pp.
Watson, J. R. (1923). Synopsis and catalogue of the Thysanoptera of North America. *Tech. Bull. Univ. Florida* **168**, 98 pp.

California

Bailey, S. F. (1957). The thrips of California, Part I: suborder Terebrantia. *Bull. Calif. Insect Surv.* **4**, 143–220.
Cott, H. E. (1956). Systematics of the suborder Tubulifera (Thysanoptera) in California. *Univ. Calif. Publs Ent.* **13**, 216 pp.

Hawaii

Zimmerman, E. C. (1948). *Insects of Hawaii* **2**, *Apterygota to Thysanoptera*. 425 pp. University of Hawaii Press, Honolulu.

Eastern U.S.A.

Stannard, L. J. (1968). The thrips or Thysanoptera of Illinois. *Bull. Ill. St. nat. Hist. Surv.* **29**, 215–552.

New Jersey

Comegys, G. R. and Schmitt, J. B. (1966). A list of the Thysanoptera or thrips of New Jersey. *Jl N.Y. ent. Soc.* **73**, 195–222.

Central and South America

Argentina

De Santis, C. (1957–1964). Various papers in *Notas del Museo Univ. Nacional de la Plata* and in *Rev. del Museo de la Plata*.

Brazil

Hood, J. D. (1955). Brazilian Thysanoptera VI. *Revta bras. Ent.* **4**, 51–160.

Chile

Bailey, S. F. and Campos, L. E. (1962). The Thysanoptera of Chile. *Publ. Inst. Estud. Entomol. Univ. Chile* **4**, 19–26.

Mexico

Crawford, D. C. (1909). Some Thysanoptera of Mexico and the south. I. *Pomona Coll. J. Ent.* **1**, 109–119.
Crawford, D. C. (1910). Thysanoptera of Mexico and the south. II. *Ibid.* **2**, 153–170.

Panama

Hood, J. D. (1934). New Thysanoptera from Panama. *Jl N.Y. ent. Soc.* **41**, 407–434.
Hood, J. D. (1935). Eleven new Thripidae (Thysanoptera) from Panama. *Ibid.* **43**, 143–170.

Africa

Congo

Priesner, H. (1932b) Thysanopteren aus dem Belgischen Congo. *Rev. zool. Bot. afr.* **22**, 192–221.
Priesner, H. (1932c) Thysanopteren aus dem Belgischen Congo. *Ibid.* **22**, 320–344.

Egypt

Priesner, H. (1964a). A monograph of the Thysanoptera of the Egyptian deserts. *Publs Inst. Désert Égypte* (1964) **13**, 549 pp.

Morocco

Strassen, R. zur (1968). Okologisiche und zoogeographische Studien über die Fransenflügler-Fauna (Ins., Thysanoptera) des südlichen Marokko. *Abh. senckenb. naturforsch Ges.* **515**, 125 pp.

South Africa

Strassen, R. zur (1960). Catalogue of the known species of South African Thysanoptera. *J. ent. Soc. S. Afr.* **23**, 321–367.

West Africa

Pitkin, B. R. and Mound, L. A. (1973). Check list of West African Thysanoptera. *Bull. Inst. franç. Afr. noire* (in press).

Indo-Malaya

India

Ananthakrishnan, T. N. (1969a). *Indian Thysanoptera*. CSIR Zool. Monograph No. 1. 171 pp. New Delhi.

Indonesia and Malaya

Priesner, H. (1929). Indomalayische Thysanopteren. I. *Treubia* **10**, 447–462.
Priesner, H. (1930). Indomalayische Thysanopteren. II. *Ibid.* **11**, 357–371.
Priesner, H. (1932a). Indomalayische Thysanopteren. IV. *Konowia* **13**, 193–202; **14**, 58–67; 159–174; 241–255; 323–339.

Australia and New Zealand

Australia

Kelly, R. and Mayne, R. J. B. (1934). *The Australian Thrips*. 81 pp. Australian Medical Publishing Co. Sydney.

Mound, L. A. (1968b). A taxonomic revision of the Australian Aeolothripidae (Thysanoptera). *Bull. Br. Mus. nat. Hist.* (Ent.) **20**, 41–74.

Mound, L. A. (1971a). Gall-forming thrips and allied species (Thysanoptera: Phlaeothripinae) from *Acacia* trees in Australia. *Bull. Br. Mus. nat. Hist.* (Ent.) **25**, 389–466.

Mound, L. A. and Pitkin, B. R. (1969–1972). Various papers in *J. Aust. ent. Soc.* and *Aust. J. Zool.*

New Zealand

Spiller, D. (1951). Notes on Thysanoptera. *N.Z. Jl Sci. Technol.* (B) **33**, 142–143.

Pacific and Oceania

Fiji

Moulton, D. (1944). Thysanoptera of Fiji. *Occ. Pop. Bernice P. Bishop Mus.* **17**, 268–311.

Japan

Kurosawa, M. (1968). Thysanoptera of Japan. *Insecta Matsum. Suppl.* **4**, 95 pp.

Solomon Is.

Mound, L. A. (1970a). Thysanoptera from the Solomon Islands. *Bull. Br. Mus. nat. Hist.* (Ent.) **24**, 85–126.

World

Jacot-Guillarmod, C. F. (1970). Catalogue of the Thysanoptera of the world (Pt 1). *Ann. Cape Prov. Mus.* (*Nat. Hist.*) **7**, 1–126.

Jacot-Guillarmod, C. F. (1971). Catalogue of the Thysanoptera of the world (Pt 2). *Ann. Cape Prov. Mus.* (*Nat. Hist.*) **7**, 217–515.

Priesner, H. (1949) Genera Thysanopterorum. *Bull. Soc. Fouad I Ent.* **33**, 31–157.

Appendix 2

METHODS OF PRESERVING AND MOUNTING THRIPS
FOR MICROSCOPIC EXAMINATION AND PERMANENT STORAGE

Satisfactory killing, preserving and permanent mounting methods should retain the natural colour of the insects, not distort their shape, make all cuticular structures including colourless sensory organs visible, and preserve these features for an indefinite period when slides are stored in the dark. Methods recommended by Hood (1940), Priesner (1964a) and Mound and Pitkin (1972) are incorporated in this summary. The original papers of Hood and Priesner should be consulted before embarking on elaborate dissections or preparations for sectioning.

Collection and temporary storage

Specimens should be collected directly into a mixture of 60% ethyl alcohol (10 pts), glycerine (1 pt) and glacial acetic acid (1 pt). In this fluid they remain relaxed but sometimes distend; this is often an advantage when tergal and sternal characters need examining. Thrips can be kept in this fluid for several months but 60% alcohol is preferable for longer periods of storage; a drop of glycerine added to the alcohol will prevent damage to the insects should the alcohol evaporate.

Mounting

Specimens should be removed from the collecting or storage fluid and placed in 60% alcohol for at least 24 h, then dehydrated through a series of alcohols as rapidly as effective, but with no great changes in concentration. The time spent in each concentration depends largely on the size of specimens and the thickness of their cuticle, but generally 1 h in 70% alcohol, 20 min in 80%, 10 min in 95% and 5 min in absolute, followed by another 5 min in absolute, is recommended. Finally the specimens are cleared in clove oil for $\frac{1}{2}$ to 10 min. It helps to dehydrate and clear specimens if their bodies are pierced with a very fine needle through the sternal or pleural membranes. The appendages should be spread before and during dehydration and finally spread when in clove oil and Canada balsam. Alternatively, specimens may be cleared in carboxylol (1 pt phenol and 3 pts xylol) for $\frac{1}{2}$ to 1 min.

Specimens are best mounted singly, ventral side uppermost, on a cover slip in a drop of balsam; the amount depends on the size of the thrips but should be sufficient to support the cover slip when the microscope slide is lowered on to it, without crushing the head. With practice the slide may be tilted and gently pressed to spread the wings and arrange the specimen. The Canada balsam should be dilute enough to flow slowly but easily. Small bubbles around the insect will eventually be absorbed by the balsam.

Labelling

Accurate labelling is essential. Two labels, on good-quality paper and written in permanent Indian ink, should be stuck to each slide. One label should include information on:

(i) The locality in which the specimen was found.
(ii) The date of collection.
(iii) The name of collector.
(iv) A concise description of the habitat and plant.
(v) A reference number to further notes.
(vi) The name of the mountant.

On the other label should be written:

(i) The scientific name of the specimen, including the authority.
(ii) The name of the person responsible for the determination.
(iii) Further taxonomic and structural details where necessary.

Drying

Dry slides in a horizontal position for about a week near a radiator, or in a slide oven.

Storing

Store slides in a horizontal position in the dark, preferably in wooden slide boxes.

Temporary mounts

For a quick laboratory check on field identifications, specimens can be displayed adequately by putting them on to a slide in a drop of Hoyers of Berlèse fluid, covering them with a cover slip and warming briefly until they are partially cleared.

Appendix 3a

THRIPS AND THEIR INSECT PARASITES IN DIFFERENT PLACES

Compiled from Thompson (1950) and Ferrière (1958) with amendments. (All parasites listed are in the family Eulophidae: Hymn unless otherwise stated)

Thrips	Parasite	Place
Caliothrips fasciatus	*Camptoptera pulla* Gir. (Mymaridae: Hymn)	North America
	Thripoctenus russelli Crwf.	U.S.A.
Cryptothrips rectangularis	*Thripoctenus nubilipennis* Wms	U.S.A.
Dinurothrips hookeri	*Dasyscapus parvipennis* Gah.	Trinidad
Frankliniella lilivora	*Megaphragma longiciliatum* Subba Rao (Trichogrammatidae: Hymn)	India
Frankiniella occidentalis	*Thripoctenus americensis* Gir.	Canada
Frankliniella schultzei	*Ceranisus rosilloi* De Santis	Argentina
	Ceranisus nigrifemora De Santis	Argentina
Frankliniella tritici	*Thripoctenus russelli* Crwf.	England
Gynaikothrips ficorum	*Tetrastichus gentilei* Del G.	France
Gynaikothrips uzeli	*Tetrastichus thripophonus* Wtstn	Brazil
	Tetrastichus gentilei Del G.	Puerto Rico
Heliothrips haemorrhoidalis	*Dasyscapus parvipennis* Gah.	Java
	Megaphragma mymaripennis Timb. (Trichogrammatidae: Hymn)	Hawaii
Kakothrips robustus	*Thripoctenus* sp.	Switzerland
	Thripoctenus brui Vuil.	Germany
		France
	Thripoctenus kutteri Ferr.	Switzerland
Leucothrips sp.	*Megaphragma mymaripennis* Timb. (Trichogrammatidae: Hymn)	Haiti
		U.S.A.
		Hawaii
Liothrips floridensis	*Tetrastichus* sp.	U.S.A.
Liothrips oleae	*Adelgimyza tripidiperda* Del G. (Cecidomyiidae: Dipt)	Italy
	Tetrastichus gentilei Del G.	Italy
		France
Liothrips setinodis	*Thripoctenoides gaussi* Ferr.	Germany
	Tetrastichus atratulus Nees	Germany
Liothrips urichi	*Tetrastichus thripophonus* Wtstn	Trinidad
Megalothrips spinosus	*Thripoctenus nublipennis* Wms	U.S.A.
Panchaetothrips noxius	*Megaphragma ghesquieri* Novicky (Trichogrammatidae: Hymn)	Congo

Thrips	Parasite	Place
Retithrips aegypticus	*Thripoctenus* sp.	Tanzania
	Thripobius hirticornis Ferr.	Tanzania
	Megaphragma priesneri Kryger	Egypt
	(Trichogrammatidae: Hymn)	
Rhipiphorothrips cruentatus	*Ceranisus maculatus* Wtstn	India
Selenothrips rubrocinctus	*Dasyscapus parvipennis* Gah.	Puerto Rico
		Trinidad
		Ghana
	Baryconus sp.	Brazil
	(Scelionidae: Hymn)	
Taeniothrips alliorum	*Thripoctenus brui* Vuil.	Japan
Taeniothrips atratus	*Thripoctenus russelli* Crwf.	England
Taeniothrips distalis	*Thripoctenus vinctus* Gah.	Philippines
Taeniothrips inconsequens	*Thripoctenus russelli* Crwf.	U.S.A.
Thrips tabaci	*Dasyscapus parvipennis* Gah.	Java
		Trinidad
		Puerto Rico
		U.S.A.
	Thripoctenus brui Vuil.	Japan
		Java
		San Domingo
	Thripoctenus russelli Crwf.	England
		U.S.A.
		Hawaii
	Ceranisus menes Walker	Japan
		Hawaii
	Ceranisus rosilloi De Santis	Argentina
	Ceranisus nigrifemora De Santis	Argentina
Thrips sp.	*Thripoctenus bicoloratus* Ishii	Japan

Appendix 3b

THRIPS AND THEIR INSECT AND MITE PREDATORS IN DIFFERENT PLACES

Compiled from Thompson and Simmonds (1964) and Herting and Simmonds (1971) with amendments

Thrips	Predator	Place
Aeolothrips fasciatus	*Nabis alternatus* Parshley (Nabidae: Hem)	Laboratory
Anaphothrips obscurus	*Spilomena vagans* Blüthgen (Sphecidae: Hymn)	Finland
Anaphothrips orchidaceus	*Franklinothrips vespiformis* (Thysanoptera)	Puerto Rico
Aptinothrips rufus	*Trombidium* sp. (Acari)	Scotland
Bregmatothrips iridis	*Orius insidiosus* (Say) (Anthocoridae: Hem)	
Caliothrips fasciatus	*Aeolothrips fasciatus* (Thysanoptera)	U.S.A.
	Chrysopa plorabunda Fitch (Chrysopidae: Neur)	U.S.A.
	Hippodamia convergens (Guér.) (Coccinellidae: Col)	U.S.A.
	Orius tristicolor (White) (Anthocoridae: Hem)	U.S.A.
	Sphaerophoria sulphuripes (Thom.) (Syrphidae: Dipt)	U.S.A.
Caliothrips indicus	*Cheilomenes vicina* Muls. (Coccinellidae: Col)	Sudan
	Chrysopa sp. (Chrysopidae: Neur)	Sudan
	Ischiodon aegypticus Wied (Syrphidae: Dipt)	Sudan
	Orius sp.	Sudan
	Orius tantilus (Motsch.) (Anthocordiae: Hem)	India
Caliothrips insularis	*Franklinothrips vespiformis* (Thysanoptera)	Trinidad
	Termatophylidea maculosa Usinger (Miridae: Hem)	Trinidad
Dinurothrips hookeri	*Franklinothrips vespiformis*	Trinidad

Thrips	Predator	Place
Drepanothrips reuteri	*Leptothrips mali* (Thysanoptera)	U.S.A.
	Orius minutus (L.) (Anthocoridae: Hem)	Switzerland
Frankliniella minuta	*Orius tristicolor* (White) (Anthocoridae: Hem)	U.S.A.
Frankliniella moultoni	*Aeolothrips fasciatus*	U.S.A.
	Aeolothrips kuwanai	U.S.A.
	Leptothrips mali (Thysanoptera)	U.S.A.
	Nabis alternatus Parshley (Nabidae: Hem)	Laboratory
	Orius insidiosus	U.S.A.
	Orius tristicolor (Anthocordiae: Hem)	U.S.A.
Frankliniella occidentalis	*Aeolothrips fasciatus* (Thysanoptera)	Canada
	Orius insidiosus (Say)	U.S.A.
	Orius tristicolor (White) (Anthocoridae: Hem)	U.S.A.
Frankliniella tenuicornis	*Spilomena troglodytes* Lind. (Sphecidae: Hymn)	Finland
Frankliniella tritici	*Chrysopa plorabunda* Fitch (Chrysopidae: Neur)	U.S.A.
Gynaikothrips ficorum	*Adactylidium* sp. (Acari)	Egypt
	Baccha livida Schiner (Syrphidae: Dipt)	Brazil
	Chrysopa carnea Steph. (Chrysopidae: Neur)	Egypt
	Haplothrips cahirensis (Thysanoptera)	Egypt
	Ectemnus sp.	Israel
	Macrotracheliella laevis Champ.	Brazil
	Montandoniola moraguesi Put. (Anthocoridae: Hem)	Sicily
		Spain
		Egypt
		Algeria
	Orius sp. (Anthocoridae: Hem)	Egypt
		Brazil
	Termatophylum insigne Reuter (Miridae: Hem)	Egypt
Haplothrips aculeatus	*Aeolothrips fasciatus* (Thysanoptera)	Russia
	Orius minutus (L.)	Italy
	Orius niger (Wolff) (Anthocoridae: Hem)	Russia
Haplothrips sorghicola	*Hexagonia terminalis* Genom & Harold (Carabidae: Col)	West Africa
Haplothrips tritici	*Adonia variegata* Goeze (Coccinellidae: Col)	U.S.S.R.

Thrips	Predator	Place
Haplothrips tritici—contd.	*Aeolothrips fasciatus* (Thysanoptera)	U.S.S.R.
	Chrysopa sp. (Chrysopidae: Neur)	U.S.S.R.
	Malachius viridis F. (Malachiidae: Col)	U.S.S.R.
	Orius niger (Wolff) (Anthocoridae: Hem)	U.S.S.R.
Haplothrips verbasci	*Orius tristicolor* (White) (Anthocoridae: Hem)	U.S.A.
Heliothrips haemorrhoidalis	*Aeolothrips fasciatus* (Thysanoptera)	Italy
	Cardiastethus consors White	New Zealand
	Cardiastethus poweri White (Anthocoridae: Hem)	New Zealand
	Chrysopa carnea Steph. (Chrysopidae: Neur)	Israel
	Franklinothrips tenuicornis	Trinidad
	Franklinothrips vespiformis	Trinidad
	Franklinothrips myrmicaeformis (Thysanoptera)	Israel
Kakothrips robustus	*Aeolothrips fasciatus* (Thysanoptera)	Great Britain
	Undet. (Cecidomycidae: Dipt)	Germany
	Undet. (Syrphidae: Dipt)	Switzerland
Limothrips cerealium	*Medetera ambigua* Zett.	Great Britain
	Medetera dendrobaena Kowarz	Great Britain
	Medetera jaculus Fall.	Great Britain
	Medetera truncorum Meig. (Dolichopodidae: Dipt)	Great Britain
Limothrips denticornis	*Medetera truncorum* Meig. (Dolichopodidae: Dipt)	Great Britain
Limothripis schmutzi	*Orius minutus* (L.) (Anthocoridae: Hem)	Italy
Liothrips africanus	*Montandoniola moraguesi* Put. (Anthocoridae: Hem)	West Africa
Liothrips floridensis	*Anthocoris* sp. (Anthocoridae: Hem)	U.S.A.
	Chrysopa oculata Say (Chrysopidae: Neur)	U.S.A.
Liothrips fluggae	*Montandoniola moraguesi* Put. (Anthocoridae: Hem)	Tibet
Liothrips oleae	*Adelgimyza tripiperda* Del G. (Cecidomyiidae: Dipt)	Italy
	Cheyletus sp. (Acari)	Yugoslavia
	Ectemnus reduvinus (H.–S.)	Italy
	Montandoniola moraguesi Put. (Anthocoridae: Hem)	France
Liothrips setinodis	Undet. (Anystidae: Acari)	Germany

Thrips	Predator	Place
	Adalia bipunctata (L.)	Germany
	Adalia conglomerata (L.)	Germany
	Anatis ocellata (L.)	Germany
	Aphidecta obliterata (L.)	Germany
	Exochomus quadripustulatus (L.)	Germany
	Propylaea quatuordecimpunctata (L.)	Germany
	(Coccinellidae: Col)	
	Hemerobius sp.	Germany
	(Hemerobiidae: Neur)	
	Syrphus corollae F.	Germany
	(Syrphidae: Dipt)	
Liothrips urichi	*Thripsobremia liothripis* Barnes	Trinidad
	(Cecidomyiidae: Dipt)	
Microcephalothrips abdominalis	*Orius tristicolor* (White)	U.S.A.
	(Anthocoridae: Hem)	
Odontothrips loti	*Orius tristicolor* (White)	U.S.A.
	(Anthocoridae: Hem)	
Phlaeothrips sycamorensis	*Adalia bipunctata* (L.)	California
	(Coccinellidae: Col)	
	Cephalothrips sp.	California
	(Thysanoptera)	
	Chilocorus stigma Say	California
	Lindorus lophanthae (Blais.)	California
	(Coccinellidae: Col)	
	Xylocoris sp.	California
	(Anthocoridae: Hem)	
Prosopothrips cognatus	*Chrysopa oculata* Say	U.S.A.
	(Chrysopidae: Neur)	
	Orius insidiosus (Say)	U.S.A.
	(Anthocoridae: Hem)	
Retithrips aegypticus	*Franklinothrips myrmicaeformis*	Libya
Retithrips syriacus	*Franklinothrips myrmicaeformis*	Israel
	(Thysanoptera)	
Scirtothrips aurantii	*Anystis baccarum* (L.)	South Africa
	(Acari)	
	Exochomus flavipes (Thunb.)	South Africa
	(Coccinellidae: Col)	
	Haplothrips bedfordi	South Africa
	(Thysanoptera)	
	Orius thripoborus (Hesse)	South Africa
	(Anthocoridae: Hem)	Transvaal
	Scymnus trepidulus Weise	South Africa
	(Coccinellidae: Col)	
Scirtothrips longipennis	*Franklinothrips vespiformis*	Puerto Rico
	(Thysanoptera)	
Selenothrips rubrocinctus	*Chrysopa alobana* Banks	Trinidad
	Chrysopa arioles Banks	Trinidad
	Chrysopa claveri Navas	Trinidad
	Chrysopa iona Banks	Trinidad
	Leucochrysa varia Schneider	Trinidad
	(Chrysopidae: Neur)	

Thrips	Predator	Place
Selenothrips rubrocinctus—contd.	*Franklinothrips tenuicornis*	Trinidad
		Surinam
	Franklinothrips vespiformis	Trinidad
	(Thysanoptera)	Surinam
	Ninyas torvus Dist.	Trinidad
	(Lygaeidae: Hem)	
	Paracarnus sp.	West Indies
	Termatophylidea maculata Usinger	Trinidad
	(Miridae: Hem)	Jamaica
	Wasmannia auropunctata Roger	Trinidad
	(Formicidae: Hymn)	
Sericothrips variabilis	*Aeolothrips fasciatus*	U.S.A.
	Aeolothrips kuwanai	U.S.A.
	(Thysanoptera)	
	Spilomena pusilla (Say)	U.S.A.
	(Sphecidae: Hymn)	
Stenothrips graminum	*Aeolothrips fasciatus*	Russia
	(Thysanoptera)	
Taeniothrips distalis	*Psallus* sp.	India
	(Miridae: Hem)	
Taeniothrips inconsequens	*Chrysopa plorabunda* Fitch	U.S.A.
	(Chrysopidae: Neur)	
	Gyrophaena manca Er.	U.S.A.
	(Staphylinidae: Col)	
	Hemerobius californicus Banks	U.S.A.
	Hemerobius pacificus Banks	U.S.A.
	(Hemerobiidae: Neur)	
	Hippodamia convergens (Guér)	U.S.A.
	(Coccinellidae: Col)	
	Orius insidiosus (Say)	U.S.A.
	Orius tristicolor (White)	U.S.A.
	(Anthocoridae: Hem)	
	Scymnus ater Kug	U.S.A.
	(Coccinellidae: Col)	
Taeniothrips laricivorus	*Aeolothrips fasciatus*	Germany
	Aeolothrips vittatus	Germany
	(Thysanoptera)	Czechoslovakia
	Chrysopa sp.	Germany
	(Chrysopidae: Neur)	
	Tetraphleps bicuspis (H.–S.)	Germany
	Adalia bipunctata (L.)	Germany
	Adalia decempunctata (L.)	Germany
	Anatis ocellata (L.)	Germany
	Aphidecta obliterata (L.)	Germany
	Coccinella septempunctata L.	Germany
	Exochomus quadripustulatus (L.)	Germany
	Neomysia oblongoguttata (L.)	Germany
	(Coccinellidae: Col)	
	Undet. (Syrphidae: Dipt)	

Thrips	Predator	Place
Taeniothrips pini	*Scolothrips sexmaculatus* (Thysanoptera)	Germany
	Undet. (Cecidomyiidae: Dipt)	Germany
Taeniothrips simplex	*Aeolothrips fasciatus* (Thysanoptera)	Austria
	Coleomegilla maculata floridana Leng.	U.S.A.
	Cycloneda sanguinea (L.) (Coccinellidae: Col)	U.S.A.
	Orius insidiosus (Say)	U.S.A.
	Orius tristicolor (White)	U.S.A.
	Orius niger (Wolff) (Anthocoridae: Hem)	Austria
Thrips flavus	*Orius minutus* (L.) (Anthocoridae: Hem)	Switzerland
Thrips fuscipennis	*Orius minutus* (L.) (Anthocoridae: Hem)	Switzerland
Thrips linarius	*Aeolothrips fasciatus* (Thysanoptera)	Russia
	Hauptmannia brevicollis Oud. (Acari)	Russia
Thrips tabaci	*Aeolothrips fasciatus*	U.S.A.
	Aeolothrips kuwanai (Thysanoptera)	U.S.A.
	Coccinella novemnotata Hbst.	U.S.A.
	Coccinella repanda Thunb.	Java
	Coccinella undecimpunctata L.	Egypt
	Coleomegilla maculata (Deg.)	San Domingo
	Hippodamia convergens (Guér.) (Coccinellidae: Col)	U.S.A.
	Chrysopa carnea Steph. (Chrysopidae: Neur)	Egypt U.S.A.
	Nabis alternatus Parshley (Nabidae: Hem)	U.S.A. Laboratory
	Oecanthus longicauda Mats	Japan
	Oecanthus turanicusc Uv. (Gryllidae: Orth)	Egypt
	Orius albidipennis Reut.	Egypt Ukraine
	Orius insidiosus (Say)	U.S.A. Bermuda
	Orius laevigatus (Fieb.)	Egypt
	Orius tristicolor (White) (Anthocoridae: Hem)	U.S.A.
	Mesograpta marginata (Say)	Bermuda
	Syrphus corollae F. (Syrphidae: Dipt)	Egypt
	Typhlodromus cucumeris Oud. (Acari)	Great Britain

Thrips	Predator	Place
Thrips sp.	*Ammoplanus perrisi* Giraud (Sphecidae: Hymn)	France
	Medetera ambigua Zett. (Dolichopodidae: Dipt)	Great Britain
Vuilletia houardi	*Montandoniola moraguesi* Put. (Anthocoridae: Hem)	Senegal Sudan

PREDATORY THRIPS AND THEIR PREY IN DIFFERENT PLACES

Compiled from Thompson and Simmonds (1965) with amendments

Thrips	Prey	Place
Aeolothrips fasciatus	*Anaphothrips sudanensis* (Thysanoptera)	Russia
	Anthonomus pomorum (L.) (Curculionidae: Col)	Russia
	Aphis crataegi Kalt.	Russia
	Aphis rumicis L. (Aphidae: Hem)	Russia
	Caliothrips fasciatus (Thysanoptera)	U.S.A.
	Chlorops taeniopus Meig. (Chloropidae: Dipt)	Russia
	Cicadula sexnotata Fall. (Cicadellidae: Hem)	Russia
	Frankliniella occidentalis (Thysanoptera)	Canada
	Habrocytus cioni Thoms. (Pteromalidae: Hymn)	Russia
	Haplothrips aculeatus	Russia
	Kakothrips robustus	Great Britain Canada
	Limothrips denticornis (Thysanoptera)	Russia
	Psammotettix striatus (L.) (Jassidae: Hem)	Russia
	Sciaphobus squalidus (Gyll.)	Russia
	Sitona sp.	Russia
	Sitona lineatus L. (Curculionidae: Col)	Germany
	Stenothrips graminum (Thysanoptera)	Russia
	Sympiesis viridula (Thoms.) (Eulophidae: Hymn)	Italy
Aleurodothrips fasciapennis	*Aonidiella aurantii* (Mask.)	China
	Aspidiotus destructor Sign.	Fiji

Thrips	Prey	Place
Aleurodothrips fasciapennis— contd.	*Comstockiella sabalis* (Comst.)	Bermuda
	Lepidosaphes beckii (Newm.)	China
	Lepidosaphes gloveri (Pack.)	China
	Chrysomphalus aonidum L.	U.S.A.
	Parlatoria pergandii Comst. (Coccidae: Hem)	U.S.A.
	Eotetranychus sexmaculatus (Ril.) (Acari)	U.S.A.
Anaphothrips obscurus	*Taeniothrips inconsequens* (Thysanoptera)	U.S.A.
Diceratothrips harti	*Icerya purchasi* Mask. (Coccidae: Hem)	U.S.A.
Frankliniella fusa	*Tetranychus telarius* (L.) (Acari)	U.S.A.
Frankliniella occidentalis	*Tetranychus telarius* (L.) (Acari)	U.S.A.
Franklinothrips sp.	*Selenothrips rubrocinctus* (Thysanoptera)	Trinidad
Franklinothrips myrmicaeformis	*Heliothrips haemorrhoidalis*	Israel
	Retithrips aegyptiacus (Thysanoptera)	Libya
Franklinothrips tenuicornis	*Heliothrips haemorrhoidalis*	U.S.A.
	Selenothrips rubrocinctus	Surinam
	Caliothrips insularis	Trinidad
	Dinurothrips hookeri	Trinidad
Franklinothrips vespiformis	*Heliothrips haemorrhoidalis*	Trinidad
	Selenothrips rubrocinctus (Thysanoptera)	Surinam
Haplothrips sp.	*Panonychus ulmi* (Koch) (Acari)	Canada U.S.A.
	Tomaspis flavilatera Urich	Guyana
	Tomaspis saccharina Dist. (Cercopidae: Hem)	Trinidad
	Typhlodromus pomi (Parr.) (Acari)	Canada
Haplothrips aculeatus	*Pegomyia betae* (Curt.) (Anthomyiidae: Dipt)	Germany
Haplothrips cahirensis	*Chrysomphalus ficus* Ashm. (Coccidae: Hem)	Egypt
Haplothrips faurei	*Bryobia rubrioculus* (Scheuten) (Acari)	Canada
	Cydia molesta Busk.	Canada
	Cydia pomonella (L.) (Tortricidae: Lep)	Canada
	Panonychus ulmi (Koch) (Acari)	Canada
	Spilonota ocellana (Schiff.) (Olethreutidae: Lep)	Canada
Haplothrips inquilinus	*Gynaikothrips kuwanai*	Formosa
	Gynaikothrips uzeli	Formosa

Thrips	Prey	Place
	Mesothrips claripennis (Thysanoptera)	Formosa
Haplothrips kurdjumovi	*Aspidiotus ostraeformis* (Curt.) (Coccidae: Hem)	Russia
	Cydia pomonella (L.) (Tortricidae: Lep)	Russia
	Polydrusus mollis (Stroem.)	Russia
	Sciaphobus squalidus (Gyll.) (Curculionidae: Col)	Russia
	Stephanitis pyri (F.) (Tingidae: Hem)	Russia
Haplothrips tritici	*Anthonomus pomorum* (L.) (Curculionidae: Col)	Russia
	Brachycolus noxius Mordv. (Aphidae: Hem)	Russia
Karnyothrips flavipes	*Asterolecanium* spp.	Mediterranean
	Parlatoria spp.	Mediterranean
	Pseudaonidia duplex (Ckll.)	U.S.A.
	Saissetia spp. (Coccidae: Hem)	Mediterranean
Karnyothrips melaleucus	*Ceroplastes cirripediformis* Comst. (Coccidae: Hem)	U.S.A.
Leptothrips mali	*Anarsia lineatella* Zell. (Gelechiidae: Lep)	U.S.A.
	Bryobia praetiosa Koch	U.S.A.
	Caliptrimerus baileyi Keifer (Acari)	U.S.A.
	Drepanothrips reuteri	U.S.A.
	Frankliniella moultoni (Thysanoptera)	U.S.A.
	Panonychus ulmi (Koch)	U.S.A.
	Tetranychus telarius (L.)	Central, North and South America
	Tetranychus yothersi McG. (Acari)	U.S.A.
Limothrips cerealium	*Brachycolus noxius* Mordv. (Aphidae: Hem)	Russia
Scolothrips acariphagus	*Tetranychus telarius* (L.) (Acari)	Russia
Scolothrips indicus	*Panonychus ulmi* (Koch)	India
Scolothrips longicornis	*Tetranychus telarius* (L.) (Acari)	Russia
Scolothrips sexmaculatus	*Bryobia praetiosa* Koch	U.S.A.
	Panonychus ulmi Koch	U.S.A.
	Paratetranychus indicus Hirst	India
	Tetranychus pacificus McG.	U.S.A.
	Tetranychus telarius (L.)	Russia
		U.S.A.

Thrips	Prey	Place
Scolothrips sexmaculatus—contd.	*Tetranychus yothersi* McG. (Acari)	U.S.A.
Thrips sp.	*Aspidiotus destructor* Sign. (Coccidae: Hem)	Fiji
	Phytomyza atricornis Meig. (Agromyzidae: Dipt)	Italy
Thrips tabaci	*Rhadinoceraea reitteri* Kon. (Tenthredinidae: Hymn)	Rumania
Trichinothrips breviceps	*Archipsocus* sp. (Psocoptera)	India

Appendix 4

SPECIES ASSOCIATED WITH DIFFERENT TYPES OF VEGETATION IN SWEDEN

(von Oettingen, 1954)

Sand dune

Anaphothrips silvarum
Frankliniella pallida
Cephalothrips monilicornis
Haplothrips acanthoscelis
Haplothrips armeriae
Haplothrips hukkineni

Dwarf shrub and heathland

Sericothrips gracilicornis
Taeniothrips ericae

Woodland

Aeolothrips melaleucus
Aeolothrips vittatus
Aptinothrips stylifer
Oxythrips ajugae
Oxythrips brevistylis
Taeniothrips inconsequens
Taeniothrips picipes
Taeniothrips pini
Thrips minutissimus
Hoplothrips corticis
Hoplothrips ulmi
Haplothrips subtilissimus
Leptothrips pini
Phlaeothrips coriaceus
Megathrips lativentris

Meadows
Grass and weeds generally damp places

Chirothrips manicatus
Limothrips angulicornis
Limothrips denticornis
Aptinothrips rufus
Aptinothrips stylifer
Sericothrips gracilicornis
Anaphothrips obscurus
Odontothrips phaleratus
Odontothrips loti
Kakothrips robustus
Frankliniella intonsa
Frankliniella tenuicornis
Taeniothrips atratus
Taeniothrips vulgatissimus
Thrips fuscipennis
Thrips hukkineni
Thrips major
Thrips physapus
Thrips validus
Haplothrips aculeatus
Haplothrips distinguendus
Haplothrips leucanthemi
Nesothrips dentipes

"Steppes"
Exposed expanses of grass in dry places

Chirothrips manicatus
Anaphothrips silvarum
Belothrips acuminatus
Frankliniella pallida
Thrips angusticeps
Thrips physapus
Haplothrips acanthoscelis
Haplothrips angusticornis
Nesothrips icarus

Ubiquitous

Aeolothrips fasciatus
Thrips tabaci

SUMMARY OF PRINCIPAL INSECTICIDES, FORMULATIONS AND DOSAGES USED
AGAINST THRIPS

(Compiled from recorded field trials and practices)

Common name (with alternatives)	*Chemical name*	*Use*	*Formulation and dose*
Chlorinated hydrocarbons			
Aldrin	1,2,3,4,10,10-hexachloro-1,4,4a,5,8,8a-hexahydro-exo-1,4-endo-5,8-dimethanonaphthalene	FS SI	0·75 lb a.i./ac Granules 2–3 lb a.i./ac
BHC Benzene hexachloride	1,2,3,4,5,6-hexachlorocyclohexane, mixed isomers and a specified percentage of gamma	SI	12% Dust, 45 lb/ac
DDT	1,1,1-trichloro-2,2-di-(4-chlorophenyl) ethane	FS FD GA, GSK GD Bulbs in store	0·5–1/lb a.i./ac 5–10% Dust, 18–30 lb/ac 3·5 g/1,000 ft³ 5% Dust, 15–20 lb/ac 5% Dust
Dieldrin	1,2,3,4,10,10-hexachloro-6,7-epoxy-1,4,4a,5,6,7,8,8a-octahydroexo-1,4-endo-5,8-dimethanonaphthalene	FS FD	0·3 lb a.i./ac 2·5% Dust, 30 lb/ac
Endrin	1,2,3,4,10,10-hexachloro-6,7-epoxy-1,4,4a,5,6,7,8,8a-octahydro-exo-1,4-exo-5,8-dimethanonaphthalene	FS	0·5–2 lb a.i./ac
Heptachlor	1,4,5,6,7,8,8-heptachloro-3a,4,7,7a-tetrahydro-4,7-methonoindene	FS FD	0·25 lb. a.i/ac 5% Dust, 20 lb/ac

Name	Chemical		
Lindane (gamma BHC)	1,2,3,4,5,6-hexachlorocyclohexane, 99% or more gamma isomer	GD GA, GSK Bulbs in store	0·6% Dust, 15–20 lb/ac 0·7 g/1,000 ft³ 0·6% Dust
Toxaphene	Chlorinated camphene containing 67-69% chlorine	FS	0·5–2 lb a.i./ac

Derivatives of phosphorus compounds

Name	Chemical		
Carbophenothion (Thrithion)	O,O-diethyl S-p-chlorophenylthiomethyl phosphorothiolo thionate	FS FD	1 lb a.i./ac 2·5% Dust, 40 lb/ac
Chlorthion	O,O-dimethyl O-3-chloro-4-nitrophenyl phosphorothioate	FS	0·5 lb a.i./ac
Demeton (Systox)	Mixture of O,O-diethyl S- (and O)-2-(ethylthio)ethyl phosphorothioates	FS SD	0·2 lb a.i./ac 1 lb a.i./ac
Diazinon	O,O-diethyl O-2-isopropyl-4-methyl-6-pyrimidinyl phosphorothioate	FD GS GA GA	4% Dust, 20 lb/ac 0·2 lb a.i./ac 1·8 g/1,000 ft³ 1 g/1,000 ft³
Dichlorvos (Vapona)	dimethyl 2,2-dichlorovinyl phosphate		
Dicrotophos (Bidrin) (Carbicron)	cis-dimethyl 1-dimethylcarbamoyl-prop-1-en-2-yl phosphate cis-3-(dimethoxyphosphinyloxy)-N,N-dimethyl crotonamide	FS	0·13–0·25 lb a.i./ac
Dimethoate (Cygon, Rogor)	O,O-dimethyl S-methylcarbamoylmethyl phosphorodithioate	FS SI	1·5–2 lb a.i./ac Granules 2 lb a.i./ac
Ethion	tetra-O-ethyl S,S'-methylene bisdithiophosphate O,O,O', "O"-tetraethyl S,S'-methylene bisphosphorodithioate	FS FD	0·5 lb a.i./ac 4% Dust, 25 lb/ac
Fenitrothion (Folithion)	O,O-dimethyl O-3-methyl-4-nitrophenyl phosphorothionate	FS	0·5 lb a.i./ac
Malathion	O,O-dimethyl S-[1,2-di(ethoxycarbonyl) ethyl] phosphorodithioate	FS GA	0·5–2 lb a.i./ac 2·5 g/1,000 ft³
Methyl demeton (Metasystox)	Mixture of O,O-dimethyl S- (and O)-2- (ethylthio) ethyl phosphorothioate	TI	5% soln, 250 ml/tree 10% soln, sprayed on trunks
Methyl trithion	O,O-dimethyl S-(4-chlorophenylthiomethyl) phosphoro-thiolothionate	FS	1 lb a.i./ac

Common name (with alternatives)	Chemical name	Use	Formulation and dose
Naled (Dibrom)	dimethyl 1,2-dibromo-2,2-dichloroethyl phosphate	GA	1·5 g/1,000 ft³
Parathion	O,O-diethyl O-p-nitrophenyl phosphorothioate	FS	0·2–0·4 lb a.i./ac
		FD	2% Dust, 15–20 lb/ac
		GSK	0·5 g/1,000 ft³
Phorate (Thimet)	O,O-diethyl S-ethylthiomethyl phosphorodithioate	SI	Granules 2 lb a.i./ac
		SD	1 lb a.i./ac
Trichlorphon (Dipterex)	dimethyl 2,2,2-trichloro-1-hydroxyethyl phosphonate	FD	5% Dust, 9 lb/ac
Carbamates			
Aldicarb (Temik)	2 methyl-2-(methylthio) propionaldehyde O-(methylcarbamoyl)oxime	SI	Granules 1 lb/ac
Carbaryl (Sevin)	1-naphthyl methylcarbamate	FS	0·5–1 lb a.i./ac
Plant derivatives			
Derris	(rotenone)	FS	3% soln
Nicotine	(-3-(1-methyl-2-pyrrolidyl) pyridine)	FS	3% soln
Pyrethrum	Mixed esters of pyrethrolone and cinerolone with chrysan- themic and pyrethric acids	FS	0·015% soln
Fumigants			
Calcium cyanide	Ca(CN)₂		0·13 lb/100 ft³
Napthalene	C₁₀H₈		1 lb/2,000 corms

FS—Foliar Spray; FD—Foliar Dust; SD—Seed Dressing; SI—Soil Insecticide; TI—Tree Injection; GA—Glasshouse Aerosol; GSK—Glasshouse Smoke; GS—Glasshouse Spray; GD—Glasshouse Dust. 1 lb/ac ≡ 1·1 kg/ha; 1 g/1,000 ft³ ≡ 3·5 g/100 m³.

(Nomenclature after *Rev. appl. Ent.* (*A.*) **59**, 1–12)

Appendix 6

LIST OF THRIPS SPECIES MENTIONED WITH AUTHORITIES, OFTEN-USED SYNONYMS AND COMMON NAMES

Scientific Name	Common Name
Acaciothrips ebneri (Karny)	
Acanthothrips nodicornis (Reuter)	
Aeolothrips Haliday	
Aeolothrips albicinctus Haliday	
Aeolothrips bicolor Hinds	
Aeolothrips ericae Bagnall	
Aeolothrips fasciatus (L.)	
Aeolothrips intermedius Bagnall	
Aeolothrips kuwanii Moulton	
Aeolothrips melaleucus Haliday	
Aeolothrips tenuicornis Bagnall	
Aeolothrips versicolor Uzel	
Aeolothrips vittatus Haliday	
Aleurodothrips fasciapennis (Franklin)	White fly thrips
Amynothrips andersoni O'Neill	Alligatorweed thrips
Anactinothrips Bagnall	
Anathothrips Uzel	
Anaphothrips cameroni (Bagnall)	
Anaphothrips obscurus (Müller)	American grass thrips
≡ *Anaphothrips striatus* Osborne	
Anaphothrips orchidaceus Bagnall	Yellow orchid or Yellow thrips
≡ *Neophysapus orchidaceus* Bagnall	
Anaphothrips sandersoni Stannard	
Anaphothrips secticornis (Trybom)	
≡ *Sericothrips apteris* Daniel	
Anaphothrips silvarum Priesner	
Anaphothrips sudanensis Trybom	
≡ *Anaphothrips flavicinctus* (Karny)	
≡ *Anaphothrips alternans* Bagnall	
Androthrips Karny	
Androthrips ramanchandrai Karny	
Aneurothrips Karny	
Apelaunothrips Karny	
Aptinothrips Haliday	
Aptinothrips elegans Priesner	
Aptinothrips rufus (Gmelin)	Grass thrips
Aptinothrips stylifer Trybom	Grass thrips

Scientific Name	Common Name
Arachisothrips Stannard	
Arachisothrips millsi Stannard	Peanut-winged thrips
Austrothrips cochinchinensis Karny	
Azaleothrips amabilis Ananthakrishnan	
Baenothrips Crawford	
Bagnalliella yuccae (Hinds)	
Belothrips acuminatus Haliday	
Bhattithrips frontalis (Bagnall)	
Bolacothrips jordani Uzel	
Bregmatothrips iridis Watson	
Caliothrips Daniel	
Caliothrips fasciatus (Pergande)	Bean thrips
≡ *Heliothrips fasciatus* Pergande	
Caliothrips impurus (Priesner)	
Caliothrips indicus Bagnall	
≡ *Heliothrips indicus* Bagnall	
Caliothrips insularis	
Caliothrips phaseola Hood ≡ *Heliothrips phaseola* Hood	
Caliothrips sudanesis (Bagnall & Cameron)	Cotton leaf thrips
Caudothrips buffai Karny	
Cephalothrips Reuter	
Cephalothrips monilicornis (Reuter)	
Chaetanaphothrips orchidii (Moulton)	Banana rust or Orchid thrips
Chaetanaphothrips signipennis (Bagnall)	Banana thrips
≡ *Scirtothrips signipennis* (Bagnall)	
Chirothrips Haliday	
Chirothrips crassus Hinds	
Chirothrips falsus Priesner	
Chirothrips hamatus Trybom	Meadow foxtail thrips
Chirothrips manicatus Haliday	Timothy thrips
Chirothrips mexicanus Crawford	
Chirothrips pallidicornis Priesner	Cocksfoot thrips
Chirothrips ruptipennis Priesner	
≡ *Chirothrips molestus* Priesner	
Chloethrips oryzae Williams ≡ *Thrips oryzae* Williams	Paddy thrips
Corycidothrips inquilinus Ananthakrishnan	
Cryptothrips nigripes (Reuter)	
Cryptothrips rectangularis Hood	
Ctenothrips bridwelli Franklin	
Dendrothrips Uzel	
Dendrothrips ornatus (Jablonowski)	Privet thrips
Desmothrips reedi Mound	
Diarthrothrips Williams	
Diarthrothrips coffeae Williams	Coffee thrips
Diceratothrips Bagnall	
Diceratothrips harti Hood	
Dictyothrips betae Uzel ≡ *Anaphothrips omissus* Priesner	

Scientific Name	Common Name
Dinothrips sumatrensis Bagnall	
Dinurothrips hookeri Hood	
Dixothrips onerosus Ananthakrishnan	
Dolicholepta inquilinus Ananthakrishnan	
≡ *Dolichothrips inquilinus* Ananthakrishnan	
Drepanothrips reuteri Uzel	Grape or Vine thrips
Elaphrothrips Buffa	
Elaphrothrips brevicornis (Bagnall)	
Elaphrothrips tuberculatus (Hood)	
Eothrips coimbatorensis Ramakrishna	
Eothrips aswamukha Ramakrishna	
Euchaetothrips kroli (Schille)	
Eugynothrips intorquens (Karny)	
Euphysothrips Bagnall	
Euphysothrips minozii Bagnall	
≡ *Euphysothrips fungivora* (Ramakrishna)	
Frankliniella Karny	
Frankliniella fusca (Hinds)	Tobacco thrips
Frankliniella intonsa (Trybom)	Flower thrips
Frankliniella lilivora Takahashi	
Frankliniella minuta Moulton	
Frankliniella moultoni Hood	
Frankliniella occidentalis (Pergande)	Western flower thrips
Frankliniella pallida (Uzel)	
Frankliniella parvula Hood	Banana flower thrips
Frankliniella schultzei (Trybom)	Common blossom or Cotton bud
≡ *Frankliniella sulphurea* Schmutz	thrips
Frankliniella tenuicornis (Uzel)	
Frankliniella tritici (Fitch)	Eastern flower thrips
≡ *Euthrips tritici* (Fitch)	
Frankliniella vaccinii Morgan	Blueberry thrips
Franklinothrips Back	
Franklinothrips myrmicaeformis Zanon	
Franklinothrips tenuicornis Hood	
Franklinothrips vespiformis (Crawford)	
Gynaikothrips Karny	
Gynaikothrips chavicae (Zimmermann)	
Gynaikothrips chavicae s.sp. *heptapleuri* Karny	
Gynaikothrips crassipes Karny	
Gynaikothrips ficorum (Marchal)	Cuban laurel thrips
≡ *Phlaeothrips ficorum* Marchal	
Gynaikothrips kuwanai Moulton	
Gynaikothrips liliaceae Moulton	
Gynaikothrips pallipes Karny	
Gynaikothrips takahashii Takahashi	
Gynaikothrips uzeli (Zimmermann)	
Haplothrips Amyot & Serville	

Scientific Name	Common Name
Haplothrips acanthoscelis (Karny)	
Haplothrips aculeatus (Fabricius)	
Haplothrips angusticornis Priesner	
Haplothrips arenarius Priesner	
Haplothrips armeriae Maltbaeck	
Haplothrips befordi Jacot-Guillarmond	
Haplothrips caespitis Priesner	
Haplothrips cahirensis (Trybom)	
Haplothrips cottei (Vuillet)	
Haplothrips dianthinus Priesner	
Haplothrips distalis Hood	
Haplothrips distinguendus Uzel	
Haplothrips faurei Hood	
Haplothrips flavipes (Jones)	
Haplothrips gowdeyi (Franklin)	
Haplothrips graminis Hood	
Haplothrips hukkineni Priesner	
Haplothrips inquilinus Priesner	
Haplothrips kurdjumovi Karny	
Haplothrips leucanthemi (Schrank)	Daisy, Black clover, Red clover
≡ *Haplothrips niger* (Osborne)	or Statices thrips
Haplothrips limoniastri Priesner	
Haplothrips ochradeni Priesner	
Haplothrips reichardti Priesner	
Haplothrips reuteri (Karny)	
Haplothrips setiger Priesner	
Haplothrips sorghicola Bagnall	
Haplothrips statices Haliday	
Haplothrips subtilissimus (Haliday)	
Haplothrips tritici (Kurdjumov)	Wheat thrips
Haplothrips verbasci (Osborne)	Mullein thrips
≡ *Neoheegeria verbasci* (Osborne)	
Helionothrips errans (Williams)	
Heliothrips haemorrhoidalis (Bouché)	Glasshouse thrips
Hercinothrips bicinctus (Bagnall)	Smilax thrips
Hercinothrips femoralis (Reuter)	Banded greenhouse or Sugar beet thrips
Heterothrips azaleae Hood	Azalea thrips
Heterothrips salicis Shull	
Hoplandrothrips Hood	
Hoplandrothrips annulipes (Reuter)	
Hoplandrothrips bidens (Bagnall)	
Hoplandrothrips pergandei (Hinds)	
Hoplothrips Serville	
Hoplothrips corticis (DeGeer)	
≡ *Phlaeothrips corticis* (DeGeer)	
Hoplothrips flumenellus Hood	
≡ *Haplothrips flumenellus* Hood	
Hoplothrips fungi (Zetterstedt)	
≡ *Phlaeothrips fungi* (Zetterstedt)	
Hoplothrips pedicularius (Haliday)	

Scientific Name	Common Name

Hoplothrips ulmi (Fabricius)
  ≡ *Phlaeothrips ulmi* (Fabricius)

Idolothrips spectrum Haliday
Iridothrips mariae Pelikan
Isoneurothrips australis Bagnall   Gum tree thrips

Kakothrips Williams
Kakothrips robustus (Uzel)   Pea thrips
Karnyothrips flavipes (Jones)
Karnyothrips melaleucus Bagnall
Kladothrips Froggatt
Kladothrips augonsaxxos Moulton   Bubble leaf-gall thrips
Kleothrips gigans Schmutz
Kurtomathrips morrilli Moulton

Leptogastrothrips ≡ *Oedalothrips* Hood
Leptothrips mali (Fitch) ≡ *Haplothrips mali* (Fitch)   Black hunter
Leptothrips pini (Watson) ≡ *Haplothrips pini*
  (Watson)
Leucothrips Morgan
Leucothrips nigripennis (Reuter)   Fern thrips
Limothrips Haliday
Limothrips angulicornis (Jablonowski)
Limothrips cerealium Haliday   Corn or Grain thrips
Limothrips consimilis Priesner
Limothrips denticornis Haliday   Barley thrips
Limothrips schmutzi Priesner
Liothrips Uzel
Liothrips africanus Vuillet
Liothrips associatus Ananthakrishnan & Jagadish
Liothrips caryae (Fitch)
Liothrips castaneae Hood
Liothrips citricornis (Hood)   Hickory thrips
Liothrips floridensis (Watson) ≡ *Cryptothrips*
  floridensis Watson
Liothrips fluggeae Bournier
Liothrips oleae (Costa)   Olive thrips
Liothrips ramakrishnai Ananthakrishnan &
  Jagadish
Liothrips setinodis (Reuter) ≡ *Liothrips hradecensis*
  Uzel
Liothrips urichi Karny
Liothrips vaneekei Priesner   Lily thrips

Maderothrips longisetis (Bagnall)
Megalothrips spinosus Hood
Megaphysothrips subramanii Ramakrishna
  & Margabandha
Megathrips Targioni-Tozzetti
Megathrips lativentris (Heeger)

Scientific Name	Common Name
Megathrips nobilis Bagnall	
Melanthrips Haliday	
Melanthrips fuscus (Sulzer)	
Mesothrips claripennis Moulton	
Mesothrips jordani Zimmermann	
Microcephalothrips abdominalis (Crawford)	Composite thrips
≡ *Thrips abdominalis* Crawford	
Mycterothrips longirostrum (Jones)	
Nesothrips dentipes (Reuter)	
≡ *Bolothrips dentipes* (Reuter)	
Nesothrips icarus (Uzel) ≡ *Bolothrips icarus* (Uzel)	
Neurothrips indicus Ananthakrishnan	
Neurothrips magnafemoralis (Hinds)	
Odontothrips Amyot & Serville	
Odontothrips confusus Priesner	
Odontothrips cytisi Morison	Broom thrips
Odontothrips loti (Haliday) ≡ *Odontothrips uzeli* Bagnall	Birdsfoot trefoil thrips
Odontothrips meridionalis Priesner	
Odontothrips phaleratus (Haliday)	
Odontothrips ulicis (Haliday)	Gorse thrips
Organothrips bianchi Hood	Taro thrips
Oxythrips Uzel	
Oxythrips ajugae Uzel	
Oxythrips brevistylus (Trybom)	
Oxythrips ulmifoliorum (Haliday)	
Panchaetothrips noxius Priesner	
Parafrankliniella verbasci Priesner	
Parthenothrips dracaenae (Heeger)	Palm thrips
Phlaeothrips Haliday	
Phlaeothrips coriaceus Haliday	
Phlaeothrips sycamorensis Mason	
Physothrips consociatus Targioni-Tozzetti	
≡ *Taeniothrips consociatus* Targioni-Tozzetti	
Platythrips tunicatus (Haliday in Walker)	
Podothrips Hood	
Podothrips aegyptiacus Priesner	
Podothrips graminum Priesner	
Prosopothrips cognatus Hood	Wheat thrips
Pseudothrips inaequalis (Beach)	
Ramakrishnothrips cardomomi (Ramakrishna)	
Retithrips Marchal	
Retithrips aegyptiacus (Marchal)	
Retithrips syriacus (Mayet)	Castor thrips
Rhipidothrips elegans Priesner	

Scientific Name	Common Name
Rhipiphorothrips cruentatus Hood	Grapevine thrips
Rhopalandrothrips annulicornis (Uzel)	
Schedothrips tumidus Ananthakrishnan	
Scirtothrips Shull	
Scirtothrips aurantii Faure	South African citrus thrips
Scirtothrips citri Moulton	Citrus or Orange thrips
Scirtothrips dorsalis Hood	Assam or Chillie thrips
Scirtothrips longipennis (Bagnall)	Begonia or Long-winged thrips
Scirtothrips manihoti Bondar	
Scolothrips Hinds	
Scolothrips acariphagus Yakhontov	
Scolothrips indicus Priesner	
Scolothrips longicornis Priesner	
Scolothrips sexmaculatus (Pergande)	Six-spotted thrips
Selenothrips rubrocinctus (Giard) ≡ *Heliothrips rubrocinctus* Giard	Cacao or Red-banded thrips
Sericothrips Haliday	
Sericothrips bicornis (Karny)	
Sericothrips cingulatus Hood	
Sericothrips gracilicornis Williams	
Sericothrips variabalis (Beach)	
Smerinthothrips Schmutz	
Sorgothrips Priesner	
Stenothrips Uzel	
Stenothrips graminum Uzel	Oats thrips
Stomatothrips flavus Hood	
Symphyothrips concordiensis (Liberman & Gemignani)	
Taeniothrips Amyot & Serville	
Taeniothrips alliorum Priesner	
Taeniothrips atratus (Haliday)	Carnation thrips
Taeniothrips croceicollis (Costa)	
Taeniothrips dianthi Priesner	Dianthus thrips
Taeniothrips distalis Karny	
Taeniothrips ericae (Haliday) ≡ *Amblythrips ericae* (Haliday)	Heather thrips
Taeniothrips firmus (Uzel)	
Taeniothrips inconsequens (Uzel) ≡ *Taeniothrips pyri* Daniel	Pear thrips
Taeniothrips laricivorus Kratochvil	Larch thrips
Taeniothrips lefroyi (Bagnall)	
Taeniothrips orionis Treherne	
Taeniothrips picipes (Zetterstedt)	
Taeniothrips pini (Uzel)	
Taeniothrips simplex (Morison)	Gladiolus thrips
Taeniothrips sjostedti (Trybom)	
Taeniothrips vulgatissimus (Haliday)	
Thaumatothrips froggatti Karny	
Thrips Linnaeus	
Thrips angusticeps Uzel	Cabbage, Field or Flax thrips

Scientific Name	Common Name
Thrips flavus Schrank	Honeysuckle thrips
Thrips fulvipes Bagnall	
Thrips fuscipennis Haliday	
Thrips hawaiiensis (Morgan)	
Thrips hukkineni Priesner	
Thrips imaginis Bagnall	Apple blossom or Plague thrips
Thrips linarius Uzel	Flax thrips
Thrips madroni Moulton	
Thrips major Uzel	Rose thrips
Thrips minutissimus Linnaeus	
Thrips nigropilosus Uzel	Chrysanthemum thrips
Thrips physapus Linnaeus	Dandelion thrips
Thrips praetermissus Priesner	
Thrips tabaci Lindeman	Onion or Potato thrips
Thrips validus Uzel	
Trichinothrips breviceps Bagnall	
Trichromothrips bellus Priesner	
Veerabahuthrips bambusae Ramakrishna Ayyar	
Vuilletia houardi (Vuillet)	
Xylaplothrips Priesner	

AUTHOR INDEX

Numbers in italic indicate the pages on which references are listed in full

GENERAL INDEX

*Insects (other than thrips) listed only in Appendix 3 are omitted. Numbers in italics
refer to page on which a relevant figure appears*

A

Abdomen,
  deformation of, 72
  flexing, *134*
  structure, *3*, *4*, 5, 9
Abundance,
  agricultural practice, effect of, 196
  bark, in, 172
  crops, on, 185
  emigration, 197, 199
  fire, effect of, 197
  flowers, on, 47, 185
  habitat, effect of, 158
  immigration, 197, 199
  key factors, 253
  life tables, 189
  local crops, effect of, 172
  long-term changes, 191
  losses during dispersal, 198
  methods of analysis, 191, 192, 193, 197,
    200
  natural enemies, effect of, 190, 251
  shelter effects, 161
  short-term changes, 187
  soil, in, 169
  tree, on, 186, 199
  weather, effect of, 187, 191, 192
    et seq.
Acacia, 41, 52, 56, 61, 207
Acaciothrips ebneri, 41, 307
Acanthothrips nodicornis, 196, 307
Accidental introductions, 132, 226
Acclimatization, 177
Achillea, 207
Achillea millefolium (*see* Yarrow)
Activity recorder, 92
Adactylidium, 71

Adalia bipunctata, 69
Adaptive colour, 58, 61
Adaptive form, 58, *59*, 212
Aeolothripidae, 5, 6
  basking, 191
  eggs, 15
  flight height, 152
  flight periodicity, 150
  larvae, 24, 60
  mating, 13, *14*
  mouthparts, 38, 65
  natural enemies, 71
Aeolothripinae, 2, 6
Aeolothrips,
  adaptive colour, 61
  aerial population, *119*, 143
  birds' nests, in, 78
  feeding, 39, 41, 292 *et seq.*
  habitats, 212, 213, 217, 303
  nomenclature, 307
  silk, 27
  walking-speed, 131
  wings, *2*, 61
Aeolothrips albicinctus, 61, 154, 182, 212,
  307
Aeolothrips bicolor, 61, 307
Aeolothrips ericae, 191, 307
Aeolothrips fasciatus, *14*, 27, 48, 64, 79,
  139, 140, 154, 212, 217, 292, 293, 294,
  296, 297, 299, 303, 307
Aeolothrips intermedius, 64, 214, 216, 307
Aeolothrips kuwanii, 27, 56, 79, 293, 296,
  297, 307
Aeolothrips melaleucus, 27, 63, 303, 307
Aeolothrips tenuicornis, 139, 140, 154, 167,
  307
Aeolothrips versicolor, 65, 307
Aeolothrips vittatus, 217, 296, 303, 307